Statistics for Political Analysis

*To my dad, M. P. Marchant, who loved statistics
during his life and who has been my guardian angel since then.*

Statistics for Political Analysis

Understanding the Numbers

Theresa Marchant-Shapiro
Southern Connecticut State University

Los Angeles | London | New Delhi
Singapore | Washington DC

Los Angeles | London | New Delhi
Singapore | Washington DC

FOR INFORMATION:

SAGE Publications, Inc.
2455 Teller Road
Thousand Oaks, California 91320
E-mail: order@sagepub.com

SAGE Publications Ltd.
1 Oliver's Yard
55 City Road
London EC1Y 1SP
United Kingdom

SAGE Publications India Pvt. Ltd.
B 1/I 1 Mohan Cooperative Industrial Area
Mathura Road, New Delhi 110 044
India

SAGE Publications Asia-Pacific Pte. Ltd.
3 Church Street
#10-04 Samsung Hub
Singapore 049483

Publisher: Charisse Kiino
Associate Editor: Nancy Matuszak
Editorial Assistant: Davia Grant
Production Editor: Olivia Weber-Stenis
Copy Editor: Liann Lech
Typesetter: C&M Digitals (P) Ltd.
Proofreader: Dennis W. Webb
Indexer: Judy Hunt
Cover Designer: Rose Storey
Marketing Manager: Erica DeLuca

Copyright © 2015 by SAGE Publications, Inc.

All rights reserved. No part of this book may be reproduced or utilized in any form or by any means, electronic or mechanical, including photocopying, recording, or by any information storage and retrieval system, without permission in writing from the publisher.

Printed in the United States of America

ISBN 978-1-4522-5865-2

Cataloging-in Publication data is available from the Library of Congress.

This book is printed on acid-free paper.

14 15 16 17 18 10 9 8 7 6 5 4 3 2 1

Brief Contents

Preface	**xiii**
About the Author	**xvii**
1 The Political Use of Numbers: Lies and Statistics	1
2 Measurement: Counting the Biggel-Balls	17
3 Measures of Central Tendency: That's Some Mean Baseball	51
4 Measures of Dispersion: Missing the Mark	76
5 Continuous Probability: So What's Normal Anyway?	111
6 Means Testing: Sampling a Population	145
7 Hypothesis Testing: Examining Relationships	179
8 Describing the Pattern: What Do You See?	212
9 Chi-Square and Cramer's V: What Do You Expect?	245
10 Measures of Association: Making Connections	273
11 Multivariate Relationships: Taking Control	319
12 Bivariate Regression: Putting Your Ducks in a Line	343
13 Multiple Regression: The Final Frontier	377
14 Understanding the Numbers: Knowing What Counts	417
Appendixes	430
Notes	449
Glossary	455
Index	459

Detailed Contents

Preface xiii

About the Author xvii

1 The Political Use of Numbers: Lies and Statistics 1

The Power of Numbers 2
The Science of Politics 3
Introductory Statistics: An Overview 7
Removing the Barriers to Understanding How Statistics Works 9
The Importance of Statistics: This Book's Approach 11
Using Data to Answer a Question 11
A Political Application: Indoctrination U. 13
YOUR TURN: USING STATISTICS 15
APPLY IT YOURSELF: ASSESS GRANTS TO POLITICAL SCIENTISTS 16
Key Terms 16

2 Measurement: Counting the Biggel-Balls 17

Finding Your Cases 18
Measure an Attribute 20
    Distinguish the Conceptual and Operational Differences 20
    Articulate the Operational Measure 21
Evaluate the Conceptual and Operational Definitions 25
    Valid Measures 25
    Measuring Variables Reliably 26
Translate Information in Numbers: Coding Your Data 29
    Create a Coding Sheet 29
    Code Your Data 30
Get a Frequency Distribution 31
Summarizing the Process: Measurement 33
Use SPSS to Answer a Question with Measurement 34
    Transfer Your Data to an Electronic Database 34
    Clean Your Data 36
    Get a Frequency Table 38
    An SPSS Application: Europe and the European Union 39

Your Turn: Measurement   48
Apply It Yourself: Measure the Norm for Chief Justice Appointments   49
Key Terms   50

3   Measures of Central Tendency: That's Some Mean Baseball   51

Measures of Central Tendency   53
   Mean   54
   Median   58
   Mode   62
Summarizing the Math: Averages   64
   Calculate Averages from Raw Data   64
   Calculate Averages from Tables   66
Use SPSS to Answer a Question with Averages   68
   An SPSS Application: Income in Poor Places   69
Your Turn: Measures of Central Tendency   72
Apply It Yourself: Calculate the Percent of Earned Income   74
Key Terms   75

4   Measures of Dispersion: Missing the Mark   76

Ranges   78
Distance from Mean   82
   Calculate the Standard Deviation from Raw Data   82
   Calculate the Standard Deviation from Tabular Data   84
Summarizing the Math: Dispersion   90
   Calculate the Range, Interquartile Range, and Five-Point Summary   90
   Calculate the Standard Deviation from Raw Data   92
   Calculate the Standard Deviation from Tabular Data   95
   Calculate the Standard Deviation for a Dichotomous Variable   97
Use SPSS to Answer a Question with Measures of Dispersion   98
   An SPSS Application: Law School Tuition   101
Your Turn: Measures of Dispersion   106
Apply It Yourself: Evaluate Graduates' Salaries   108
Key Terms   110

5   Continuous Probability: So What's Normal Anyway?   111

The Normal Curve   112
z-Scores   114
   Using a z Table   115
   Find the Probability of Negative z-Scores   117
   Find the Probability of Different Ranges: One-Tailed   117
   Find the Probability of Different Ranges: Two-Tailed   120

Finding a z-Score   122
Use Probability to Calculate z-Scores   123
Summarizing the Math: Probabilities of Normally Distributed Events   126
    Calculate z-Scores   126
    Find the Probability of a z-Score   128
    Use a Probability to Calculate a Value   132
Use SPSS to Answer a Question with Continuous Probability   135
    An SPSS Application: Crime Data   137
YOUR TURN: CONTINUOUS PROBABILITY   142
APPLY IT YOURSELF: EVALUATE THE MURDER RATE   143
Key Terms   144

6   Means Testing: Sampling a Population   145

Type I and Type II Errors   146
Means Testing   148
    Compare Populations with Samples   148
    Standard Error and t-Tests   152
    Proportions as a Special Case: Showing Employment Discrimination   157
Confidence Intervals: Two-Tailed Distributions   159
    Confidence Intervals with Proportions   162
Choose a Sample Size   163
Summarizing the Math: Sampling a Population   164
    Means Testing   165
    Means Testing with Proportions   167
    Confidence Intervals   169
    Choosing a Sample Size   170
Use SPSS to Answer a Question with Means Testing   171
    A Political Example: Civil War and Infant Mortality   173
YOUR TURN: MEANS TESTING   176
APPLY IT YOURSELF: ASSESS MATERNAL MORTALITY RATE INCREASES   177
Key Terms   178

7   Hypothesis Testing: Examining Relationships   179

Hypothesis Testing   180
    Hypotheses   181
    The Null Hypothesis   183
    Analysis of Variance   185
Summarizing the Math: Hypothesis Testing and ANOVA   194
    Hypothesis Testing   194
    ANOVA   195
Use SPSS to Answer a Question with ANOVA   199
    A Political Example: Partisanship and Support for the President   202

Your Turn: Hypothesis Testing   208
Apply It Yourself: Examine Partisanship's Effect on Feelings toward the Democratic Party   208
Key Terms   211

8 Describing the Pattern: What Do You See?   212

Choosing the Appropriate Form of Presentation   213
Graphs: Relationships and Scales   213
Visualizing a Relationship: Contingency Tables   220
    Set up a Contingency Table   221
    Describe the Pattern in a Contingency Table   224
    Collapse Data for a Contingency Table   225
Summarizing the Math: Graphs and Contingency Tables   228
    Graphs   228
    Contingency Tables   231
Use SPSS to Answer a Question Using a Contingency Table   235
    An SPSS Application: Life Expectancy and Female Literacy   237
Your Turn: Describing the Pattern   242
Apply It Yourself: Determine Stability across Legislative Systems   243
Key Terms   244

9 Chi-Square and Cramer's V: What Do You Expect?   245

The Probability of Discrete Events   246
    Probabilities of Independent Events   247
    Probabilities of Contingent Events   249
Chi-Square   252
    What We Expect to See   253
    Comparing What We Expect with What We Actually Observe   254
Cramer's V   258
Summarizing the Math: Chi-Square and Cramer's V   260
    Chi-Square   260
    Cramer's V   261
Use SPSS to Answer a Question with Chi-Square and Cramer's V   264
    An SPSS Application: Cross Tabulation of Religious Freedom   265
Your Turn: Chi-Square and Cramer's V   269
Apply It Yourself: Analyze Data by Type   272
Key Terms   272

10 Measures of Association: Making Connections   273

Basic Principles of Measures of Association   274
Pearson's r   278
    Get a Visual Intuition for Correlation   278

      Setting up the Work Table for Pearson's r 281
      Working through an Example: Political Rights and Civil Liberties 283
      Gamma 285
      Positive and Negative Relationships in Contingency Tables 286
      Gamma for Dichotomous Variables 287
      Gamma for More than Two Categories 290
      Kendall's Tau 296
Lambda 297
      Calculating Lambda 297
      Working through an Example: Generations of Partisanship 298
      Limitations of Lambda 300
Summarizing the Math: Measures of Association 302
      Pearson's r for Two Interval-Level Variables 303
      Gamma for Two Ordinal-Level Variables 306
      Lambda for Nominal-Level Variables 308
Use SPSS to Answer a Question with Measures of Association 310
      An SPSS Application: The Internet and Political Instability 312
YOUR TURN: MEASURES OF ASSOCIATION 314
APPLY IT YOURSELF: MEASURE POOR STUDENT GRADUATION RATES 317
Key Terms 318

11   Multivariate Relationships: Taking Control   319

Spurious Relationships 320
      Antecedent Variables 321
      Intervening Variables 321
      Spurious Non-Relationships 322
Interaction Effects 323
Three-Way Contingency Tables 324
Summarizing the Process: Setting up Three-Way Contingency Tables 329
      A Political Example: Abortion and Religious Fundamentalism 330
Use SPSS to Answer a Question with a Three-Way Contingency Table 332
      An SPSS Application: Views on Legal Abortion for Rape among Catholics 333
YOUR TURN: MULTIVARIATE RELATIONSHIPS 339
APPLY IT YOURSELF: ANALYZE DATA ON RACE FOR PARTISANSHIP AND INCOME 340
Key Terms 342

12   Bivariate Regression: Putting Your Ducks in a Line   343

Graph a Relationship 344
      Plot the Data 344
      Find a Line 346
Fit the Data with the Ordinary Least Squares Estimate of the Line 346
Find the Statistical Significance 350

Find the Strength of the Relationship   353
Use Regressions with Time Series Data   354
    A Political Example: Murder Rates   354
Interpret Regressions with Dichotomous Independent Variables   356
    A Political Example: Income and Gender   356
Summarizing the Math: Regression   357
    Describe the Pattern by Estimating the Line   357
    Identify the Statistical Significance with the Standard Error   358
    Evaluate the Strength of the Association with R^2   359
    A Political Example: Birth Rates and Abortion Rates   359
Use SPSS to Answer a Question with Bivariate Regression   363
    A Political Example: State Expenditures and Population   366
YOUR TURN: BIVARIATE REGRESSION   371
APPLY IT YOURSELF: ANALYZE INFLUENCES ON CORRUPTION   374
Key Terms   376

13 Multiple Regression: The Final Frontier 377

Using Regression to Control for Other Variables   378
The Assumptions of Regression   381
    Gauss-Markov Assumption 1: Interval-Level Variables   382
    Gauss-Markov Assumption 2: Linear Relationship   384
    Gauss-Markov Assumption 3: Model Correctly Specified   393
    Gauss-Markov Assumption 4: Non-Collinear   394
    Gauss-Markov Assumption 5: Errors Have a Mean of Zero   397
    Gauss-Markov Assumption 6: Errors are Homoscedastic   397
Summarizing the Process: Multiple Regression   398
    Analyzing Multiple Regression   398
    The Assumptions of Regression   399
    A Political Example: Teen Pregnancy   400
Use SPSS to Answer a Question with Multiple Regression   401
    Dummy Variables   401
    Bivariate Scatter Plots   402
    Residual Scatter Plots   403
    An SPSS Application: The Relationship of Multiple Factors on State Instability   404
YOUR TURN: MULTIPLE REGRESSION   412
APPLY IT YOURSELF: EVALUATE THE IMPACT OF MULTIPLE FACTORS ON THE 2012
  PRESIDENTIAL ELECTION   414
Key Terms   416

14 Understanding the Numbers: Knowing What Counts 417

Measurement   418
Univariate Statistics   419

Multivariate Statistics   421
    Hypotheses   421
    Describe the Pattern   422
    Identify the Statistical Significance   423
    Evaluate the Substantive Significance   424
Keeping the Numbers Meaningful   427
Embracing the Uncertainty   428

Appendix 1: Tips for Professional Writing   430

Appendix 2: How to Use SPSS   436

Appendix 3: z Table   443

Appendix 4: t Table   445

Appendix 5: Chi-Square Table   447

Notes   449

Glossary   455

Index   459

Preface

I wrote *Statistics for Political Analysis: Understanding the Numbers* at the request of my students. Every semester, when the representatives for different publishers visited my office, I asked whether they had a statistics textbook designed for political science undergraduates. Although they diligently sent me political science research methods textbooks and sociology statistics textbooks, they never had quite what I was looking for. I had two primary criteria in choosing a textbook. First, I wanted it to teach the math in a step-by-step way that my math-phobic students could follow easily. Second, I wanted it designed for political science students so that it contained all of the statistics that my students needed to learn, but not much more.

I had settled on a public administration textbook as the best fit for those two requirements. But over the years, it morphed into a graduate-level textbook that contained much I didn't need to cover. Unfortunately, for continuity purposes, cutting sections out was difficult. And I was still getting comments from my students asking, "Why can't the authors just explain it the way you do in class?" Finally, one day, a book rep, seeing my frustration, asked, "Have you ever thought about writing your own textbook?" My students pounced on the idea, and so I prepared to write a chapter a week over the next semester to post online for my students to use. The final result is *Statistics for Political Analysis*, a book that my students list on their evaluations as one of the best parts of the course.

UNDERSTANDING THE NUMBERS: STEP-BY-STEP

My top priority is to make statistics accessible. In teaching the math in this book, I include the appropriate equations in the book, but I know that my students find the equations terrifying. So immediately after giving an equation, I take the students through a step-by-step process of what they need to do to solve the equation. Usually, this means completing a work table in which each column contains the next mathematical operation. This allows students the needed practice to master the process without making the process intimidating.

In addition to teaching the math in an accessible way, I limit the topics to the ones that I see as most important for political science students to master. The book covers the basic univariate statistics and the basic measures of association, as well as both bivariate and multivariate regression. It also includes a few statistical concepts that are not core for political scientists: z-scores, means testing, and analysis of variance. I include these for pedagogical reasons. I remember as an undergraduate having my mind blown by the concept of statistical significance: It took several exposures before the concept really sunk in.

I see the same confusion in my students. As a result, the book covers means testing and analysis of variance because I see them as helpful in building the cognitive structure that students need to master the notion of statistical significance. Students are still a bit uncertain about the concept, but by the end of the semester, most of them are able to explain statistical significance on the final exam correctly.

Organization

Although my minimum requirements in looking for a textbook were accessibility and coverage, in writing my own I was able to structure it to fit my philosophy of teaching and learning. Pedagogically, each chapter of the book works to build a cognitive framework for the student to learn a particular concept. Every chapter begins with a cultural example that depicts the concept in a way that is familiar to the student. As it works through the math, it explains why statisticians take each step. The steps are then summarized briefly, followed by an example in which I model doing the math. At the end of each chapter are "Your Turn" exercises for students to actually apply what they've learned. The modeling in the chapter makes completing the exercises very doable.

Math courses are active by nature: You can't learn math from lectures; you have to do problems. But I think that statistics needs to be even more active than that. Learning the math is an important step in the process of understanding what assumptions statistical tests make and what the results mean. But although it is vital to develop those intuitions, the math is only a means to the more important end of analyzing data so that we can better understand the world. As a result, I have always taught my students both how to calculate statistics mathematically and how to get statistics using SPSS. Thus, after teaching the math, each chapter proceeds to teach the equivalent SPSS commands. Once again, I build cognitive frameworks by first modeling the process of using SPSS to answer a question and then giving a similar assignment in the "Apply It Yourself" section. In each problem-based learning situation, students answer a real-world question by writing a memo based on their statistical results. Each memo serves as something of a capstone for the chapter, requiring students to both analyze data and explain their results. This practical approach allows students to see the professional relevance of the material.

In 2005, the American Statistical Association recommended using real data so that students can learn about the messiness of real data and engage in the learning process,[1] and I use real-world political data throughout the book. Students major in political science because they find politics interesting; many of them hate math. Using political data for these majors certainly does spice up the otherwise boring math. But more than that, using real data shows how understanding statistics is relevant to students' education. I believe that, as a result, this is as much a political science textbook as it is a statistics textbook.

In Chapter 1, I discuss how important statistics are to the student professionally, academically, and civically. In one of the "Your Turn" exercises in that chapter, I ask them to look for an example of statistics being used in the news. My goal is to make students conscious of how often statistics are used in the political world. To reinforce that notion, each chapter contains a "Numbers in the News" feature. Each of these depicts a time in which a concept from the chapter was relevant to something that happened politically. My hope is that these features will make students more sensitive to how often events depicted in the news are dependent on the statistical concepts they have learned in this course.

The goal of this book is to give students a vision of how to apply the principles of statistics to analyze real-world data. Hopefully, the accessibility of the math instructions will enable students to get beyond their fears and develop intuitions about what the numbers mean. The political discussions should keep students interested long enough to realize that this course isn't just another requirement. By the end, they should realize that they have learned valuable skills that are relevant to their studies and their professional futures.

ACKNOWLEDGMENTS

Thanks go to John Kelly for asking that fateful day, "Have you ever thought about writing your own textbook?" Even more thanks go to Michael Tweedie for his enthusiasm for the idea, without which I am not sure I would have had the courage to undertake the project. I am very appreciative for the time he spent with me during the next semester, brainstorming topics that students would find interesting as examples and then finding the data that made the examples come to life.

I'm grateful for the support of all of my students and for all of their encouragement. They were unfailingly patient with the process. In particular, Michele Kuck, Kyle Tamulevich, and Paul Vitale were very helpful in finding errors in earlier drafts.

I'm blessed with supportive colleagues at Southern Connecticut State University. When I started teaching Quantitative Analysis, many students so dreaded it that they procrastinated taking it until their last semester. I'm very grateful that my colleagues agreed to refocus their student advising so that most students now finish the course well before their senior year. This contributes immeasurably to having a positive learning environment. Thanks in particular to Art Paulson and Kevin Buterbaugh for their consistent reminders to students that what they learn in this course matters both academically and professionally.

A project of this magnitude always requires family support. Thanks to my husband Andy for always having his eye out for interesting numbers in the news. Thanks to my youngest son Abram for his patience during those evenings and Saturdays when I was focused on my computer instead of him. This book has a special connection to my third child. During the fall of 2012, I spent each Saturday morning revising a different chapter at East Haven beach while my son Isaac and his fellow future soldiers did physical training in preparation for serving their country. The process reinforced to me what an incredible generation this is. Many of my students are veterans and so have already shown their commitment to service, but my other students are equally committed to serving their communities. I feel privileged to be able to spend my days helping to prepare them to reach their goals.

I'm grateful to all those who have taught me. Stan Taylor and Dave Magleby introduced me to all these concepts in the two methods classes I took as an undergraduate. I was fortunate that my graduate years at Chicago overlapped with some excellent methodologists: Henry Brady, Art Miller, and Lutz Erbring. I owe the most to my mentor, Chris Achen, who taught me both the Gauss-Markov assumptions and that OLS is robust. I never saw him do statistics for statistics' sake. He taught me that statistics is an incredibly valuable tool for learning about the political world.

This book benefited from an excellent editorial team at SAGE/CQ Press. Publisher Charisse Kiino and Development Editor Nancy Matuszak have been full of good suggestions and encouragement. In addition to being a great copy editor, Liann Lech has a statistical mojo that she contributed to this book. The perspectives of the reviewers were very helpful in reminding me what concepts are most important: David Damore, University of Nevada, Las Vegas; Teri Fair Platt, Suffolk University; Sean Gailmard, University of California, Berkeley; and Steve B. Lem, Kutztown University of Pennsylvania. Because of their suggestions, the book has a much clearer and tighter structure. Christopher Lawrence, Middle Georgia State College, served as the technical reviewer for the book—I am indebted to his careful work. Considering the magnitude of assistance I received from this team, I have to take sole credit for any mistakes and weaknesses that remain.

---

[1] Guidelines for Assessment and Instruction in Statistics Education (GAISE), "College Report," *American Statistical Association*, 2005. Accessed 13 July 2013. http://www.amstat.org/education/gaise/.

About the Author

Theresa Marchant-Shapiro received her PhD from the University of Chicago. Throughout her career, she has been just as interested in teaching as she is in research, and she has participated in various NSF and APSA-sponsored programs for teaching statistics and research methods. She currently teaches the methods classes at Southern Connecticut State University, where she normally receives high student evaluations—an unusual occurrence in methods classes. When not busy teaching measurement with a freshly baked loaf of bread and standard errors with M&Ms, Tess teaches classes related to mass political behavior, such as Race and Ethnicity in American Politics and Political Participation. Her research focuses on various aspects of decision making, particularly in an electoral context.

CHAPTER 1

The Political Use of Numbers

Lies and Statistics

In 1904, Mark Twain was busy writing his last book, an autobiography, in which he reminisced about his life. At one point, as he recalled writing *Innocents Abroad* in 1868, he became a bit mournful as he remembered that while writing it, he had been able to write 3,000 words a day. He began to feel his age as he remembered that twenty-nine years later, in 1897, he had slowed down from producing 3,000 to only 1,800 words a day, and seven years later, in 1904, his output had declined by an additional 400 words to only 1,400 words a day. But as he began to despair over the limitations of age, Twain realized that what had declined in those years was not how fast he wrote, but how much time he spent writing. He had begun by writing seven to nine hours a day, but ended with only four to five. After he realized that the number of pages had halved when the number of hours spent writing had also halved, he understood that the change in writing was not that his mind had gotten any less sharp; rather, he just didn't work as much. Twain concluded that numbers do not always speak for themselves. And then he repeated a well-known saying of the day, attributing it to Benjamin Disraeli: "There are three kinds of lies: lies, damned lies, and statistics." For Twain,[1] the aphorism was a reminder to be careful about the use of numbers.

In 1954, well-known "How to" author Darrell Huff[2] repeated the refrain when he wrote his best-selling statistics textbook *How to Lie with Statistics*. In it, Huff regales readers with examples of statistics used to delude the public. The interesting thing about his examples is that the lies are perpetrated by the violation of principles taught in all introductory texts. He begins the work in the same place as most statistics texts, with measurement. The ability to perform statistical analyses depends on our ability to attach numbers to concepts. How well we measure those concepts determines how accurately we can describe our world. Huff then proceeds to work his way through all the major topics addressed in any introductory statistics textbook: measures of central tendency, measures of dispersion, measures of association, probability, and control variables. Although his ostensible purpose is to teach the reader how to lie, in reality, the book shows how knowing the basic principles of statistics empowers us to be skilled consumers of data. To that end, he concludes the book with Chapter 10, "How to Talk Back to a Statistic." More than half a century later, that principle still holds: Understanding statistics is a fundamental life skill for being an active citizen in a democracy.

THE POWER OF NUMBERS

The ability to, in Huff's words, "look a phony statistic in the eye and face it down"[3] has societal benefits. Sir Francis Bacon said that knowledge is power. I normally think about that quote at the personal level: The more I know, the more I am able to accomplish. The concept is certainly a valid reason to get an education. But it also applies at the national level. Those political actors with more knowledge are more politically powerful.

Political entities can use knowledge in ways that can be helpful. I'm sure that doctors probably get irritated with filling out the forms required by the Centers for Disease Control and Prevention (CDC). But the CDC's efforts to document the occurrence of various diseases have incredible societal benefits. In 2006, because of requirements that doctors report cases of *E. coli,* the CDC was able to initially recognize that the seventy-one cases in the northeastern United States were not random and so constituted an outbreak with a single cause. Next, they were able to identify the commonality of the affected people having eaten at Taco Bell. They then narrowed the source down to lettuce and so recalled all potentially contaminated produce. The next step was to identify the source of the contamination at the farm of origin. Finally, the CDC was able to enter the policy-making arena in making suggestions about what policies would prevent future outbreaks. A similar process is replicated in all areas of public policy. This careful collection and analysis of data make government agencies very powerful actors in public health as well as in other public policy areas.

Political entities can also use knowledge in harmful ways. Information was one of the major sources of power for the Nazi regime. The Nazis were able to collect detailed information about individuals living within their borders. They kept such good records that genealogists today have a wealth of information that is not available in most countries. The Nazis' goal, however, was not to aid family historians. Rather, they used their knowledge to identify neighborhoods with higher Jewish populations and then enclosed these ghettos in preparation for transporting the residents to concentration camps. They were also able to use records documenting family ties to extend their control over those of mixed family origin. Perhaps it is this type of misuse of knowledge that led Chinese philosopher Lao-Tzu to say, "People are difficult to govern because they have too much knowledge."[4] A monopoly of information is a very powerful resource for a government intent on keeping control of its citizens.

But the power of statistics is a two-way street. In one direction, it allows politicians to gather information to make decisions. Indeed, the word "statistics" originally referred to the political state in the context of governments collecting data to help them govern better.

But access to information can be just as powerful to citizens as it is to policymakers. Statistics allow citizens to see the causes and effects of political decisions more clearly. A wealth of information is available that allows us to keep tabs on the government and potentially check its misdeeds. If you go to the website of virtually any government agency, you can find readily accessible data pertaining to the mission of that agency. Max Weber[5] identified the reliance on written documents as the third basic function of bureaucracies. One of the points of all that paperwork is to provide transparency into the workings of government. And with the advent of the Internet, all of that information is easily available to the citizen who knows enough to use it.

The collection of statistics has expanded dramatically since governments first began collecting demographic data through censuses. In this book, you will see data from a wide variety of sources. From government sources, you will see discrimination data collected by the Equal Employment Opportunity Commission (EEOC), crime data from the FBI's Uniform Crime Reports and demographic data from the Census Bureau. From research organizations, you will see survey data from the Pew Research Center, the National Opinion Research Center, and the American National Election Studies. You will see international data collected by the World Bank, the Center for Systematic Peace, and the United Nations. All of these organizations collect the data for their own purposes, but they are also bureaucracies that rely on written information. The benefit to us is that we have access to all that information. It is up to you to learn how to use it. With it, you have the power to be an effective citizen. And as you share what you learn, you can make the powers that be more responsive to this nation of citizens who gave them the power to begin with.

THE SCIENCE OF POLITICS

The power of information rests on the scientific use of it. I can't count the number of times people have laughed when I told them I am a political scientist: "What's so scientific about politics?" Their confusion is a result of misunderstanding the nature of science. Sometimes, we mistakenly confuse science with technology—test tubes and mass spectrometers. Sometimes, we confuse science with the areas of study that use those technologies—biology and chemistry. But the nature of science is found in neither the tools nor the subject matter. Rather, it is found in the method used to find out how the world works. Depending on the subject matter, we may call it "hard science" or "social science," but the process is the same.

The scientific method begins with a question of how the world works. We call this our research question. Sometimes, we ask why a phenomenon occurred the way it did. Sometimes, we ask how an event affects another phenomenon. Depending on our worldview, we may posit different answers to our research question. This allows us to develop a model of how we think the world works.

Independent Variable → Dependent Variable

In regular English, we would say that a cause leads to an effect. But in science-speak, we say that the independent variable affects the dependent variable. Model in hand, we are able to articulate a hypothesis detailing the precise causal relationship between the dependent and independent variables. But not all hypotheses are scientific. To earn that appellation, the hypothesis needs to be subject to an empirical test that is capable of evaluating whether the evidence confirms or disconfirms the hypothesis.

Notice that nowhere in the description of the scientific method do we limit what aspect of the world is appropriate to study. We could use this approach to study living organisms, in which case we would call ourselves biologists. But we can also use this approach to study political interactions, in which case we call ourselves political scientists.

The description of the scientific method also does not rigidly limit the nature or method of collecting the evidence we use to test our hypotheses. Sometimes, political science uses the same data collection techniques as the hard sciences: experiments, observation, or public records. But due to the nature of our subject, we have additional sources of data not necessarily useful for the hard sciences: content analysis and survey data.

Sometimes, we collect data through experiments, or controlled tests where we vary a possible cause in order to measure the resulting change in the effect. Political scientists Daniel Butler and David Broockman[6] decided to use an experimental research design to answer the question of whether politicians discriminate against constituents due to race. Although their experiment did not take place in a laboratory using test tubes, the structure of their experiment is not dissimilar to experiments used by scientists studying the physical world. They decided to write to state legislators using letters that differed only in the names of the senders. Controlling for the party and race of the legislator, they found that the purported race of the constituent did make a difference on whether legislators responded to the letter.

Whereas some scientists collect data by observing the natural world, political scientists can collect data by observing political interactions. Both hard scientists and social scientists can participate in scientific observation by noting and recording phenomena in a systematic way. Political scientist Richard Fenno[7] wondered how members of Congress interact with their constituents. By following several representatives around, he was able to observe how they behave in their home districts, as opposed to in Washington, D.C. Fenno found that each representative builds trust with his or her constituency by explaining who he or she is and how that relates to the needs of the district. On the basis of that trust, representatives are able to justify their activities in Washington and so get reelected.

We not only expect scientists to collect original data through experiments and observation, we also expect them to use public data archived for public use. In the same way that scientists who study public health use data collected by the CDC and the World Health Organization, political scientists have access to huge amounts of political data found in archived public records, which would include any information gathered and maintained by a governmental body that is openly available. Political scientist Gary Jacobson[8] wondered how campaign spending affects elections. He used campaign spending data collected by the Federal Election Commission in conjunction with publicly collected election data to find that, at least in congressional elections, incumbents who spend the most win by the smallest margin. (Ponder on that. In Chapter 11, I'll be asking you to come up with a possible explanation!)

Although some political scientists use data collection techniques similar to other scientists, others use techniques that are not part of the hard science repertoire. Because much political data are contained in documents, political scientists study political communications systematically using a technique called **content analysis**. Political scientist Daniel Coffey[9] wondered how polarized the two major political parties are in the United States. He collected state party platforms for both the Republican and Democratic parties between 2000 and 2004 and analyzed the difference between the number of liberal and conservative sentences contained in them. Through this quantitative content analysis, he was able to conclude that the state parties are very polarized.

Finally, although popular opinion does not determine scientific reality, public opinion is an important political factor. As a result, **surveys** (questionnaires used to collect the self-reported attitudes of people) can be a very useful tool for political scientists. For example, political scientist Robert Putnam[10] wondered whether Americans have the kind of social resources they need to be good citizens in a democracy. He used data from various sources to measure how social capital in the United States has changed over time. By using the General Social Survey conducted yearly by the National Opinion Research Center, he found that one aspect of social capital, group membership, has been declining.

Experiments, observation, public records, content analysis, and surveys are also used by political decision makers in order to gather the information they need to make informed decisions. For example, in 1954, when the Supreme Court was deliberating *Brown v. Board of Education* and the segregation of races in public schools, it received an amicus brief containing the results of an experiment conducted at Columbia University. In it, psychologists Kenneth and Mamie Clark wondered what influence race had on young children. In the experiment, they showed two dolls, one white and one black, to a series of children. Although they expected to find that white children would be biased, they were surprised to find that black children were also more likely to associate negative attributes with the black doll. The Supreme Court was equally surprised with the results of the experiment and found it sufficient grounds to conclude that separate could never be equal in an educational context.

During the Cold War, data about Soviet internal politics were sorely lacking. As a result, a form of content analysis called Kremlinology was carefully developed in order to draw conclusions from what little data were available—public speeches and photographs of public events. During the Cuban missile crisis, this technique was used to analyze two contradictory messages that had been received, both of which were purportedly signed by Soviet premier Nikita Khrushchev. Because the content of the two letters was so different, U.S. President John F. Kennedy asked who had written each message. Using content analysis, the Kremlinologists were able to conclude that Premier Khrushchev had written the first, more conciliatory message, whereas the second had been written by hardliners in the Kremlin who were trying to take control. President Kennedy was able to use this information to resolve the conflict successfully.

Much more recently, President Obama charged the Department of Defense with studying the feasibility of discontinuing the "Don't Ask, Don't Tell" (DADT) policy toward gay members of the military. One way in which the Department of Defense chose to study the issue was with a survey of military personnel. The survey found that an overwhelming majority of soldiers had no problem serving with gay comrades. On the basis of the survey, Secretary of Defense Robert Gates advised President Obama to repeal DADT and allow gays to serve openly in the military.

It is not an accident that political scientists and political decision makers use the same data-gathering techniques as other scientists. We are all interested in figuring out how the world works. As we collect data and analyze them systematically to see whether they support or disconfirm our hypotheses, we are better able to understand what is happening around us. Sometimes we do so in a quantitative way, collecting and analyzing information as numbers, as in Coffey's study of political parties and the Department of Defense's study of military personnel. Sometimes we do so in a qualitative way, collecting information as descriptions and explanations rather than as numbers, as in Fenno's study of members of Congress and the Kremlinologists' study of the Khrushchev letters. Regardless of whether the data are numerical or not, the scientific method requires that they be analyzed systematically. Those researchers who rely on qualitative data have developed various specialized techniques to ensure the validity of their results. Those of us who rely on quantitative data use statistical analysis to ensure the validity of our results.

BOX 1.1   Numbers in the News

Since 1939, when FDR first made a request that George Gallup measure public opinion about potential U.S. involvement in the war in Europe, presidents have used public opinion polls in various ways. Late in 2001, in the warm bipartisan aftermath of 9/11, President George W. Bush's press secretary, Ari Fleischer, invited a slew of prior press secretaries to the White House for lunch. At one point, the conversation turned to whether, in the face of vague terrorist threats, it is better to stay quiet or alert the public. Dee Dee Myers, who had been press secretary in the Clinton administration, asked what the polls said on the subject. President Bush replied, "In this White House, Dee Dee, we don't poll on something as important as national security."[11]

Although Bush consciously distanced himself from Clinton's practice of molding policy to fit public opinion, he still availed himself of the valuable information that polls contain. For President Bush, polls were valuable when times were rough. He primarily used them to determine how to frame unpopular policies to seem more appealing to the public. Whether presidents use polls to find popular opinion or to figure out how to mold it, that use reflects their belief that for policies to succeed, they need (at least some) public support.

The purpose of this textbook is to teach you the basics of statistical analysis. This will be useful in many ways to you as a political science student. It will undoubtedly help you as you read articles assigned in your classes because a lion's share of political science research is quantitative. In addition, if your department requires you to conduct original research, understanding statistics will help you in that process. Or, if you choose to pursue graduate studies, this course could form the foundation of your future study of statistical techniques.

But if you are like my students, you are less likely to go to graduate school than you are to get a job in the public arena. In that case, the skills you learn in this course might be the most profitable skills you acquire in college. Whether you work for the government, a political organization, or a public interest group, you will need to know how to both collect and analyze data. Your potential employers will all be impressed when they see you list "SPSS" as one of the computer programs with which you are proficient because all of them need employees who know how to answer questions by systematically analyzing data.

INTRODUCTORY STATISTICS: AN OVERVIEW

Statistics are normally divided into two types: descriptive statistics and inferential statistics. By definition, the difference between them relies on who is being studied, a population or a sample. If you have data about all of the cases you are trying to describe, you are measuring the population and so your analysis would be called descriptive statistics. You are actually describing the phenomenon. If you can measure only a sample or subset of the population, your analysis would be called inferential statistics. You are inferring characteristics about the population from the sample.

Frequently, though, we use the term "descriptive statistics" to refer to the techniques with which we analyze population characteristics. Most frequently, we want to describe univariate statistics, or statistics that analyze a single variable, such as measures of central tendency and measures of dispersion. By "measures of central tendency" we mean the different statistical ways to measure the average for the single variable in question—the mean, median, and mode. By "measures of dispersion" we mean the way all of the data are distributed around the average. The tighter the distribution, the better the average describes the variable. These measures of dispersion can include the range, the interquartile range, the variance, or (most often used for statistical purposes) the standard deviation.

But because any analysis of a population is by definition descriptive, some relationships can be analyzed using bivariate statistics, which would be classified as descriptive in spite of the fact that they analyze the relationship between two variables. In particular, cross tabulations can be considered descriptive statistics if they compare the relative frequency of variables for the full population. Although statistics as a technique for analyzing data began with census data and the process of counting people (a univariate question), it developed as political leaders realized that these numbers could be useful in answering questions about the differences in distributions either between groups or over time (both of which are bivariate questions). The first systematic census taking began in Sweden in 1749 with the Tabellverket. This census did not just count the people of Sweden, it also gathered detailed information about them. As a result, the government learned that many women died in childbirth and that many children died young. By looking at the distributions, the government was able to ask why this was happening. And as it found the answers, it was able to take action. At the heart of answering the questions of why some groups have different distributions than others, or why some factors change over time, are principles of probability. Of course, there will be fluctuations in whatever concept is being measured. The question is whether these fluctuations have a cause or are merely random effects of time and place.

Remember Mark Twain's warning that numbers don't speak for themselves—they need to be placed in context. Probability allows us to find that context. For example, Progressive

Insurance did a study of car accidents and found that 77 percent of policy holders who had been involved in an accident were less than fifteen miles from home. Fifty-two percent of them were less than five miles from home. Car safety advocates uniformly responded by advising drivers to be particularly careful when driving on short trips.[12] This advice reflected two unsupported assumptions: (1) You are more likely to get in an accident if you are driving close to home than if you are on a longer trip, and (2) there must be something about driving close to home that makes it more dangerous. So should we only drive on long-distance trips? Or would we be safer if we parked our car five miles from home and then walked or took mass transit to get to it? What the reported statistics fail to do is put the numbers into context: What percent of driving occurs close to home? If more than 52 percent of the driving occurs within five miles of home, then we might actually be safer, statistically speaking, driving closer rather than further from home. Before we can begin to analyze why a particular event occurs, we must first answer the question of whether anything actually even happened. Perhaps there is no event to explain; perhaps what we thought we observed is only a random product of underlying probabilities. The first half of this textbook will address descriptive statistics and probability.

Our uncertainty about the reality of a causal relationship can be exacerbated by how many cases we can measure. Although the goal of the United States' decennial census is to count all individuals residing within our borders, the Census Bureau obtains only limited information at that time. In off years, however, it gathers a wealth of information about Americans and American businesses through surveys. Although it is able to obtain this information in valid and reliable ways, the process of generalizing from a sample to a population inserts uncertainty into our measures. The result is an increased level of complexity in our statistical analysis of the data.

Our understanding about underlying probabilities allows us to make the jump from descriptive statistics to inferential statistics. With this basic understanding of probability and sampling in hand, we are able to begin thinking about the world in terms of cause and effect. Recall that with the scientific method, we hypothesize that a change in what we think of as a cause (our independent variable) has an impact on the effect (the dependent variable) we are interested in studying. Once we begin to use inferential statistics, we can begin the process of testing hypotheses in the ways required by the scientific method. With descriptive statistics, we simply described the distribution of the population. With inferential statistics, we will analyze data in a three-step process. In the first step, we describe the pattern that we see in a way similar to the process we used for descriptive statistics. But then we'll add on two more steps.

After describing the pattern of the relationship between our dependent variable (the effect) and independent variable (the cause), we will then ask how confident we are that there is actually a relationship between them. This is where probability comes in. With probability, we can describe what we expect to see if there is no relationship (or if what statisticians call the null hypothesis is correct). Of course, with data from a sample we will not necessarily see that precise pattern. Simply due to the random process of the sample selection, we could see a pattern very different from what the null hypothesis would predict. The second step of our analysis, then, is to find the probability that the relationship we thought we observed is only the result of random error. If we find that it is very unlikely to have occurred due to random error, we reject the null hypothesis and say that the pattern we saw is statistically significant. Notice that this use of the word "significant" is different from what you expect. When a non-statistician hears the word "significant," he or she usually interprets it to mean that a strong relationship exists. Not so here. A statistician says a relationship is significant if he or she is confident that a relationship exists, however weak it might be. We'll come back to the non-statistical use of the word "significant" in the third step of the analysis when we examine how strong the relationship appears to be. But for now, you need to be very conscious of the fact that when a statistician uses the word "significant," contrary to your expectation, he or she usually means the relationship is statistically significant, meaning that we can be fairly confident that a relationship exists. Another thing to keep in mind is that the use of probability dictates that we never know with certainty. We may be confident, but we are never sure. The scientific method never proves anything. Hopefully, by the end of this semester you will have broken yourself of the habit of using the dreaded "p" word. Good statisticians (and scientists) do not claim to prove their hypotheses. We speak in terms of confidence and probabilities.

Once we've concluded that there probably is a relationship between the two variables, we can move to the third step in the process of analyzing data: We ask how strong that relationship probably is. I call the strength of the relationship the substantive significance to line up this third step with what non-statisticians mean when they say a relationship is significant: an important or strong relationship between two variables. In this third step of the analysis, we choose the appropriate measure of association to describe how strong the relationship appears to be. There are various measures of substantive significance that we can use to answer that question. Which measure of association we choose depends on the ways the variables are measured. The second half of the book will address several different measures of association and when to use each.

In parallel with our discussion of the various measures of association will be a discussion of how to control for other factors. One of the problems that dogs statisticians is the reality that correlation is not the same as causation. It is possible for two variables to be correlated even if there is no causal relationship between them. No matter how strong the measure of association may be, it is always possible that there is an alternative explanation. In particular, sometimes a third variable is connected to both the independent and dependent variables that is the actual causal factor. If there is a third factor, we call the relationship we originally saw *spurious*, meaning an apparent relationship that is actually false. Although it is always good to begin an analysis by looking at the raw relationship between our dependent and independent variables, it is also important to control for alternative explanatory variables. We will discuss how to do this as we proceed through the second half of the text.

REMOVING THE BARRIERS TO UNDERSTANDING HOW STATISTICS WORKS

Huff's tongue-in-cheek title *How to Lie with Statistics* plays off the common assumption that the math involved in statistical analysis is so esoteric that statisticians can manipulate the numbers with impunity. I believe the opposite is actually true. Most of the statistics that affect politics and the public actually use fairly basic mathematics. After all, policymakers are not usually professional statisticians. If policymakers can understand the numbers well enough to make a decision, as a citizen you should be able to understand the numbers well enough to call them on their decisions. And in fact, if you can do the most basic of algebra, you can learn to do the introductory statistics that make up the preponderance of evidence used in policy making.

I have structured this book to remove the barriers standing between you and a clear understanding of how statistics works. The structure of most statistics textbooks gets in the way of communicating what should be straightforward concepts. Unfortunately, most statistics textbooks are written by mathematicians and so teach the techniques using an equation-centered format. I know from experience that most political science majors take one look at an equation and are paralyzed with the fear of interpreting what looks like a foreign language (and actually is filled with an awful lot of Greek letters). My approach is to take the equations and interpret them into English. In doing that, I clearly describe, in a step-by-step way, how to do the math. This process should make doing the math very straightforward. So when you see an equation, don't panic. That equation will immediately be followed by instructions on how to do the math.

My goal, not unlike Huff's, is to write a "How to" manual, but for me, it is "How to Keep the Numbers Honest." In the concluding chapter of *How to Lie with Statistics,* "How to Talk Back to a Statistic," Huff drops his literary device of teaching you "how to lie" and suggests five questions to ask in order to distinguish good statistics from bad statistics.

1. Who says so?
2. How does he know?
3. What's missing?
4. Did somebody change the subject?
5. Does it make sense?[13]

All of his suggestions are excellent for a novice, but you can do even better. Your best bet is to actually understand the language of statistics, what the different statistical techniques are, and what assumptions they make. All of that is easier if you know the math that is used to calculate the different statistical measures. So in this book, my first goal is to teach you how to calculate the numbers.

A second barrier most students have to learning statistics is that the equation format is very dry. Students decide to major in political science because they love talking, reading, and learning about politics. Most statistics textbooks set the students up for failure because they fail to show how the techniques are relevant to understanding interesting topics. Frequently, their problem sets rely on made-up data instead of the widely available wealth of real-world data. My second goal is to make the study of statistics interesting by using the techniques I want you to learn to analyze real-world data.

A third barrier most students have is in seeing how they will ever use what they learn. I know my students are registered for the course only because it is required. But I firmly believe that of all the classes they take in the Political Science Department, this is the one that teaches them the most marketable skills. This book is set up using an active learning paradigm in order to show the professional relevance of knowing statistics. To that end, I will model how to use statistics and real-world data to answer hypothetical professional assignments. Although the assignment might be hypothetical, the historical situation giving rise to the political question is not. After I model how to answer the question, I will give similar historical assignments that will require the students to use SPSS (the most commonly used statistical package) to analyze real-world data in order to write a memo answering a question. Thus, my third goal is that students will learn to use statistics to answer realistic professional questions.

THE IMPORTANCE OF STATISTICS: THIS BOOK'S APPROACH

Although statistics are, on occasion, used to deceive and manipulate, the honest use of them can make a great deal of difference politically. For political decision makers, the systematic collection and analysis of data can make all the difference in identifying and solving problems. For citizens, the open dissemination of statistics can keep government in line. For political scientists, statistical analysis allows us to better describe how politics works and why.

My goal is to help you become better political scientists, better citizens, and better political decision makers by teaching you the basic statistics that are relevant to understanding politics. This book covers many of the same topics covered in any introductory statistics class. First, you'll learn about descriptive statistics, which describe the average and distribution of data from a population. Because the conclusions we draw are always uncertain, you will then learn some basic principles of probability. We will then turn to inferential statistics, with which we draw conclusions about data based on a sample.

Unlike most introductory statistics textbooks, I will be modeling the process of calculating the various statistical measures using real-world political data. Each chapter will begin with a cultural example of the relevant concept to help you build an understanding about how it works. I will then work through the details of the concept, usually giving you detailed instructions of how to do the appropriate math in a work table. After covering the concept, I'll summarize it, modeling one last time how to do the math. At the end of the chapter, you'll find an exercise section titled "Your Turn," which will have problems structured very similarly to the ones in the chapter. By mirroring the steps I take in the chapter, you can complete the work. Take the time to do these exercises because doing the math really does help you understand what the numbers mean.

USING DATA TO ANSWER A QUESTION

Although doing the math helps you understand the numbers better, in the real world, you will not be doing statistics by hand. In the real world, you will use a statistical package to get a computer to calculate the numbers for you. In the political world, this statistical package will usually be the Statistical Package for the Social Sciences, or SPSS. As a result, although the bulk of each chapter focuses on building your intuition about how to interpret numbers by teaching you the math, after the "Summarizing the Math" section, you'll find a final section describing how to use the concept in order to answer a question using SPSS.

Each "How to Answer a Question" section (with the exception of the current chapter) will begin by describing how to get the relevant statistics using SPSS. The steps will be outlined in summary tables, which you can find both in the chapter and at the end of the book. Following the description of how to use SPSS will come a section with the heading "A Political Application." This section will describe a real-world situation in which policymakers had to analyze data to answer a question. I place myself in that situation as a hypothetical actor who is asked to answer the question. I then model for you how to use SPSS to answer it. I will include screen shots to show how I navigate SPSS's drop-down menus as well as what the resulting output tables look like. Finally, I write a memo summarizing my findings.

Although most of your academic career will be spent writing papers, in your professional career, you are much more likely to be required to write memos. This is your chance to become proficient at that. Although memos have introductions, bodies, and conclusions in common with term papers, they differ in one key respect. Whereas the purpose of writing a term paper is to convince your professor you know a lot, the purpose of writing a memo is to give your boss an answer to a question. As a result, memos need to be brief and clearly written, and give the answer right up front.

The memos you write in this class will have four parts: the heading, the introduction, the body, and the conclusion. You can find a template for writing a memo in Microsoft Word. If you click on "File" and "New," one of the template options is "Memos." There are several options available for you to download. As you begin to write the memo, the heading should include who it is to, who it is from, the date, and the subject. The introduction should state the problem that you have been asked to solve, describe the data you will use, and give a short answer to the question. The body should give evidence to answer the question. The conclusion should summarize the findings, state how they answer the original question, and indicate the broader implications those findings have for the decision maker.

Box 1.2 provides a quick reference on how to write a memo. It is also contained in Appendix 1, "Tips for Professional Writing." The focus here is answering a practical question in a clear and professional way. You do not want to write a long-winded memo—supervisors give assignments to save themselves time. Stay focused on answering the specific question using only relevant information.

BOX 1.2 How to Write a Memorandum

A. The Heading

　1. To: Name and title

　2. From: Your name

　3. Date: Date sent

　4. Subject: Keyword the nature of the assignment

B. The Introduction
   1. Describe the assignment briefly
   2. Describe the data you will use
   3. Summarize your results in one sentence
C. The Body
   Analyze the data
D. The Conclusion
   1. Summarize the data
   2. Explain how it answers the original question
   3. Describe any broader implications

My goal in writing each "Political Application" section is to model for you how to use SPSS to analyze data in a real situation. Following the "Your Turn" exercises at the end of each chapter, you will find a final assignment to "Apply It Yourself." This will place you (hypothetically) in a historical situation where you need to use SPSS to answer a question using real-world data. The datasets are found on the textbook's website, http://college.cqpress.com/sites/statspa. You will use SPSS to analyze the data and then write a memo answering the question. We'll start using SPSS in the next chapter. For now, let me walk you through how to write a memo.

A POLITICAL APPLICATION: INDOCTRINATION U.

It is February 2012 and the Republican presidential primaries are in full force. Candidate Rick Santorum has been attacking President Obama as a snob for thinking that all Americans should be able to attend college. Santorum not only argues that there are lots of great Americans who never attended college, but also accuses Obama of only wanting youth to attend college so that liberal professors can indoctrinate them. You are interning in the presidential campaign offices for former governor Mitt Romney, who, like President Obama, attended Harvard. Your boss is Ryan Williams, the press secretary for the campaign. He is afraid that Santorum might make similar attacks on Romney and asks you to take a look at an op-ed piece on Santorum's comments in the *New York Times*[14] and summarize the statistical argument in a memo.

1. Set up the heading.
   The joy of writing a memo is that you do not need to think about how to start. Because memos have a structured heading, I know that I begin by addressing the

memo to its recipient, in this case, Ryan Williams. The memo is from me and dated today. The subject line usually begins with "Re," which is short for "regarding." Because my boss is likely to have given me multiple assignments, I need to be clear here in identifying what this memo is about. I choose the three key words "Liberal University Indoctrination" in order to focus on this specific topic.

2. Write an introduction.
   I begin the introduction by summarizing the question I was asked to answer: Does college make students more liberal? I indicate that the data I am going to use come from the four studies described in the op-ed piece. Finally, I answer the question. It is very important to have a clear answer at the end of the introduction because frequently, bosses will stop reading at this point.

3. Analyze the data in the body of the memo.
   Here is where I give the evidence that I used to draw my conclusion. I see four relevant points. First, young adults do become more liberal. Second, professors do tend to be more liberal than the American public. Third, among young adults, college students are actually less liberal than non-college students. Fourth, college students are actually more likely to be religious than non-college students.

4. Conclude the memo.
   In the conclusion, it is important to return to the original question so that your boss understands how the data answer it. Because students are not more liberal than non-students, I can conclude that young adults are not more liberal because they have been indoctrinated by their professors. But I need to do more than answer the question. After having done my research, I now know more about this particular topic than my boss. What does my boss need to know about the implications of my results? The underlying fear that motivated this question was that Santorum might make similar attacks against Romney, so I choose to address that fear in the conclusion. Although the finding that college students are more religious isn't actually relevant to answering the question, it might be a good hook if Santorum does end up attacking Romney for being a Harvard elitist.

Memo

To: Ryan Williams, Press Secretary
From: T. Marchant-Shapiro, intern
Date: May 30, 2012
Re: Liberal University Indoctrination

You asked me to summarize the statistical arguments made regarding former Senator Santorum's accusation that universities indoctrinate students to become liberal. The *New York Times* op-ed piece[15] refers to four different studies relevant to the statement. These studies find that although professors do tend to be more liberal and less religious than the general population, if you compare students to non-students of the same age, young adults typically become more liberal regardless of education.

To a certain degree, Santorum is correct in his description of university professors. Whereas only 20 percent of Americans consider themselves liberals, fully 50 percent of university professors do. Similarly, professors are less likely to believe in God than the general population: 20 percent of professors are atheists as compared to 4 percent of Americans more broadly. He is also correct that students tend to become more liberal while in school. But the relationship is not causal. If you compare students to non-students, you find that the change observed among students is simply a result of their age. Students do not become any more liberal than non-students of the same age and actually are more likely to retain their religious views than non-students.

Former Senator Santorum is incorrect when he says that universities indoctrinate students. Although professors do tend to be more liberal and less religious than the general public, their beliefs do not rub off on their students. Governor Romney would be justified in extolling the virtue of a college education both for the economic benefits of being better prepared for the job market and for its tendency to build religious faith in its students.

Your Turn: Using Statistics

YT 1.1 Get a copy of a major newspaper or news magazine and thumb through it looking for statistics. Lots of times, the numbers are missing, but you can look for decisions being made based on some study or another. Even if the news story doesn't report the numbers, the numbers contained in the original study will have led to that particular decision. Find one article that reports something about a study in which statistics were used. Usually, such articles will be based on a news release by whoever did the study—track it down. Now that you've done your research, answer Huff's five questions:

1. Who says so?
2. How does he know?
3. What's missing?
4. Did somebody change the subject?
5. Does it make sense?

Evaluate the statistics in your news story on the basis of those five questions to conclude whether it is a good or bad use of statistics. (Be sure to attach a copy of the article!)

YT 1.2 Watch "The Joy of Statistics" online. How are statistics important for politics? Give three examples from the movie. You can find the movie at www.gapminder.org/videos/the-joy-of-stats/.

YT 1.3 In April 2012, the U.S. Supreme Court spent three days hearing oral argument on the constitutionality of the Affordable Care Act. Connected to that argument, the justices had read amicus curiae briefs, many of which contained empirical studies relevant to their decision. A few weeks later, Justice Stephen Breyer spoke at the Midwest Political Science Association meetings and indicated how helpful the empirical evidence was. He also talked about a decision that the justices had made at the time to not allow television cameras in the court. Their fear was that the ability of news reporters to edit the statements of justices would lead viewers to have less respect for the Supreme Court. In this case, no empirical data had been available. Justice Breyer said that he wished that the justices had had empirical evidence of how decisions by states to televise court proceedings had influenced the public views of those courts. As an appeals court, the Supreme Court does not hear new empirical evidence about the facts of the case. Why would empirical studies about the possible impact of the Court's decisions be useful as justices deliberate?

Apply It Yourself: Assess Grants to Political Scientists

It is May 2012. Arizona representative Jeff Flake has successfully passed a bill through the House of Representatives banning the National Science Foundation (NSF) from funding research by political scientists. Flake's accusation is that studying how politics works is a waste of taxpayer money. In response, Rick Wilson, the editor of the *American Journal of Political Science*, one of the premier journals in the field, blogged about some of the articles published in his journal using data funded by the NSF. You are interning at the National Science Foundation in the Division of Legislative Affairs. Because your boss, Anthony Gibson, is going to need to testify in the Senate when it takes up the appropriations bill (S.2323), the speech director, Lee Herring, asks you to write a memo describing how prior grants to political scientists have been used. Choose two of the studies described in Wilson's blog to summarize and explain why the results are useful. Conclude by identifying which of the two studies would best support the argument that political science research is not a waste of taxpayer money. The blog is found at http://themonkeycage.org/blog/2012/05/15/what-has-the-nsf-wrought/.

Key Terms

Bivariate statistics (p. 7)
Content analysis (p. 5)
Descriptive statistics (p. 7)
Empirical (p. 4)
Experiment (p. 4)
Inferential statistics (p. 7)
Observation (p. 4)
Population (p. 7)

Public record (p. 4)
Qualitative data (p. 6)
Quantitative data (p. 6)
Sample (p. 7)
Scientific method (p. 3)
Survey (p. 5)
Univariate statistics (p. 7)

CHAPTER 2

Measurement

Counting the Biggel-Balls

Whenever someone tells me "I love statistics," my gasp of surprise is inevitably followed by the discovery that the individual is a baseball fan. I can understand the enthusiasm. I have fond memories of sitting in Wrigley Field on a warm, sunny afternoon, scorecard in hand, keeping track of the stats for that day's game. But I've never been enough of a devotee to tabulate the numbers that most fans follow during the season: home runs, runs batted in, batting average, slugging average. Apparently, collecting statistics originated with a reporter, Henry Chadwick, who gathered data on outs, runs, home runs, and strikeouts in order to measure how valuable different players were to their teams. It was natural that he would begin collecting the data he calculated. One of his major concerns was that data be collected in a uniform way so that baseball players could be compared legitimately.[1]

Like Chadwick, as a statistician, my question is always "Where do the numbers come from?" In baseball, like most sports, the record keeping is done in accordance with a prescribed set of rules. A runner will never be able to claim the record for the fastest mile on an informal neighborhood sprint. The mile needs to be officially measured, the speed has to be observed by an unbiased timer, and the run has to take place in an approved race. Similarly, baseball statistics are collected in official ways during Major League games using a form very similar to the one devised by Chadwick a hundred years ago.

Although we usually hear numbers reported with an attitude of acceptance that they "just are," actually collecting the information is not always such an obvious process. This was brought home to me when my children were little and I ended each day by reading them *Dr. Seuss's Sleep Book*. I interspersed my reading with many fake yawns because, as everyone knows, "A yawn is quite catching, you see. Like a cough. It just takes one yawn to start other yawns off." As a tired mother, I hoped that as my children "caught" my yawns, they would also assimilate my exhaustion and fall asleep. But as a statistician, the page that always caught my attention pictured an enormous counting device:

Counting up sleepers . . . ?

Just how do we do it . . . ?

> Really quite simple. There's nothing much to it.
>
> We find out how many, we learn the amount
>
> By an Audio-Telly-o-Tally-o Count.
>
> On a mountain, halfway between Reno and Rome,
>
> We have a machine in a plexiglass dome
>
> Which listens and looks into everyone's home.
>
> And whenever it sees a new sleeper go flop,
>
> It jiggles and lets a new Biggel-Ball drop.
>
> Our chap counts these balls as they plup in a cup.
>
> And that's how we know who is down and who's up.[2]

As I read that page each night, I'm not sure which I was most amazed by: the device that could listen and look into everyone's home, or the chap who could count up to ninety-nine zillion nine trillion and three each night by the time my children had fallen asleep. But those are precisely the initial tasks of any statistician: first, identifying the cases, and second, counting them.

In this chapter, we will begin our exploration into statistical analysis by looking at measurement. The first step is identifying our cases—who (or what) is it we are actually studying? Second, we need to pinpoint the attribute of those cases that we are interested in studying—how can we measure it? Third, we need to evaluate our measurement of the attribute to make sure that it reflects what we are trying to study. Fourth, we record that measurement before we finally enter it into a database preparatory to analysis. As with all chapters in this book, this chapter will conclude with a description of how to use this concept to answer a political question.

FINDING YOUR CASES

Like the Biggel-Ball chap, the U.S. Census Bureau is tasked with two Herculean feats every ten years: The Constitution requires first that they find every person living on U.S. soil and, second, that they count them. It's really not simple, and there's quite a lot to it.[3] The twofold process of finding and counting U.S. residents has evolved and expanded for the past two centuries. The first stage of the census requires enumerators to identify all U.S. residents. When the first census was taken in 1790, U.S. marshals were sent around their districts, tasked with visiting every home and recording the name of each head of household along with the number of members in the household. They wrote down this information on whatever paper was available to them, made two copies, one of which they sent on to Washington for counting and one of which they posted in a public place in their jurisdiction for all to inspect. It wasn't until 1830 that the government provided the enumerators with standard forms on which to record the required information. The U.S. marshals retained responsibility for filling out these forms until 1880, when enumerators were hired specifically for the purpose of completing the census. Their task of finding each individual became easier in 1890, when the enumerators were issued maps of the roads they were assigned to cover. It became easier still in 1970 after U.S. postal workers were tasked with compiling an address register of all addresses in the United States. That same year, the U.S. Post Office also eased the workload of the enumerators with the institution of mailed questionnaires rather than personal visits. The task has eased further in the past two censuses with the availability of Internet and phone responses. Of course, the increased clarity of the process still doesn't change the fact that the population of the United States has increased from 3.9 million counted in the 1790 census to 308.7 million counted in the 2010 census. That is a lot of people to find, even with simplified processes. Somehow, in 2010, they were able to complete this first part of the process in two months.

The second part of the process, counting all the people identified during the enumeration, is even more time consuming. Originally, all those handwritten lists had to be tallied by hand. As the population of the United States increased, the time it took to count it increased as well. Although standardizing the forms helped the counting process, by 1870, counting the 39.8 million inhabitants by hand proved to be impractical. As a result, the chief clerk of the Census Office invented a basic counting machine to ease the problem. But even with the device, the magnitude of counting to 50.2 million in 1880 took seven months to complete. One of the employees of the 1880 census, Herman Hollerith, left that job to invent a more sophisticated device that could not only count, but also cross tabulate variables. His invention used cards based on those used by the Jacquard loom. Each individual identified in the census had a corresponding card with various holes punched in it to identify characteristics about the individual—age, gender, and so on. In the long run, Hollerith cards, with their eighty-column format, went on to become the basis of modern computers, and Hollerith's company went on to become IBM. But in the short run, Hollerith rented his counting machines to the U.S. Census Bureau for the 1890 census, allowing them to complete their marathon count up to 62.9 million in a record six weeks. The use of the first non-military computer in 1950 further helped the process. And the 1960 advent of optical scanners connected to bubble-coded responses helped further. But even with the sophisticated technology, the 2010 census took nine months to count.

As with the Biggel-Ball chap and the census, the first step of any analysis is to identify the cases in question and to find a way to measure them. For example, in the waning days of the 2010 lame duck Congress, President Barack Obama was able to get the Senate to ratify the New START Treaty with Russia. The issues raised in it echoed the issues raised by arms control treaties negotiated beginning in 1969. On the surface, it was relatively easy to negotiate a conceptual agreement that both countries be limited to the same number of nuclear weapons. In practice, though, measuring the number of nuclear weapons proved difficult. Much of the difficulty originated in the Soviet preference for weapons with multiple warheads, in contrast to the U.S. preference for single-headed missiles. As a result, the Soviets had fewer weapons but more total warheads than the United States. Before the two nations could agree on a treaty, negotiators had to agree on a definition of what constitutes a single nuclear weapon. Similarly, as we study political events, we need to clearly define our unit of analysis—what or who we are studying.

MEASURE AN ATTRIBUTE

Rarely are we interested only in counting the number of cases. Normally, we want to find the cases in order to measure the variation that those individuals exhibit on a particular attribute. For example, the Constitution mandated the decennial census because two key functions of government were dependent on the population of the states: representation and taxation. The marshals were charged not simply with counting the number of people in each state, but also with distinguishing between the status of those people. In particular, the Three-Fifths Compromise written into the Constitution required that they distinguish between "free Persons," those "bound to Service for a Term of Years," "Indians," and "all other Persons." The authors of this portion of the Constitution found it easier to have a residual ("all other persons") category than to define the concept of slavery, even though that was what they meant. But this concept of slavery had to be measured by the census. So the enabling legislation for the 1790 census instructed the marshals to draw a chart with six columns with the headings: "Names of heads of families"; "Free white males of 16 years and upwards, including heads of families"; "Free white males under 16 years"; "Free white females, including heads of families"; "All other free persons"; and "Slaves." The marshals then operationalized the concept of slavery by asking each head of household how many slaves he or she owned. Based on that operational definition, the marshals were able to identify a total of 694,280 slaves in the United States.[4] The point is that even the Census Bureau, which ostensibly simply wants to count people, also wants to measure attributes about individuals.

Distinguish the Conceptual and Operational Definitions

In statistical analysis, it is essential to identify both the conceptual definition of an attribute and its operational definition. The conceptual definition is something like the dictionary definition of a concept: What do we visualize when we use a term? The operational definition is the process by which we translate our observations of reality into a measurement. Although we want both definitions to be closely connected, we need to clearly define the concept first so that when we decide how we are going to measure it, we can actually evaluate how well our measure connects to the concept.

If we fail to specify both definitions in a systematic way, we will end up being confusing. In *Through the Looking-Glass,* Alice meets Humpty Dumpty, who begins to use words in ways that don't make sense to Alice. And so Alice challenges him on the definitions. "'When *I* use a word,' Humpty Dumpty said in rather a scornful tone, 'it means just what I choose it to mean—neither more nor less.'"[5] Conflating conceptual and operational definitions is akin to Humpty Dumpty's arrogance. A political example of the failure to distinguish between conceptual and operational definitions can be found in court debate over the definition of pornography. Justice Potter Stewart made many people uncomfortable in his suggestion of how to identify pornography: "I know it when I see it."[6] Not every community involved in a dispute over whether something is pornographic can run it by Justice Stewart for evaluation. So in 1973, the Supreme Court detailed the Miller Test to determine if obscene materials are protected.[7] The Miller Test leaves the definition of obscenity to the state, although it requires that the state's measure be clear. It then says that obscenity is not protected if it fails the SLAPS test—if it lacks Serious Literary, Artistic, Political, or Scientific value. Critics of Miller fear that basing the definition of obscenity in community standards makes too encompassing a definition. But in practice, the equal breadth of the SLAPS test has increased the umbrella of what is protected. The failure to define how to measure what is and what is not obscene has led to the inability of communities to limit it. In general, any time we fail to give clear conceptual and operational definitions, we limit the usefulness of our work. In the end, we have measured what we have measured, but everyone is left confused about what it means.

The meaning of statistical results relies on the use of the scientific method. This requires that we describe our procedures in advance in order to prevent us from being tempted to mold the data to support our theories about the world. Take the example of British doctor Andrew Wakefield, who believes that childhood vaccinations cause autism. He conducted a case study of twelve children, finding that eight of them developed "regressive autism" after receiving an MMR vaccine. A recent study in the *British Medical Journal* suggested, however, that Wakefield was so invested in his theory that he failed to fully define the indicators of autism in advance. The current study reexamined Wakefield's original data and found that several of the children he reported getting autism after an inoculation do not actually fit into the commonly accepted definition of the affliction, whereas others actually were exhibiting symptoms before they received the shot.[8] Proper scientific method requires that we define our concepts and measures clearly so that we can be equally clear about the difference between what it would look like if we are right and if we are wrong. This means that before we ever start collecting data, we need to clearly define the concepts contained in our theory.

Articulate the Operational Measure

A clear measurement requires a description of not only what the attribute in which we are interested looks like, but what the alternatives look like as well. When the United States and the Soviet Union negotiated arms limitation treaties, they not only discussed the definition of a nuclear weapon, they also discussed verification. This meant that there had to be monitors to verify the count of weapons. But the monitors had to be able to examine places where weapons were *not* in order to verify that they were not there. Years later, the role of monitors was key in the invasion of Iraq: How could monitors know if there were no weapons of mass destruction? If that question was not answered before the monitoring ever began, there was no way for Iraq to prove to the United States' satisfaction that our expectations were being met.

Assignment of a Number. In Chapter 1, we learned that the difference between quantitative and qualitative analysis is that quantitative analysis looks at variables numerically and analyzes them on the basis of those numbers. As a result, an essential component of the process of operationalizing a variable entails assigning numbers to the different values of the variable. Sometimes, the numbers seems obvious. For example, if we are discussing government spending for programs, it seems obvious to quantify the variable in terms of the dollars spent. Similarly, if we are talking about age, it seems obvious to quantify it in years.

But for other variables, the process of quantifying the values is less obvious. Particularly in the social sciences, some of our most important concepts do not have physical manifestations that are easy to count. We reify concepts like democracy—discussing them as if they are real—when they are actually abstractions. Conceptually, we think about democracy as a system of government in which the people rule. But that conceptualization can include multiple factors: competitive elections, political alternatives articulated by multiple parties, social and political freedoms, and even the absence of corruption. Trying to take all those dimensions into consideration while measuring which countries are more (or less) democratic is very difficult. In practical terms, the common practice is to have individuals who are experts about a particular country answer a series of questions about whether the country embodies each of the constituent indicators, or characteristics reflecting different aspects of a concept. Each "yes" response increases the final index of democracy by an additional point. This method of measuring democracy is not perfect,[9] but it does allow a number to be assigned to an abstract concept.

Levels of Measurement. Depending on the underlying concept, the numbers associated with its operationalization may have more or less meaning. In statistics, we address three facets of the measure (categorization, order, and scale) and place them in a hierarchy. The more of these dimensions reflected in the numbers associated with a variable, the higher the level of measurement. We label the levels of measurement (in order from lowest to highest) nominal, ordinal, interval, and ratio. Figure 2.1 summarizes the characteristics of the different levels of measurement.

The level of measurement about which we have the least amount of information is nominal. Nominal-level data have only categories—they do not have a natural order. Some common examples of nominal-level data are race, gender, hair color, and astrological sign. For each of these variables, any individual can be placed in a group, but the variable isn't really an attribute about which an individual can have more or less. You really cannot be more or less gender; you are either a man or a woman. Our computers require us to associate numbers with the categories, but there is no logical reason why I would code men as "0" and women as "1" rather than the reverse. Nominal variables identify only membership in groups.

The second level of measurement is ordinal. In this case, the categories actually do have an order from less to more. But the numbers we associate with those categories do not reflect a uniform amount of more. Some common examples are class, educational degrees, and socioeconomic status. We can rank degrees in order from high school diploma, associate's, bachelor's, master's, to PhD and know that we have gone from less to more education. But the transition between each category does not reflect an equal amount of education. I can assign each of them a number from zero to five, but the number only puts them in order; it does not reflect a uniform amount of education.

Figure 2.1 Levels of Measurement

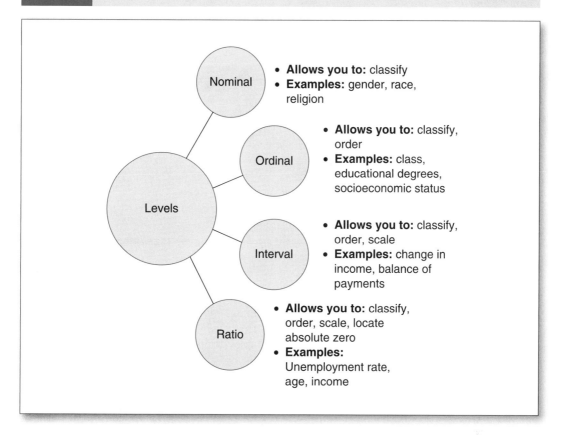

The third and fourth levels of measurements are interval and ratio. For both of these, the units not only reflect an order, they also measure a uniform amount. You can say that the numbers have a measurable distance on a scale. As the value of the variable increases by one point, the attribute increases by the same amount regardless of where you are on the scale. If the description of the variable uses the word "scale" or "index," you can be confident it is either interval or ratio. The difference between interval and ratio levels of measurement is that interval-level measures can have negative values whereas ratio-level measures have an absolute zero. A ratio-level measure is a special kind of scale that can have only a positive value. Because zero means a total absence of the attribute being measured, two ratio values can be compared proportionally: If one number is twice as big as another, we can conclude that it has twice as much of that attribute. If the variable is presented either as a percentage or physical measure (size, weight, quantity, or amount), it is probably ratio level.

Figure 2.2 Decision Tree to Determine Level of Measurement

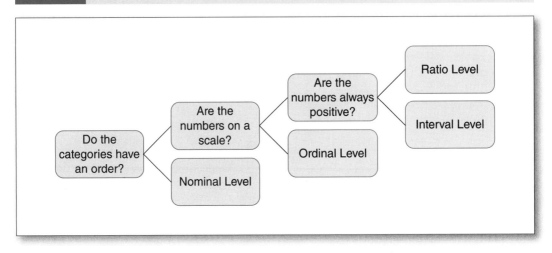

Because the different levels of measurement contain different amounts of information, what you can do with them statistically varies. Statistically, you can conclude more from variables with higher levels of measurement. That is why we spend the second half of this textbook learning about different measures of association. The higher the level of measurement of the variables you are studying, the more you can do in measuring their association. (Except that for statistical purposes, interval- and ratio-level variables get treated the same way mathematically.)

Because identifying the appropriate measure of association depends on the variable's level of measurement, it is essential that you learn to identify the level of measurement correctly. I suggest using a series of three questions to identify the level of measurement. These questions are presented as a decision tree in Figure 2.2. First, ask whether the categories have an order. If they don't, it is a nominal-level variable and you can stop there. If there is an order, then ask if the numbers are on a scale. Does each unit equal the same amount of the attribute you are measuring? If the answer is no, the numbers don't tell you anything more than the order of the categories, then it is an ordinal-level variable and you can stop there. If the answer is yes, then the last question you ask is whether there is an absolute zero, which indicates the total absence of the attribute in question. If the answer is no, you can have negative values, then the variable is interval level. If the answer is yes, the variable can have only positive values, then the variable is ratio level.

Conclusion. After identifying our cases (who we are studying), we need to define what attribute we are studying both conceptually and operationally. The conceptual definition paints a word picture of what we are talking about—along the lines of a dictionary definition. Only after defining the concept do we consider how best to measure it. And that operational definition needs to be clear enough that anyone else could use it and get exactly the same results. Because it is easier statistically to find associations between variables with higher levels of measurement, make that a priority as you are operationalizing your variable.

EVALUATE THE CONCEPTUAL AND OPERATIONAL DEFINITIONS

Once we have defined both our concept and how we are going to measure it, we can evaluate how well the two definitions connect. In practice, no measure is going to quantify a concept perfectly. The 1790 census erred in its count of slaves for at least two reasons. First, it is off by at least sixteen because the sixteen "colored" residents of Vermont were all free but for some reason were included in the final count as slaves.[10] It is also off because in this first census, the relationship between the results and taxes was unclear. As a result, many respondents were cautious about revealing information about their families and "property." As social scientists, we usually look at two aspects in order to evaluate the quality of a measure: validity and reliability.

Valid Measures

Validity means that it makes logical sense to measure our concept in this way. The question here is, How close does the measure come to getting at the underlying concept? When we argue that a particular measure is valid because it makes sense, we call it face validity. After the Civil War, the census operationalized race by the identification of the enumerator, but more recently, the census has asked individuals how they identify themselves. It has been valuable for social scientists to realize that physical characteristics are not synonymous with race. Furthermore, this self-identification has led to the realization that "black" and "Hispanic" are not mutually exclusive categories, as was assumed by earlier censuses. As a result, since 2000, the census has used two different questions, the first to identify those of Hispanic descent and the second, those who think of themselves as "black." We can now identify at least some of those who think of themselves as multi-racial. The face validity of self-identification even led to including "negro" as an option in the 2010 census because some older Americans identify themselves that way.[11]

Sometimes, social scientists claim validity because a measure has widespread use—we call this consensual validity. Unfortunately, everyone agreeing does not mean that there is a perfect connection between a concept and its measure. Consider, for example, the concept of unemployment. Conceptually, when we use the term "unemployment rate," what we mean is the proportion of the workforce that is without a job at any given point in time. Operationally, though, in the United States unemployment is measured by the Census Bureau, which takes a survey and asks a series of questions without ever using the word "unemployment." Someone in Washington takes those responses and excludes anyone who is jobless but is either not looking for a job or not available for work at that time. Conversely, it counts anyone as employed who can name an employer even if they worked only a minimal amount. And some people who didn't work at all or earn anything are included in the "employed" category if they were ill, on family leave, or involved in an industrial dispute.[12] An alternative method of measuring unemployment would be to use the number of people filing unemployment claims. But that would mean excluding anyone who didn't file for unemployment because, for example, they were being supported by a spouse. It also would exclude all those for whom benefits had expired. Logically speaking, the first operationalization is probably a more valid measure than the second. It has consensual validity because, although it is not perfect, most experts agree that it is the best we can do at measuring what we think of conceptually when we use the term "unemployment."

Alternatively, we can claim associational validity if our measure is correlated with other measures connected to the concept about which we are interested. For example, when measuring the partisanship of voters, political scientists will ask two questions, a uniform question first and one of two possible follow-up questions. First, they ask all respondents if they think of themselves as a Democrat or a Republican. If any respondents refuse to associate with one of the two major parties and insist that they think of themselves as Independents, the follow-up question is, "Which party do you lean toward?" Some respondents will continue to insist they are Independents, but many will say they lean toward one of the two major parties. If, however, the respondents allied with one of the two parties in the first question, the follow-up question asks if they think of themselves as a strong or a weak partisan. The possible responses to the questions lead to seven possible categories of partisanship: Strong Democrat, Weak Democrat, Leaning Democrat, Independent, Leaning Republican, Weak Republican, and Strong Republican. In practice, many political scientists will take those seven categories and lump the three categories for each party together. For example, "Democrats" would include Strong Democrats, Weak Democrats, and Leaning Democrats. Thus, those respondents who originally identified themselves as Independents and only when pushed said they leaned toward the Democratic Party will, in the end, get labeled as Democrats. Lumping the leaners with the partisans clearly defies the face validity of self-identification. Political scientists justify it, though, by arguing that if you correlate self-identified partisanship with how these individuals vote, the leaners actually are more likely to vote in a partisan way than the weak partisans. Because we think of partisanship as including how someone votes, political scientists define leaners as partisans on the basis of associational or correlational validity.

We can also claim a measure to be valid if it is a good predictor of an effect we are trying to explain by our concept. We call this predictive validity. For example, those who complain about affirmative action say that admission to a good college or getting a job should be based on merit. For them, it is obvious that merit should be operationalized as good grades and high scores on standardized exams. Because both of those measures of merit are good predictors of success in life, opponents of affirmative action claim predictive validity for their operationalization. Face validity, consensual validity, associational validity, and predictive validity are all explanations of why it makes sense to measure a concept in a particular way.

Measuring Variables Reliably

In contrast to validity, reliability means that the measurement is consistent—the results don't change with time or researcher. To a large degree, reliability is dependent on how well you define your measure. If you are coding information, you need to be clear from the outset how it will be coded (hence the difference between Dr. Wakefield and his critics in coding "autism"). Usually, this will mean having at least two people coding the same data and reporting the correlation between those two codings as "inter-coder reliability." It would also mean publishing your methods so that other researchers can replicate your findings.

Unclear operationalizations can lead to problems in reliability. For example, if the questions used in surveys don't communicate clearly what they are asking, respondents will tend to respond in random or inconsistent ways. One problem in question wording is being overly general. If you were to ask "Do you support world peace?" any viewer of talent pageants knows that the answer will always be "Yes." Left unmeasured is what policy measures the supporters of peace actually advocate. Do they support intervention in Darfur? Oppose recognition of Myanmar? Support an autonomous Palestinian state? More specific questions would yield more reliable results.

At the other extreme, overly specific question wordings can also yield unreliable results if respondents don't know what the question means. Most Americans could answer reliably whether they support or oppose President Obama's health care act, but it is possible that calling the law by its official name would change the results of the survey: "Do you support or oppose the Patient Protection and Affordable Care Act?" Similarly, getting views on specific aspects of it could be more difficult because of lack of information. "Do you support or oppose the provision that gradually eliminates the donut hole in prescription care coverage?" Someone who is retired would be well informed on the topic and able to take a discerning stand. In contrast, many college students do not know what the prescription donut hole is. Frequently, survey designers will increase the reliability of overly specific questions by giving background information. "Prior to the current health care act, Medicare would cover the first $300 of prescription care, but wouldn't cover any more medicine until the elderly person had paid $6,000 out of pocket for prescriptions. Do you oppose or support the health care act's elimination of this 'donut hole' in prescription coverage?" Respondents need to know what the question means before they can answer in a consistent way.

The honesty of respondents can affect the reliability of data as well. As mentioned earlier, historians of the 1790 census had some reason to fear that property owners were less than forthcoming in their reports because of their fear of taxes. We need to avoid framing our questions in a way that might provoke an emotional response from those answering the questions (e.g., "Would you want an admitted homosexual to be a scout leader for your son?"). But even with neutral terminology, sometimes the political situation leads to unreliable responses. There have been a few elections where the preelection polls overestimated the vote a black candidate would get on Election Day. The "Bradley Effect" is named for the election Los Angeles mayor Tom Bradley unexpectedly lost when he ran for governor of California. The theory is that whites may not report their racial bias when asked a survey question, but that bias erupts in the polling booth. As a result, respondents end up voting differently from how they answered on the survey. If the Bradley Effect is correct, then in elections with one black candidate, survey questions about voting intentions are unreliable indicators of the actual vote. Along a related vein, researchers into the prevalence of workplace dishonesty found that it was less reliable to ask individuals whether they use company property (paper supplies or the company car) for personal use than to ask whether the behavior is typical of most people. In this case, the self-report may seem a more valid measure, but reported perception of peers is more reliable.[13] Sometimes, self-reports are more reliable measures of values than of behavior. For example, more Americans self-report attending church than actually show up for services in any given week. A more reliable measure comes when the survey is conducted on a Monday and the respondents are asked to detail their activities of the previous day (without any mention of church). The self-report of church attendance reflects how much Americans value church attendance. But the Monday survey is a more reliable measure of actual behavior.[14] However valid self-reports may be, the (dis)honesty of respondents can interfere with reliability.

Claims of reliability need to carefully distinguish between unreliable data and attributes that really are changing. The measure of partisanship that lumps leaning Independents with self-identified partisans is considered reliable because the proportion of Americans in each category changes slowly over time. But **cross sectional surveys** can discuss only aggregate-level changes because they interview a different group of respondents every time. When partisanship is asked about in a **panel study**, where the same people are asked the same questions at various points in time, the reliability of this measure of partisanship becomes questionable. Particularly among the self-identified Independents, their response to "toward which party they lean" can be inconsistent.[15] Categorizing the leaners as partisans may then have associational validity, but it is not necessarily a reliable measure of the long-term political attribute we were trying to measure. It could be that Independents have no long-term commitment to a party; they are really only indicating for which party they plan to vote in the next election.

Sometimes, a measure is just too volatile to be considered reliable. For example, many people follow the stock market because it has an important role both in the U.S. economy and in our personal retirement plans. But although the evening news always reports how the stock market did that day, it is rarely reported as a "leading economic indicator" because it fluctuates too much. For example, the "Flash Crash" of May 2010 highlighted the problem of computerized trading. An unusual combination of events combined with computer algorithms for selling led to an unprecedented 1,000-point drop in the stock market in just five minutes. The "crash" did not reflect real changes in the economy. Similarly, when you invest in the stock market, any financial advisor worth his or her pay will tell you to ignore what is going on in the stock market. Watching it too closely will encourage you to sell when you become concerned about your investments and buy when you are pleased that your investments are doing well. But if you want to have a comfortable nest egg when you retire, you actually want to do the reverse: Buy low, sell high. The stock market is an unreliable indicator of both the economy and your retirement because it is too volatile.

After you define the concepts you are interested in studying and carefully detail how you will measure them, you then need to evaluate how well your measure embodies your concept. This discussion should address the validity and reliability of your measure in quantifying your concept. Is your measure getting at the underlying essence of your concept? Is your measure clear enough that it will yield the same results given the same situation?

> **BOX 2.1 Numbers in the News**
>
> In the fall of 2012, the U.S. Bureau of Labor Statistics (BLS) released two contradictory numbers. First, it said that the unemployment rate had increased from 7.8 percent in September to 7.9 percent in October. Second, the BLS said that employers had hired an additional 171,000 workers. If employers were finally starting to hire new employees again, shouldn't the unemployment rate have decreased? The answer to the puzzle is found in the measures. The Bureau of Labor Statistics calculates the unemployment rate based on responses to a survey of households, whereas it bases job growth on a survey of businesses and government agencies. The two measures have two key differences. First, fewer households are surveyed than businesses, so the unemployment rate is less reliable than the estimate of job growth. Second, the estimate of job growth does not include numbers for farmers or self-employed workers, so the estimate of job growth has lower validity than the unemployment rate.[16]

TRANSLATE INFORMATION IN NUMBERS: CODING YOUR DATA

After thinking carefully about your concept and how best to measure it, you can begin the process of actually doing so. We call the process of translating information in numbers coding. This involves four steps. First, you create a coding sheet on which to record the data. Second, as you collect the information, you actually record the data on the coding sheet. Third, you transfer the data from the coding sheet to a database in preparation for data analysis. Finally, you clean any mistakes you might have made in entering your data.

Create a Coding Sheet

After you have made the decision about how to measure the variable, you then need to actually measure it. This process is easier if you have a coding sheet on which you can enter the data while you collect it. Coding sheets can be elaborate. If you are collecting a lot of data about each case, you would want to have a separate coding sheet for each case. (Remember how in the 1880 census there was a different Hollerith card for each individual?) For now, though, we are only gathering information about one variable, so we can use a single coding sheet to collect information about all of the cases. Set up the coding sheet so that each case has its own line and each variable has its own column. The first column should identify the case; the second, the attribute you are measuring.

As you are collecting the data, it is helpful to have your coding sheet as structured as possible to minimize human error. You will need to write in the identity of each case by hand, but for the attribute, it is easier to have the possibilities already identified. That will mean that you just have to check off which value applies for each case. To do this, you will have to go through the very helpful process of identifying all of the possible categories an attribute can take.

As you are identifying the possible categories that a case can have for a particular attribute, you need to obey two rules. First, the categories need to be mutually exclusive. This means that no one case can belong in more than one category. Second, the categories need to be exhaustive. This means that every case has to belong in one of the categories. You'll have problems following these rules if you try to measure two different attributes within a single variable. For example, suppose you're categorizing phenotypes and create a variable that has the alternatives "blond haired, blue eyed," "red haired, green eyed," and "brown haired, brown eyed." Even if hair color and eye color tend to be correlated, the correlation is not perfect. I have (had) red hair and brown eyes. How would you code my phenotype? One option is to include a category for each possible combination of hair and eye color. Sometimes, you will want to designate an "other" category to lump together several uncommon possibilities. Alternatively, you could include a "missing" category for the cases that don't fit into your original options as well as the cases for which you have no data. I normally use the number "–9" to indicate that I don't have data for a variable for that case, although if –9 were a possible value, I would make –99 or –999 the missing value. But in the end, every case needs to fall into one and only one of the categories of each variable.

You will not only want to set up your coding sheet to make gathering your data easier, you will also want to set it up to make using the data easier. Just as the Census Bureau can count faster using automated counting machines, in general, for quantitative data, you are going to want to use a statistical package to analyze it. That will entail entering the data electronically into a database. Although computers can keep track of words (we call that "string" or "alphanumeric" data), they can't analyze them very well. Computers (and statistics) are most useful for analyzing numerical data. So, although there isn't a problem just typing in the name of the case identifier (because you won't be analyzing that), for your attribute, you are going to want numbers rather than the labels for your categories. Your life will be much easier if you include next to each category the number you want associated with it. So if my attribute were gender, I would know that my possible categories are to be male and female. If, every time I entered another case, I had to remember to type "0" when I saw "male," I would be very likely to make mistakes along the way. If, however, I put a "0" next to "male" and "1" next to "female," then as I enter the data into a database, I easily type the number I see without risking mistake.

Code Your Data

Once you've set up your coding sheet properly, you are ready to collect your data for use. Your first column should contain the case identification. Begin by writing the name of each case in the first column. The cells of the second column should contain a list of the possible categories of the attribute you are measuring. As you find the information for each case, check off the appropriate category. As you are collecting the data, be sure to keep track of any sources you use to find the information. You will need to give them credit in the end, and it will be easier to keep track of the sources now, rather than to go back and find them later.

GET A FREQUENCY DISTRIBUTION

With only one variable and a limited number of cases, it is easy enough to just count the number of cases that have each of the possible categories of your variable. In general, though, you are going to have more cases and more variables than it is practical to count by hand. Just as the Census Bureau was able to enumerate the population in 1890 much more expeditiously with the incipient IBM computer, you'll find your job much easier if you enter your data in a spreadsheet before you try to analyze it. In the application section, I'll give instructions on how to enter the data in Excel and import it into SPSS for counting.

But even with SPSS to do the counting, the task of producing a professional-looking table is entirely yours. Box 2.2 details how to make a professional-looking frequency table. You'll notice that each of the boxes in this chapter begins with a title. The title is prefaced with an identifying number that is assigned in the order that the boxes are found. In this case, "Box 2.2" refers to the second box in Chapter 2. After labeling the table, you need to give it a descriptive title that includes the attribute you have measured.

The first row of the table should include column headings. The first column is going to contain the categories of the variable, so it should be headed with the name of the variable. The second column should be headed "Frequency" and the third, "Percent." In the second row, place the name of the first category followed by the number of cases in this category along with what percent of all the cases are in this category. In each of the subsequent rows, do the same thing for each of the other categories of the variable. After you finish entering all of the categories, there should be one row left. In the first column, place the word "Total"; in the second, the total number of cases; in the third, add up the percentages in the third column to get "100.0%," plus or minus 0.1%. (This difference is due to rounding error. If it totals less than 99.9% or greater than 100.1%, recheck your math.) If you have any cases for which you were unable to determine their values for this variable, you exclude them completely from the table.

In professional political science papers, each major section of the table is separated by a horizontal line. This is a little different than what you see in this book. For papers, the standard is to place a line at the top of the table (below the title) and at the bottom. You place a third line under the column headings. Begin by highlighting the full table (without the title). When you created the table, Word automatically put lines around all the cells of the table; you'll want to get rid of those first. If you look on the right-hand side of the "Paragraph" section of the "Home" menu of Word, you'll see a square icon—sometimes it is shaped like a box, sometimes it has a grid in it. Right now, it probably has a grid. Click on the arrow next to the icon to get the drop-down menu and choose the "Borders and Shadings" command. Within the "Borders and Shadings" window, click on the setting "None" and the lines in the "View" should vanish. A single line will already be selected as the "Style." In the "Preview," click above and below the table to insert lines there. Once you click on "OK," your table should have single lines at the top and at the bottom. Next, add the line below the headings by highlighting the first line of the table—the one containing the column labels. At this point, the "Borders and Shading" icon will have changed to a box. Click on the arrow to open the drop-down menu and choose "Bottom Border." Your table should now have the three lines: one at the top, one below the labels, and one below the table.

BOX 2.2 How to Create a Professional-Looking Frequency Table in Word

1. Give it a title describing its content.
2. After the title, insert a table of the appropriate size: columns = 3; rows = 2 + the number of categories in your variable.

   >Insert
   >>Table
   >>>Insert Table
   >>>>Number of columns = 3
   >>>>Number of rows = 2 + *number of categories of variable*

3. In the first row, label the three columns: *Variable Label*, "Frequency," "Percent."
4. In the first column, list the categories of the variable.
5. In the second column, give the frequencies of those categories.
6. In the last column, give the valid percentages—do not include the missing cases unless this is actually relevant to your analysis.
7. In the last row, give the totals for the columns: "Total," "*number of valid cases*," "100.0%."
8. Draw the appropriate lines.

   Highlight the entire table:
   >Borders and Shading *(the arrow, not the icon)*
   >>Borders and Shadings
   >>>Settings
   >>>>>None
   >>>Style
   >>>>*The single line should be highlighted.*
   >>>Preview
   >>>>*Click above and below the table to add lines there.*
   >>>>OK

   Highlight the first line (which has the column labels):
   >Borders and Shading *(the arrow, not the icon)*
   >>Bottom Border

9. Below the last line, use a smaller font to add the data source and any other clarifying information.

Finally, you can clean up the table. For example, I like to have a little extra space between the column labels in the first line and the data in the second line. To insert that space, I highlight the second line and use the "Paragraph" command in Word to add "Spacing" "Before." Similarly the "Total" line looks best separated from the data, so I highlight that line and use the "Paragraph" command to add space both before and after it. The table also looks nicer if the columns are centered and the decimals are lined up so I highlight the full table and click on the "Center Text" icon in the "Paragraph" box. In addition, I usually adjust the size of the columns. Because the three columns are stretched all the way across the page, they look too far apart. With the table highlighted, I can move my cursor across a row to find the margin of each column. If I left-click and hold, I can slide the margins over. Finally, after the concluding line, you should give any important information about your variable: where you found your information and how you measured it. If your percentages do not add up to precisely 100.0%, you can indicate here that it is due to rounding error. Use a smaller font for any information after the last line so that it doesn't merge into whatever text follows it.

SUMMARIZING THE PROCESS: MEASUREMENT

This chapter has very practical applications in answering political questions. In this section, I'll describe the general process of measuring an attribute for the purpose of answering a question. This will make more sense in the next section, where I will walk you through an example of actually answering a political question. Box 2.3 outlines the steps we've already covered in this chapter on how to measure a variable. First, you need to identify the cases you are studying. You will want to measure an attribute of those cases that varies. Define that attribute conceptually. Then define it operationally—how will you measure it? Next, evaluate the measure. How well does the measure line up with the concept? Is it valid? Is it reliable? Once you have decided how to measure your variable, you need to do so. Set up a coding sheet for entering the data and fill it out. Then copy the data into a database, cleaning the data if necessary to make sure it is correct.

BOX 2.3   How to Measure a Variable to Answer a Question

1. Identify the unit of analysis. Who or what are the cases you want to identify in order to measure their variation on an attribute?

2. Define your variable conceptually. What is the attribute you are interested in studying?

3. Define your variable operationally. How will you measure that attribute?

4. Evaluate the validity and reliability of your measure. Does it make sense? Is the measurement clear enough that anyone else would get the same result as you?

(Continued)

(Continued)

5. Create a coding sheet for collecting your data, including one column to identify your cases and one column to code your variable. List all the categories for your attribute, make sure they are mutually exclusive and exhaustive, and associate numbers with each category.

6. Code your data.

7. Enter the data into Excel.

8. Import your data into SPSS and clean them. Look for any anomalies (fix anything that looks wrong), identify the missing values, and label each of the values.

9. Request a frequency for your variable.

10. Type up a professional-looking table, indicating both the number of cases and percent for each category.

USE SPSS TO ANSWER A QUESTION WITH MEASUREMENT

If you want to answer a question by measuring a concept, you will go through all of the steps covered in this chapter. You will first define the variable conceptually. You will then translate the concept into a measurement, trying to maximize the validity and reliability of the measure. Next, you will take this operational definition and create a coding sheet, which you will then use to code this variable for each of your cases. Coding sheets in hand, you will then enter your data into an electronic database (either SPSS directly or Excel). From there, you can use SPSS to provide you with a basic frequency of the variable, which you can then use to create a professional-looking frequency for use in answering your question.

Transfer Your Data to an Electronic Database

Once you are done collecting the data, you can either enter them directly into SPSS or into an Excel worksheet. For both SPSS and Excel, the structure of this will be similar to your coding sheet: Each row will contain one case; each column, one variable. The first column will be your case identification. The second column will be the value of the attribute. I personally find it easier to enter data into Excel than SPSS, so that's what I'll describe. But you could also go into the data view window of SPSS and enter the data directly into the cells you see there. Each column is a different variable, originally named "V0001," "V0002," and so on. In each row, you take one case and enter the values of the variables in the corresponding column. If you switch to the "Variable View" of the Data Window (the tab to switch between "Data View" and "Variable View" is found in the lower left-hand corner of the window), you can change the names of the variables to be more descriptive. For example, you can change "V0001" to "Case Identifier."

The convenient aspect of Excel is that in the first row, you can insert labels for your variables. So cell "A1" will identify your unit of analysis, and "B1" will identify the attribute you have measured. When you import the file into SPSS, it is very easy to indicate that these are the names of the variables. After you have indicated the variable labels in the first row of your Excel file, you can start entering data in the second. In the first column, beginning in cell "A2," you can type in the names of your cases. In the second column, beginning in cell "B2," you will type the number associated with the category for the attribute you measured. You will end up with a spreadsheet that looks very similar to the coding sheet, except that it has only numbers in the second column. Be sure to save this file. If you haven't saved it, you won't be able to find or use it.

Although Excel can do some rudimentary statistics, in this textbook, I will describe how to do data analysis using SPSS (short for Statistical Package for the Social Sciences). This is the statistics package most commonly used by social scientists, whether they are in academic settings or in business or government agencies. As a result, learning to use SPSS can be fairly remunerative. I will focus on SPSS in order to achieve my goal of teaching you a marketable skill. Your university will probably have a site license for SPSS, so you should be able to do all of your work on campus.

Many students, though, prefer to do their work at home. Unfortunately, SPSS is not only popular, it is also expensive—expensive enough that it is normally purchased only by organizations. SPSS was purchased by IBM, which is now producing a student version you can rent for under $100. In its prior incarnation, the student version was not worth purchasing because it could analyze only a limited number of cases and variables. Those limitations no longer hold. The one drawback is that the rental license is only good for a period of time—either six or twelve months. If you search for "SPSS GradPack" with your search engine, you can find various vendors who handle the student rental process. Check all of these because the details vary by length of rental, the number of copies, whether you get disks or download electronically, and (most importantly) the price. Get version 19 or later to ensure compatibility. What they call the "Base" package is quite sufficient for your needs.

After you enter the data into an Excel file, you will need to import it into SPSS. Detailed instructions are found in Box 2.4. (You should notice that this book, focusing as it does on "How to do statistics," has a lot of "How to" boxes. These are collected at the back of the book in appendixes to help you find them later. All of the "How to Use SPSS" boxes are found in Appendix 2.) You first open SPSS by double-clicking on it, and then you open the Excel file that contains your data. There are two tricks to finding the Excel file. First, you need to be sure that you have saved it. Second, you need to be sure that the file type SPSS is looking for (at the bottom of the window) is set to Excel. By default, SPSS looks for an SPSS data file, so those are the only ones that show up. Under the window that allows you to navigate through your documents to find the data file, you will see two boxes: "File Name" and "Files of Type." The box next to "Files of Type" will be prefilled with the default file type "SPSS Statistics (*.sav)." Next to that default file type, you'll see a down arrow. Click on that to get the drop-down menu shown in Figure 2.3. Choose "Excel (*.xls, *.xlsx, *.xlsm)." Once you change the file type to "Excel," the Excel files will show up as available to you. Click on the Excel file so that its name enters the "File Name" box. Then click on "Continue."

> **BOX 2.4 How to Import an Excel Data File into SPSS**
>
> *Find "IBM SPSS Statistics" in your computer's programs and double click to open it. At this point, your first window opens called "Output [Document]." This is the window where any statistical results that you request will be reported. Superimposed on this window is a box asking if you want to open an existing data source. Since you want to import a new Excel file, you want to cancel out of this box.*
>
>   What would you like to do?
>   >Cancel
>   >File
>   >Open
>   >Data
>   *Find the appropriate folder*
>   Files of type = "Excel" (*.xls, *.xlsx, *.xlsm)
>   File Name = *younamedit.xlsx*
>   >Open
>   ✓Read variable name from first row of data
>   >OK
>
> *At this point, your second window opens called "Untitled [Dataset]." This window will show your data. But to be able to use it in the future, you will need to save the data into SPSS format. Make sure you are in the Data Window and do the following:*
>
>   >File
>   >Save As
>   File Name = *make sure you are in the right folder and give it a name*
>   >OK

    Once you have opened the data in SPSS, be sure to save it into an SPSS data file. SPSS opens two windows: the Output Window and the Data Window. To save the data, you need to save while you are in the Data Window. You'll know you are there because it will look a lot like an Excel worksheet. Once there, simply use the "File," "Save As" commands in the same way you would with Word.

Clean Your Data

    Every time you plan to analyze data, you should always make sure that it has been entered properly. We call the process of correcting any errors data cleaning. (Instructions can be

Figure 2.3 How to Find Excel Files to Open in SPSS

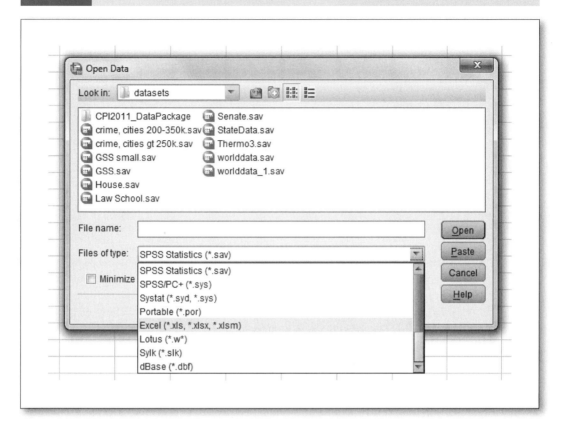

found in Box 2.5.) You can sometimes catch errors by looking over the raw data, and you definitely should do this. But you are actually more likely to find errors by looking at how the data are set up. In SPSS, go to the "Data" command and click on "Define Variable Properties." Identify the variables on which you want to work by highlighting them with a click and then clicking on the arrow. Once you have moved the variables over, click on "Continue" to process the request. You will now see a window that has a box on the left listing each of the variables you requested with the first variable highlighted. The box on the right will describe the highlighted variable by listing all the possible values and their frequencies. Look through all the possible values to see if they make sense. If there is a value that you know isn't possible for that variable, you need to fix this. Go into the Data Window, find the column for the variable, and then scan down to find the weird value. Once you find it, look to the left to find the case. With the case identifier, you can check the original coding sheet to find the correct value. Simply type that value into the appropriate cell of the Data Window.

> **BOX 2.5 Cleaning Your Data in SPSS**

>Data
>>Define Variable Properties
>>>*Click on each relevant variable to highlight it.*
>>>→Variables to Scan
>>>>Continue
>>>>>*For each variable, check that the values make sense; if not, fix in the Data Window.*
>>>>>>-9=Missing
>>>>>*Type in the value label in the box next to each number.*
>>>>OK

After assuring yourself that all the values make sense, give labels to each of the values. In the "Define Variable Properties" window, one of the columns is labeled "Missing." You'll want to click on the appropriate box to set that value to missing. If you followed my example and made "–9" missing, you'll want to click on the missing box in the row for the value of "–9." Of course, if you do not have missing data for any of your cases, "–9" will not appear as a value and so you will not be able to set a missing value. To the right of the numerical values, you can enter labels to correspond with each number. So, for the gender variable, if you coded men as "0" and women as "1," you can type "men" next to "0" and "women" next to "1." SPSS has some odd quirks with adding in variable labels. Later on, when you are working with more than one variable, you might have problems if you label from the top down. So it is probably a good idea at this point to get in the habit of assigning a label to the last category first and working your way up to the top of the categories. Once you have labeled all of the categories for the first variable, move on to the second. Identify the missing values and then start assigning labels at the bottom and work your way up. Systematically work your way through each of the variables in the box on the left, identifying the missing values and labeling each possible value.

Get a Frequency Table

After you've made your corrections, request a frequency for each of your variables to verify that everything has been fixed. A frequency is a table that shows the result of totaling the number of cases in each category of your variable. The command to analyze data is at the top of both the Data Window and the Output Window—it doesn't matter which you use. To get a frequency, the procedure is to click on "Analyze," "Descriptive Statistics," and "Frequencies." At this point, you will see a list of the variables in your dataset in a box on the left. If you click on a variable to highlight it and then click on the arrow, the variable name will move into the "Variable" box. Once you have all your desired variables in the variable box, you can click on "OK" to process your request. The frequency table will then show up in the Output Window. A summary of this process is found in Box 2.6. If everything in the frequency looks correct, go into the Data Window and save your data file. You do this in much the same way as you would in Word—find "File" in the upper left-hand corner and "Save" a file with a name and in a place you will remember in the future.

> **BOX 2.6 How to Get a Frequency of a Variable in SPSS**
>
> *After opening your data in SPSS:*
>
> >Analyze
> >>Descriptive Statistics
> >>>Frequencies
> >>>>*Click on the variable to highlight it.*
> >>>>→*To bring the variable over to the "Variable" box*
> >>>>OK
>
> *At this point, the frequency of the variable will show up in the "Output" window.*

These steps are all preparatory to actually answering a question. In order to answer a question, you'll need to translate the final (clean) frequency into a more usable form. First, you're going to want to copy the frequency table from SPSS and paste it into a Word document. Although you can save the Output Window for future reference, it is only readable by SPSS. So if you are going to want to access your frequency at home, I recommend opening a Word document at this point, copying the frequency from the Output Window, and pasting it into the Word document.

The SPSS frequency, however, is not the professional-looking table you will want to use when presenting your findings. Professional-looking tables follow certain protocols you will want to emulate so that your reader takes your results seriously. The SPSS frequency has all the information you need, but it is only a resource from which you will borrow the appropriate numbers as you are creating your own frequency table. Follow the instructions in the chapter (summarized in Box 2.2) to use Word to create a professional-looking frequency table.

Now that you have collected your data in a professional-looking table, you can write a professional memo answering your question. Chapter 1 described the elements of a memo. The box summarizing that process is found in Appendix 1. As with all writing, memos should have an introduction, a body, and a conclusion. Unlike other writing, brevity (combined with clarity) is highly valued in memos, so do your best to keep it short.

An SPSS Application: Europe and the European Union

It is November 2010. Ireland's economy has been tottering, but the government has been refusing pressure from the European Union (EU) to accept a bailout. Prime Minister Brian

Cowen, afraid of a vote of no confidence, has been claiming that an EU loan is unnecessary—that his planned austerity measures will be enough for Ireland to regain its financial footing. But on November 21, Finance Minister Brian Lenihan announces that Ireland has formally applied to the EU and the International Monetary Fund for a bailout package. You are interning in the Congressional Joint Economic Committee, and today you have been charged with dealing with constituent emails. You receive the following query:

To: Joint Economic Committee
From: Alice Concerned
Date: November 22, 2010
Re: European Bailouts

I just heard in the news that the EU is going to bail out Ireland. But last summer, they allowed Iceland to go bankrupt. Why didn't they bail out Iceland, too? I am of Scandinavian descent, and it sure sounds like the EU has a prejudice against northern climes.

When you show it to your supervisor and ask her what to do, she says, "Reply to it, of course." Noticing your look of panic, she gives you a hint: "Why don't you put together a table of how many European nations are members of the European Union?"

1. Identify the unit of analysis.
   I want to identify which countries are in Europe. I could look up Europe in a dictionary, find its boundaries on a world map, and then look for all of the countries within those boundaries. But my dictionary doesn't have an entry for Europe. (Is that weird or what?) And aren't there some European countries the size of cities that wouldn't show up on a map? I would be liable to miss some. So instead, I decide to look up "Europe" on Wikipedia—it is sure to have a list. After doing that, I am glad I went that route because it indicates that even defining the concept of "country" isn't as straightforward as I had assumed it would be. It actually has several categories: "recognized sovereign states," "partially recognized sovereign states," "unrecognized sovereign states," and "dependent territories." I choose to look only at the "recognized sovereign states" because that is the closest to how I had conceptualized a country. I might need to revise that later if it ends up that some EU nations fall into one of the other categories.

2. Conceptually define your variable: What does it means to be a member of the European Union?
   The European Union is a formal association between various countries in Europe based on a signed treaty. Each country retains its political and military autonomy, but commits to a level of economic interrelatedness. Most EU nations share a common currency, which means that each individual country's economy is dependent on the economies of all the other EU nations. As a result, there are prerequisites that a country has to achieve in order to be allowed into the organization. And once it joins, the treaty specifies certain responsibilities that each country needs to uphold.

3. Operationally define what is meant by being a member of the EU: How will you measure it?

   A member of the EU is any nation that is currently a signatory of the treaty. This is a valid measure because it is the technical definition—a country cannot be a member without signing.

   For ease of data collection, I am going to find the list of member states on Wikipedia. But because I've been told that Wikipedia is not always a reliable source of information, I decide to double-check by looking at the EU's website.

4. Evaluate the validity and reliability of your variable.

   Because both sources are publicly available, anyone can follow the same procedures and get the same results. As a result, I can be confident that my coding will be reliable.

5. Create a coding sheet for collecting your data.

Table 2.1 EU Coding Sheet, Blank

Coding Sheet	
Which European Countries are Members of the European Union?	
European Country	*Member of the European Union?*
	☐ 0 No ☐ 1 Yes
	☐ 0 No ☐ 1 Yes
	☐ 0 No ☐ 1 Yes
	☐ 0 No ☐ 1 Yes
	☐ 0 No ☐ 1 Yes
	☐ 0 No ☐ 1 Yes
	☐ 0 No ☐ 1 Yes
	☐ 0 No ☐ 1 Yes
	☐ 0 No ☐ 1 Yes

Sources: "List of Sovereign States and Dependent Territories in Europe," Wikipedia, The Free Encyclopedia. http://en.wikipedia.org/w/index.php?title=List_of_sovereign_states_and_dependent_territories_in_Europe&oldid=494486538. "Member States of the European Union," *Wikipedia, The Free Encyclopedia,* May 23, 2012. "Countries," *European Union,* 2012. http://europa.eu/about-eu/countries/index_en.htm.

Table 2.1 shows a truncated version of my coding sheet. The first column leaves space to write in the names of the European countries identified in (A). The second column contains all the possible categories for EU membership: either yes it is, or no it is not a member. Each of those categories is assigned a numerical value. So "no" it is not a member of the EU will be coded "0," and "yes" it is will be coded "1." There are check boxes next to the number so that as I am coding the data, I can just check the appropriate box. The coding sheet also includes any information a coder might need—in this case, where one should go to find the data.

6. **Code your data.**
   Table 2.2 shows a truncated version of the coding sheet with the data.

Table 2.2 EU Coding Sheet, Completed

Coding Sheet
Which European Countries Are Members of the European Union?

European Country	Member of the European Union?
Republic of Albania	☑ 0 No ☐ 1 Yes
Principality of Andorra	☑ 0 No ☐ 1 Yes
Republic of Armenia	☑ 0 No ☐ 1 Yes
Republic of Austria	☐ 0 No ☑ 1 Yes
Republic of Azerbaijan	☑ 0 No ☐ 1 Yes

Sources: "List of Sovereign States and Dependent Territories in Europe," *Wikipedia, The Free Encyclopedia,* May 28, 2012. http://en.wikipedia.org/w/index.php?title=List_of_sovereign_states_and_dependent_territories_in_Europe&oldid=494486538. "Member State of the European Union," *Wikipedia, The Free Encyclopedia,* May 23, 2012. "Countries," *European Union,* 2012. http://europa.eu/about-eu/countries/index_en.htm.

7. **Enter your data into Excel.**
   Entering data into Excel is really pretty easy. I just need to be sure to make each column corresponds to a different variable. In Excel, the first row in each column should name the variable. The first column is the case identifier; in this instance, it is the name of the European country, so this column is headed "Country." The second column is the variable about European Union membership, so it is headed "EU?" Each line in the file will look at a single case and include all the data for every variable in the appropriate column. Figure 2.4 shows what this file looks like. I am careful to save this Excel file ("File" "Save As") before trying to do anything with it.

Figure 2.4 Excel File of EU Data

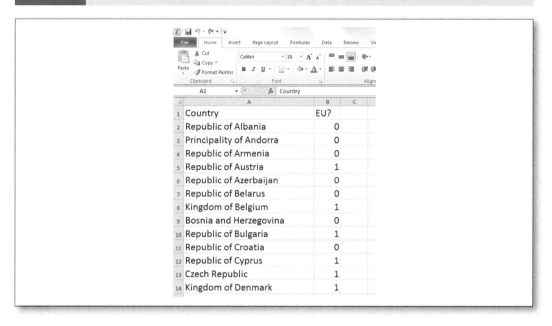

8. Import your data into SPSS and clean it. Look for any anomalies (fix anything that looks wrong), identify the missing values, and label each of the values.

Figure 2.5 Opening an Excel File in SPSS

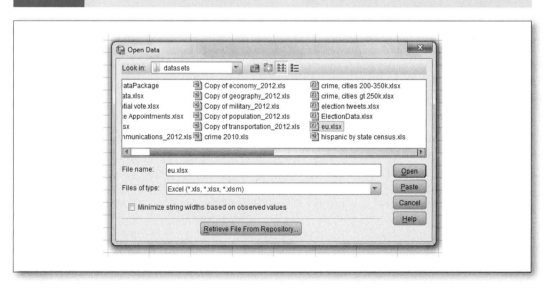

First, I open up SPSS. Once in there, I click on "File," "Open," and "Data" to open a data file. I find the location on my computer where I saved my Excel file. (Note: the file won't show up until you have saved it.) Once I got to the appropriate folder, however, only SPSS/PC+ files showed up because that is the default. At the bottom of the window, I changed the "Files of type" to indicate Excel as shown in Figure 2.5. Now Excel files show up in the "Open Data" box, so I find my file and click on it to enter it into the "File Name" box. I then click on "Open" to actually open the file in SPSS.

Figure 2.6 Indicating the First Line Contains the Variable Names

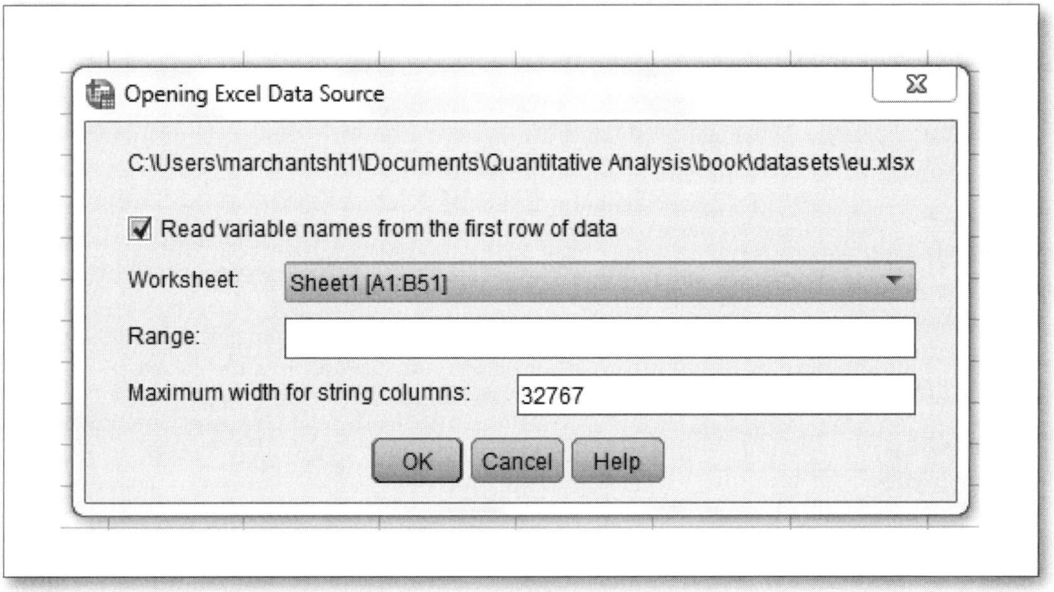

After opening the right file, a message asks if the first line includes the variable labels. Because the Excel file has the labels "European Country" and "EU?" rather than data in my first line, I leave the box checked and reply "OK" as shown in Figure 2.6. SPSS then opens two windows. In the Data Window, I see the data looking very much like my Excel file. Figure 2.7 shows what the Data Window looks like. While I am in the Data Window, I save my data in SPSS format with the standard "File," "Save As" command. (If I were to check my directory at this point I would notice that a file has appeared named "EU.sav." The ".sav" extension indicates that it is an SPSS data file.) In the Output Window, I see all of my commands along with their results. In the top of both windows are commands to get SPSS to perform statistical functions on my data.

Figure 2.7 SPSS Data Window for EU Data

	Country	EU	var	var
1	Republic of Albania	0		
2	Principality of Andorra	0		
3	Republic of Armenia	0		
4	Republic of Austria	1		
5	Republic of Azerbaijan	0		
6	Republic of Belarus	0		
7	Kingdom of Belgium	1		
8	Bosnia and Herzegovina	0		
9	Republic of Bulgaria	1		
10	Republic of Croatia	0		
11	Republic of Cyprus	1		
12	Czech Republic	1		
13	Kingdom of Denmark	1		

To clean the data, I use the "Data," "Define Variable Properties" command as shown in Figure 2.8. After designating my EU variable for scanning, I can assign missing values (which I do not have in this case) and the value labels. Figure 2.9 shows where to enter this information. I am careful to begin at the bottom when entering the value labels.

Figure 2.8 The Define Variable Properties Command, Step 1

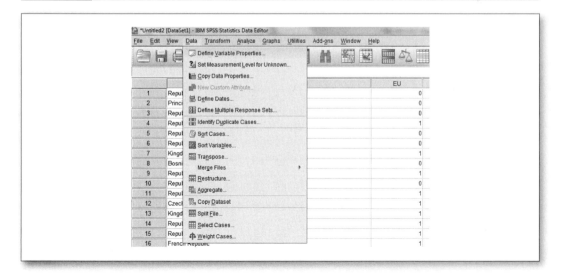

Figure 2.9 The Define Variable Properties Command, Step 2

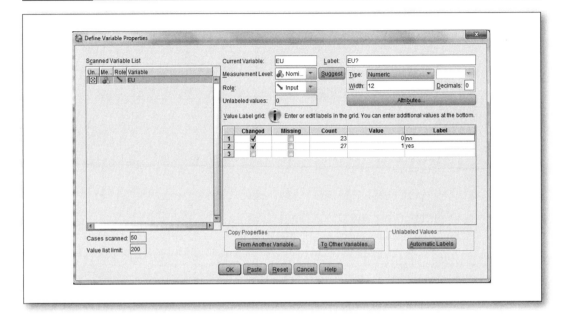

9. **Request a frequency for your variable.**
   I want a frequency, so I click on "Analyze" (in either of the windows), choose "Descriptive Statistics," and then choose "Frequencies." At this point, a window opens with two boxes and an arrow between them. The left box has a list of my variables; the right is empty. I click on "EU?" because that is the variable for which I want a frequency. I click on the arrow and "EU?" moves into the right-hand box. At this point, I click on "OK" to get my frequency. It shows up in the "Output" window. Figure 2.10 shows what this looks like.

10. **Type up a professional-looking table indicating both the number of cases and percent for each category.**
    To have access to the frequency from my computer without SPSS, I copy the components of the Output Window into a Word document. Each component is in a table format, so I copy each component and then paste it into Word. When exiting SPSS, I am sure to save the data in the Data Window as an SPSS file, but do not bother saving the output in the Output Window.

    Because it is not professional to cut and paste the printout of a statistical package into written work, I transfer the information from the SPSS output into a more professional format that looks like Table 2.3. Notice several key features of the table: (1) It has a title describing its content. (2) The first column describes the categories of the variable. (3) The second column gives the frequencies of those categories. (4) The last column gives the percentages. (5) The last row gives the total for the columns. (6) Each section of the table is divided by lines. (7) The data source and any other clarifying information are found after the last line.

Figure 2.10 SPSS Output Window of Frequency for EU Data

Frequencies

[DataSet1]

Statistics

EU?

N	Valid	50
	Missing	0

EU?

		Frequency	Percent	Valid Percent	Cumulative Percent
Valid	no	23	46.0	46.0	46.0
	yes	27	54.0	54.0	100.0
	Total	50	100.0	100.0	

Table 2.3 Frequency of Membership in the European Union by European Countries

Member of EU?	Frequency	Percent
No	23	46.0
Yes	27	54.0
Total	50	100.0

Sources: "List of Sovereign States and Dependent Territories in Europe," *Wikipedia, The Free Encyclopedia*. http://en.wikipedia.org/w/index.php?title=List_of_sovereign_states_and_dependent_territories_in_Europe&oldid=494486538. "Member State of the European Union," Wikipedia, The Free Encyclopedia, May 23, 2012. "Countries," *European Union*, 2012. http://europa.eu/about-eu/countries/index_en.htm.

11. Write a professional memo to Ms. Concerned answering her question and include a table at the end with the data you collected.

Memo

To: Alice Concerned
From: T. Marchant-Shapiro, Joint Economic Committee
Date: November 23, 2010
Subject: European Bailouts

We received your email wondering why the European Union agreed to a bailout for Ireland but allowed Iceland to go bankrupt. It is a very good question, especially in the face of the current economic crisis, which has made apparent the interrelated nature of the economies of the nations of the world. To answer your question, I collected data about which European countries are members of the European Union. The EU treats Ireland and Iceland differently because Ireland is a member of the EU and Iceland is not.

The European Union is the result of a treaty between specific European countries. In particular, these nations have forged strong economic ties and a common currency. Because of this interdependence, the EU is very careful about allowing countries to join: They must meet strict preconditions in order to join as well as make commitments about future economic behavior. As the table below shows, slightly more than half of the countries in Europe are actually members of the European Union.

That is the crux of the difference in the cases of Iceland and Ireland. Iceland is not a member of the EU and so it did not have the responsibilities of membership. And the EU did not have the responsibility to keep Iceland's economy afloat. Ireland, on the other hand, is a member of the EU and so, although its government actually did not want to accept the outside loans, it eventually buckled to pressure from its allies to accept the bailout.

You can expect that in the future, as in this instance, the European Union will bail out only its member states. Thank you for your interest in this issue. It is concerned citizens like you that keep our country strong.

Frequency of Membership in the European Union by European Countries

Member of EU?	Frequency	Percent
No	23	46.0
Yes	27	54.0
Total	50	100.0

Sources: "List of Sovereign States and Dependent Territories in Europe," *Wikipedia, The Free Encyclopedia*, May 28, 2012. http://en.wikipedia.org/w/index.php?title=List_of_sovereign_states_and_dependent_territories_in_Europe&oldid=494486538. "Member State of the European Union," Wikipedia, The Free Encyclopedia, May 23, 2012. "Countries," *European Union*, 2012. http://europa.eu/about-eu/countries/index_en.htm.

Your Turn: Measurement

YT 2.1 After the economic downturn of 2008, one aspect of the public debate was about what to call it.

1. Give a conceptual definition of an economic "recession."

2. In order to know whether or not we are in a recession, the operational definition needs to indicate both what is and what is not a recession. Find a reputable source and give its operational definition of a recession.

3. During this downturn, during what span of time was the United States in a recession according to this measure?

CHAPTER 2 Measurement: Counting the Biggel-Balls

YT 2.2

One of the questions political scientists like to ask is, "Why do countries go to war?" Before you can even begin to answer it, though, you face a sticky measurement problem. How do you decide whether a country is at war or not?

1. Identify your unit of analysis.
2. Describe a way to determine whether a country is at war or not.
3. Evaluate the quality of that measure: Is it valid? Is it reliable?

YT 2.3

Determine the level of measurement (nominal, ordinal, or interval/ratio) for each of the following variables:

1. Education (in years)
2. Level of Education (no high school diploma, high school graduate, some college, college graduate, some graduate school)
3. Graduate Degree (MA/MS, PhD, JD, MBA, MD)
4. Race (White, Black, Hispanic, Asian, Other)
5. Marital Status (married, civil union, widowed, divorced, single)
6. Abortion (number of instances in which abortion should be legal: 0–7)
7. Income (in dollars)
8. Social Class (working class, middle class, upper class)

Apply It Yourself: Measure the Norm for Chief Justice Appointments

The year is 2005. Chief Justice William Rehnquist has passed away and President George W. Bush has nominated John Roberts to replace Rehnquist both for his seat on the bench and as chief justice. You are interning in the Senate Judiciary Committee and your boss, Committee Chair Arlen Specter, knowing that the learning curve is very steep just for becoming a justice, wonders whether it is practical to become chief justice at the same time. But perhaps that's the norm for chief justice appointments? He asks you to find out how many chief justices were appointed simultaneously to being placed on the Court.

1. The unit of analysis for your assignment is Supreme Court chief justices. How will you find all your cases?

2. Your variable is a combination of two concepts: appointment to the Supreme Court and appointment as chief justice. Define these concepts.

3. How will you measure them? Think about any problems you might have and predetermine how you will resolve them. For example, will you include the first chief justice? (By necessity, he must have been appointed simultaneously.) What will be the source of the data?

4. Does this measure give you valid data? How can you check to see if the data are reliable?
5. Create a coding sheet to record your data. You should have one variable for the name of the chief justices (which can be an alphanumeric variable) and one for your appointment variable (which should be a numerical variable). Be sure to indicate the source of the data.
6. Code the data. If you realize that you need to make adjustments in your operationalization, adjust point 3 above. If you do refine your measure, be sure to review your previous codings to make sure that the refinement does not affect how they should be coded.
7. Enter the data into Excel.
8. Import the Excel file into SPSS and clean the data.
9. Get a frequency for your Chief Justice variable.
10. Translate the output into a professional-looking table. How many chief justices had already been on the Supreme Court? How many were appointed simultaneously?
11. Write a brief memo for Senator Specter answering his question. Follow the format described for writing memos, and be sure it contains a properly formatted frequency table at the end.
12. If you were actually writing a memo for an assignment from work, you would not include any SPSS printout with your memo, but this semester you should attach copies of the relevant SPSS tables to your memos for grading purposes. For this assignment, turn in your memo along with your completed coding sheet, a printout of your spreadsheet, and a printout of your SPSS frequency.

Key Terms

Absolute zero (p. 23)
Associational (correlational) validity (p. 26)
Case (p. 19)
Coding (p. 29)
Coding sheet (p. 29)
Conceptual definition (p. 20)
Consensual validity (p. 25)
Cross sectional surveys (p. 28)
Data cleaning (p. 36)
Exhaustive (p. 30)
Face validity (p. 25)
Frequency (p. 38)
Indicator (p. 22)

Interval (p. 23)
Level of measurement (p. 22)
Mutually exclusive (p. 30)
Nominal (p. 22)
Ordinal (p. 22)
Operational definition (p. 20)
Panel study (p. 28)
Predictive validity (p. 26)
Ratio (p. 23)
Reify (p. 22)
Reliability (p. 26)
Unit of analysis (p. 19)
Validity (p. 25)

CHAPTER 3

Measures of Central Tendency

That's Some Mean Baseball

In 1989, Phil Robinson directed *Field of Dreams*. Possessed of both a BA in Political Science (which is interesting, but not relevant) and a fascination for baseball (which is relevant, but not unique), Robinson had taken the storyline from the book *Shoeless Joe*[1] and adapted it for the screen. *Field of Dreams*[2] is a fantasy about the redemptive power of baseball. Although the redemption of the movie was fictional, Shoeless Joe Jackson actually was an amazing baseball player from a century ago. His career batting average of 0.356 still places him third on the list of all-time great hitters.[3] And Jackson actually may have needed forgiveness—for the sin of throwing the 1919 World Series as part of a scandal that year that caused the White Sox to be nicknamed the Black Sox.

The interesting thing is that Joe Jackson was such an amazing baseball player that it is very difficult to prove that he helped fix the World Series based on his batting statistics. If Jackson was part of the scandal, we would expect his batting record during the World Series to be worse than his career record. We already know that Shoeless Joe had a career batting average of 0.356. What was his batting average during the World Series? Table 3.1 shows the batting record for Jackson in the eight games of the 1919 World Series. In the third and fourth columns, I have separated out the number of hits and at bats for each game. We could eyeball his batting record for these games and see that he usually hit about half the time. But we want to be more precise in measuring Jackson's batting average for the series.

For statistical purposes, we need the data organized in a slightly different way. Instead of focusing on how Jackson did in each game, our unit of analysis is each time Joe Jackson was at bat. I set up my coding sheet so that each time at bat gets its own row and the variable (X) is whether or not Jackson hit the ball that time at bat. I code the variable so that "0" is assigned when Jackson did not hit the ball and "1" when he did. The data are found in Table 3.2.

Table 3.1 Batting Record for Joe Jackson, 1919 World Series

Game	Batting Record	Number of Hits	Times at Bat
1	0/4	0	4
2	3/4	3	4
3	2/3	2	3
4	1/4	1	4
5	0/4	0	4
6	2/4	2	4
7	2/4	2	4
8	2/5	2	5

Source: "1919 World Series," *Shoeless Joe Jackson's Virtual Hall of Fame*. www.blackbetsy.com/1919WorldSeries.html.

Table 3.2 Batting Record for Joe Jackson, 1919 World Series, Coding Sheet

Game	Time at Bat Case	Hit? X	Game	Time at Bat Case	Hit? X
1	1	0		18	0
	2	0		19	0
	3	0	6	20	0
	4	0		21	0
2	5	1		22	1
	6	1		23	1
	7	0	7	24	1
	8	1		25	0
3	9	1		26	1
	10	0		27	0
	11	1	8	28	0
4	12	1		29	1
	13	0		30	0
	14	0		31	1
	15	0		32	0
5	16	0		$N = 32$	$\Sigma \text{ hits} = \Sigma X = 12$
	17	0			

Source: "1919 World Series," *Shoeless Joe Jackson's Virtual Hall of Fame*. www.blackbetsy.com/1919worldseries.html.

Note: Hit = 0 if no hit; hit = 1 if yes hit.

To calculate the batting average for the entire World Series, I add up the total number of hits that Jackson made, and divide by the number of times he was at bat. In statistical terms, my variable X is whether or not Jackson hit the ball. I write the total number of hits as ΣX. (Σ is the upper case version of the Greek letter sigma. It is also called the summation sign because when you see ΣX it, you sum all of the values of X.) Jackson's being at bat is the case or unit of analysis. As statisticians, we talk about the number of cases (in this instance, the number of times Jackson was at bat) as "N." Mathematically, the process we went through of adding up the number of hits and dividing by the number of at bats is written as

$$Batting\ Average = \frac{\Sigma X}{N}$$

$$= \frac{0+0+0+0+1+1+0+1+1+0+1+1+0+0+0+0+0+0+0+0+1+1+1+0+1+0+0+1+0+1+0}{32}$$

$$= \frac{12}{32}$$

$$= 0.375$$

A baseball fan would say, "Shoeless Joe batted .375." What they mean by that is that when Joe was up to bat, he hit the ball 37.5 percent of the time. But the equation isn't just the equation for a batting average; it is the equation for any arithmetic average, although in statistics, we call it the mean.

Based on our calculations, we know that for the 1919 World Series, Shoeless Joe Jackson had a batting average of 0.375. How does that compare with how well Jackson usually played? Jackson's career batting average was 0.356, so he actually exceeded his career performance in the World Series. His batting average does not suggest that Shoeless Joe threw the 1919 World Series. But on the basis of other evidence, Joe Jackson and seven of his teammates were banned from organized baseball, and Major League Baseball established a Commissioner of Baseball who is still charged with overseeing the ethics of all league games. Based on the statistics, though, many Jackson fans still argue that Shoeless Joe was actually not a party to the 1919 fix and campaign to get him instated into the Baseball Hall of Fame.[4]

MEASURES OF CENTRAL TENDENCY

Just as Joe Jackson's batting average describes how well he usually batted, statistical averages describe the normal value of a particular variable. By definition, different cases will have different values of a variable. The question here is: What is the central tendency of that variation? On average, where do the cases usually lie? There are three different measures of central tendency: the mean, the median, and the mode. Because all three averages measure the central tendency of variables in different ways, they might look at the same data and compute a different average. Which average best describes the middle of the data will depend on the level of measurement of the variable as well as the way the data are arranged around the middle.

Mean

Like the batting average we calculated for Shoeless Joe, when we use the word "average," we are usually referring to the statistical mean. Frequently, we calculate the mean from raw data, like we did for Jackson's batting average. But as social scientists, we usually have so many cases that calculating the mean from raw data is cumbersome. So after reviewing how to calculate the mean from raw data, we'll learn how to calculate it from tabular data.

Calculating the Mean from Raw Data. Baseball fans use the same procedure to calculate a batting average that we as political scientists use to calculate the mean of political variables. In general terms, to get the mean we divide the sum of the values of our variable by the number of cases. Mathematically, we write

$$\mu = \frac{\sum X}{N}$$

You'll find throughout this book that statisticians frequently use symbols to stand in for concepts, and frequently, those symbols are Greek. In this case, the symbol that statisticians use for the mean for the data from an entire population (as opposed to a sample) is the Greek letter μ (pronounced myoo). This equation tells you to that in order to find the mean, you sum all of the values of X and then divide by the total number of cases. If you are working with raw data, the easy way to find N is simply to count the number of rows of data.

Take, for example, the death penalty. The January 2011 shooting of Representative Gabby Giffords by Jared Loughner in Arizona resurfaced discussions about the prevalence of the use of the death penalty within the different jurisdictions of the United States. Loughner was subject to prosecution in both the federal court system (for the attempted assassination of federal employees) and in the Arizona state court system (for the murders of various onlookers). At the time of the shooting, some people asked the question of how likely it was that Loughner would be executed in each of those two systems. Although we won't actually compare the state and federal execution rates, we will examine how often states usually execute prisoners. Table 3.3 shows the states that executed prisoners in 2010 along with their frequencies.

In 2010, twelve states executed inmates so $N = 12$. We get the total number of executions by adding up the number of executions in each state—this totals 46. So substituting into the equation, we get

$$\text{Mean number of executions} = \frac{\sum 5+1+1+2+1+3+8+3+17+1+3+1}{12}$$

$$= \frac{46}{12}$$

$$= 3.83$$

Among the states that executed inmates in 2010, the average rate was 3.83 executions.

Calculating the Mean from Tabular Data. Calculating statistics is relatively easy when you have only twelve cases, but gets harder when you have more. For example, if instead of being interested in the execution rate for the single year of 2010, I wanted to know the average for a decade (say 2000 through 2009), then the math would get more complicated. If I set up the raw data in a table, I get Table 3.4, where the first column identifies each case and the second column gives the number of executions performed in the first decade of the twenty-first century.

Table 3.3 State Death Penalty Executions in 2010

State Case	Number of Executions X
Alabama	5
Arizona	1
Florida	1
Georgia	2
Louisiana	1
Mississippi	3
Ohio	8
Oklahoma	3
Texas	17
Utah	1
Virginia	3
Washington	1
#rows = N = 12	Σexecutions = ΣX = 46

Source: Halperin, Rick, "Death Penalty News and Updates." January 27, 2012. http://people.smu.edu/rhalperi/.

Table 3.4 State Death Penalty Executions, 2000-2009, Raw Data

State	Number of Executions	State	Number of Executions
Alabama	25	Kentucky	1
Arizona	4	Louisiana	2
Arkansas	6	Maryland	2
California	6	Mississippi	6
Connecticut	1	Missouri	26
Delaware	4	Montana	1
Florida	24	Nevada	4
Georgia	23	New Mexico	1
Indiana	13	North Carolina	28

(Continued)

Table 3.4 (Continued)

State	Number of Executions	State	Number of Executions
Ohio	32	Texas	248
Oklahoma	72	Virginia	32
South Carolina	18	Washington	1
South Dakota	1	$N = 26$	$\Sigma X = 587$
Tennessee	6		

Source: Halperin, Rick, "Death Penalty News and Updates." January 27, 2012. http://people.smu.edu/rhalperi/.

I could use the raw data in Table 3.4 to calculate the mean by adding up the total number of executions and dividing by the number of states that performed them. The problem is that I am liable to make a mistake if I add up too many numbers. One way to cut down on the probability of making a mistake is to simplify the data by making a properly formatted frequency using the guidelines from page 32 in the previous chapter (in Box 2.2). To make this collapsing process easier, I want to reorder Table 3.4 so that it is ordered by number of executions rather than being alphabetized by state. In fact, at this point, I don't even need to know which state is connected to which execution rate. Instead, I simply want to order all the numbers of executions from least to most.

1, 1, 1, 1, 1, 1, 2, 2, 4, 4, 4, 6, 6, 6, 6, 13, 18, 23, 24, 25, 26, 28, 32, 32, 72, 248

With the values in order, it is easy enough to see that there are six ones, two twos, three fours, and so on. Because my goal is to make the data simpler to work with, I want one line for each value of my variable. I collapse all six of the states with one execution into the first line, both of the states with two executions into the second line, and so forth. If I do that, I end up with Table 3.5. Notice that I no longer have a column for the state name. Also notice the

Table 3.5 State Death Penalty Executions, 2000–2009, Collapsed

Number of Executions X	Number of States Frequency f_x
1	6
2	2
4	3
6	4
13	1
18	1
23	1
24	1
25	1
26	1
28	1
32	2
72	1
248	1

Source: Halperin, Rick, "Death Penalty News and Updates." January 27, 2012. http://people.smu.edu/rhalperi/.

CHAPTER 3 Measures of Central Tendency: That's Some Mean Baseball

statistical names for each of the columns. I always call my variable X and I call the frequency of that variable f_x.

In order to use this frequency to calculate the mean execution rate, I need to find both the total number of executions and the total number of states. Unfortunately, I can't find those two numbers from Table 3.5 using the same technique as in the previous section: I can no longer get N by counting the number of rows, and I can no longer get the total number of executions by summing the execution column. But the data are still there to calculate both of those numbers; I do this in Table 3.6. I get the total number of states by summing the "Number of States (f_x)" column; this is N. The next step is to calculate the total number of executions. To do this, I've inserted an extra column into the table that multiplies the number of executions (X, found in the first column) by the number of states that had that number of executions (f_x, found in the second column). In the original data, I had six states that executed one person. In Table 3.4, I would have added one six times to include all of those executions. But adding six ones together is the same as multiplying one by six. So in the third column of Table 3.6, I calculate $X(f_x)$ to get the total number of executions in each row. Then I can just add up the values in that column to get the total number of executions.

Table 3.6 State Death Penalty Executions, 2000–2009, Frequency

Number of Executions X	Number of States f_x	Total Executions (#Executions)(#States) $X(f_x)$
1	6	1·6 = 6
2	4	2·2 = 4
4	3	4·3 = 12
6	4	6·4 = 24
13	1	13·1 = 13
18	1	18
23	1	23
24	1	24
25	1	25
26	1	26
28	1	28
32	2	32·2 = 64
72	1	72
248	1	248
	$N = \Sigma f_x$ = 6 + 2 + 3 + 4 + 1 + 1 + 1 + 1 + 1 + 1 + 1 + 2 + 1 + 1 = 26	$\Sigma \text{executions} = \Sigma X(f_x)$ = 6 + 4 + 12 + 24 + 13 + 18 + 23 + 24 + 25 + 26 + 28 + 64 + 72 + 248 = 587

Source: Halperin, Rick, "Death Penalty News and Updates." January 27, 2012. http://people.smu.edu/rhalperi/.

With the total number of cases and the total number of executions from Table 3.6, we can now calculate the mean number of executions from the equation for a mean:

$$\text{Mean number of executions} = \frac{\Sigma \text{executions}}{N}$$

$$= \frac{587}{26}$$

$$= 22.58$$

We now know that states that executed someone between 2000 and 2009 averaged 22.58 executions in this ten-year period.

Median

The mean of 22.58 highlights one important feature of the mean as a measure of central tendency. If you look at the data, you will notice that only nine states executed more than the mean; seventeen executed less than average. If two thirds executed less than average, is that really an average? The problem here is that Texas, with its 248 executions, is dominating the math—we call that an outlier. Outliers can really skew our calculations, so if we have them, it is helpful to look at alternative measures of central tendency. An additional limitation of using the mean as a measure of central tendency is that it is meaningful only if the numbers connected to the values are on a scale. If the data are skewed or if the variable has an ordinal rather than an interval or ratio level of measurement, a more appropriate measure of central tendency is the median.

Calculating the Median from Raw Data. Intuitively, when I looked at the death penalty data, I assumed that the average should be in the middle of the data. An alternative measure of central tendency, the median, does just that. The median is the value of the middle case. George Carlin is referring to the median when he talks about the average intelligence of people: "Think of how stupid the average person is, and realize half of them are stupider than that."[5] In order to calculate the median, you take a set of data, arrange it from least to most, and find the number that is in the middle. In Table 3.3, the number of executions in each state in 2010 was

5, 1, 1, 2, 1, 3, 8, 3, 17, 1, 3, 1

If you arrange those in order from smallest to largest, you get

1, 1, 1, 1, 1, 2, 3, 3, 3, 5, 8, 17

(When you are doing this, always check to make sure that you have the same number of cases as you started out with! It is really easy to make a mistake. In this list, I still have twelve cases, so I should be all right.) To find the middle value, I can put one finger on each end of the list and slowly move to the middle to find the middle value. If I do that with these data, my fingers come together with my left finger on "2" and my right finger on "3." Because I have an even number of cases, there actually isn't a middle case. Technically, the median is the interval from 2 to 3. Normally, though, we take the mean of the two values on each side of the median: $\frac{2+3}{2}$ = 2.5. For the data in Table 3.3, 2.5 is probably a more reasonable measure of the average—once again, Texas is an outlier, so the mean of 3.83 we calculated really is a bit high to reflect how many executions occurred per state.

If you have more than a dozen cases, the two-finger approach to finding the median can get inefficient. The technical way to identify the median case is to add one to the number of cases and divide by two.

$$\text{Median Case} = \frac{N+1}{2}$$

For the 2010 executions, that would be $\frac{12+1}{2}$ = 6.5. So that formula correctly indicates that we chose a value between the values of the sixth and seventh cases on our ordered list. (Adding one to the number of cases makes sense—if you just divided 12 by 2, you would get 6. If you then counted in 6 from the left, you would get a median of 2, whereas if you counted back 6 from the right, you would get a median of 3. That just doesn't make sense. If it is the middle value, it shouldn't matter if you count left or right, the middle should be the middle. So you must add one before you divide by two so that you actually do find the middle.) Counting up to find the sixth and seventh cases, we still get values of 2 and 3. Taking the average of 2 and 3 still produces a median of 2.5.

Calculating the Median from Tabular Data. When we calculated the mean from tabular data, we had to add a column. In the same way, when working with tabular data, adding two columns helps us to find the median. Table 3.7 takes the frequency from Table 3.5 and adds the appropriate columns. Remember that the median case is given by the equation

$$\text{Median Case} = \frac{N+1}{2}$$

$$= \frac{26+1}{2}$$

$$= 13.5$$

Because an even number of states executed inmates in the 2000 decade, we're going to need to take the average between the executions of the thirteenth and fourteenth cases. Identifying the median case isn't enough. We need to find the value of our variable connected to the median case.

To find the thirteenth and fourteenth cases on Table 3.7, I add in a column that gives the cumulative number of states for each category. We begin with the six states that have

Table 3.7 State Death Penalty Executions, 2000–2009, Cumulative

Number of Executions X	Number of States f_x	Cumulative Number of States	Cases
1	6	6	1–6
2	2	2 + 6 = 8	7–8
4	3	3 + 8 = 11	9–11
6	4	4 + 11 = 15	12–15
13	1	1 + 15 = 16	16
16	1	1 + 16 = 17	17
23	1	1 + 17 = 18	18
24	1	1 + 18 = 19	19
25	1	1 + 19 = 20	20
26	1	1 + 20 = 21	21
28	1	1 + 21 = 22	22
32	2	2 + 22 = 24	23–24
72	1	1 + 24 = 25	25
248	1	1 + 25 = 26	26
	$\Sigma f_x = N = 26$		

Source: Halperin, Rick, "Death Penalty News and Updates." January 27, 2012. http://people.smu.edu/rhalperi/.

one execution. In the second row, we add that six to the two states who executed two inmates to get eight states that executed one or two inmates. For each row we add the states from that row to the cumulative total from the prior row. That cumulative number allows us to determine which cases had each value of our variable. The first row has the first case through its cumulative number. The second row begins with the next case and ends with its cumulative number, and so on. For Table 3.7, cases 1 through 6 occur in the first row; 7 through 8, in the second; 9 through 11, the third; and so on. Once you've identified which cases occur in which rows, it is easy enough to find the row that contains the middle case. On that row, you look over to the left-hand column to find the value of X. So both the thirteenth and the fourteenth cases had six executions. The median number of executions is six for the states that executed anyone between 2000 and 2009. Because it doesn't give undue weight to the outlier Texas, the median of 6 is much more reflective of the central tendency of this table than the mean of 22.58 we found earlier.

The median has the added advantage of being useful for ordinal-level data. For ordinal-level data, the mean is nonsensical because it assumes that the data are on a uniform scale. The execution rates we've been using are interval because they have order (the higher the number, the more executions) and a uniform scale (each number higher is precisely one more execution). But for ordinal-level variables, the numbers only imply order. The numbers are not associated with a constant unit. For this kind of data, you couldn't calculate a mean, and so a median is a viable alternative.

Table 3.8 Level of Political Competition, 2010

Level of Competition	Number of Countries	Cumulative Number	Cases	Percent	Cumulative Percent
1 Repressed	14	14	1–14	9.2	9.2
2 Suppressed	16	30	15–30	10.5	19.7
3 Factional	35	65	31–65	23.0	42.8
4 Transitional	51	116	66–116	33.6	76.3
5 Competitive	36	152	117–152	23.7	100.0
Total	152			100.0	

Source: Polity Project IV, "PARCOMP," *Political Regime Characteristics and Transitions, 1800-2010,* Integrated Network for Societal Conflict Research, 2010. www.systemicpeace.org/inscr/inscr.htm.

For example, the Polity IV Project collects various kinds of data about the stability of the countries of the world. One of the variables measures the level of political competition within each country—shown in Table 3.8. The categories of Repressed, Suppressed, Factional, Transitional, and Competitive clearly are ordered from less political competition to more political competition. But the numbers that I have associated with each category don't mean much beyond that ordering. Because the increase in political competition as you move from 0 (Repressed) to 1 (Suppressed) is not the same as the increase from 3 (Transitional) to 4 (Competitive), the units do not have a constant size. Without the meaning that numbers have when they are on a scale, it wouldn't make sense to calculate a mean. It would, however, make a lot of sense to talk about the median—the level of competition for the average country.

To find the median for this distribution, you could find the middle case according to the previous rule: Median Case = $\frac{N+1}{2} = \frac{152+1}{2} = 76.5$. Once again, because we have an even number of cases, we are going to need to average values of the middle two cases: 76 and 77. To find their values, you could use the method from Table 3.7; you could add up the number of cases in each category until you find the seventy-sixth and seventy-seventh cases and then look to the left to find the value of X connected to that row. The seventy-sixth and seventy-seventh cases fall in the row containing cases 66 through 116. Looking to the left, I see that the value connected to that row is "4, Transitional." Although the computer will always assign a number to all categories in order to keep track of them, for ordinal variables, the number doesn't mean anything so you report the label of the category. The median value of political competiveness in 2010 was Transitional. Although this technique works, finding the median in this way can be cumbersome. Because frequency tables do not usually include a "Cumulative Frequency" column, you have to calculate that yourself.

Frequency tables, however, do frequently give a "Cumulative Percent" column that you can use as an alternative way to calculate the median. Without even knowing the *N*, you will always know that the median value always corresponds with the fiftieth percentile. (When we say that someone is in the fiftieth percentile, we mean that 50 percent of the cases have a lower value.) So you look for the category where the cumulative percent exceeds 50 percent. If you scan Table 3.8, you can see (without doing any math) that the 50th percent is not in "Factional"—the categories up through Factional account for only 42.8 percent of the cases. But the next category, "Transitional," includes all the cases exceeding the 42.8th percentile up through the 76.3rd percentile. Thus, you can conclude that the median must be in the "Transitional" category.

Mode

The "Transitional" countries have one more attribute that reflects the central tendency of the data. Not only do half of the countries fall on each side, but that is also the category that has the most cases. The average that is the most frequently observed value is called the mode. To calculate the mode, you need only look for the category with the most cases. If you look back to Table 3.5, the mode would be one execution because six states had that number. The mode can be identified for both interval- and ordinal-level data, although it is not always useful for interval-level data. But the mode is the only measure of central tendency that can be used if your data do not have any order because it has a nominal level of measurement.

Calculating the Mode from Raw Data. Calculating the mode from raw data is fairly straightforward. Simply look for the value that occurs most often. Look at our original 2010 execution data:

$$1, 1, 1, 1, 1, 2, 3, 3, 3, 5, 8, 17$$

With the data seriatim, or in an ordered series, it is very easy to see that "1" occurs most often. Among states that executed prisoners in 2010, the modal number of executions is one. If you want to find the mode from raw data, be sure that the data are ordered. The only problem you might have is that on occasion, you will see more than one value with the same frequency. If you see two numbers that have the same frequency, we call the data "bimodal." If you have bimodal data, you report both modes.

Calculating the Mode from Tabular Data. It is actually even easier to calculate the mode from tabular data than from raw data. Because the data have already been collapsed into categories in a frequency table, you simply need to find the category with the most cases. For example, Israel has a multiparty parliamentary system. In order to form a ruling coalition, parties need to form alliances such that the prime minister gets a majority of the votes. In 2012, a new unity government was formed between the Likud and Kadima parties. Table 3.9 shows the partisan makeup of the Israeli parliament, the

Knesset, at that time. As I normally do, I have associated numbers with each of the categories just so that the computer can keep track of them easier. But this is an example of a nominal-level variable. The parties are just in alphabetical order, so the numbers on the left don't mean anything. I could just as easily have put them in order from largest to smallest or from liberal to conservative. The numbers still would have been simply standing in for the name of a category. Because the numbers are meaningless for these data, it wouldn't make sense to compute a mean. Similarly, computing a median wouldn't make sense; you couldn't put these categories into order from less to more, they are just different parties. But it is clear that the most common party is Kadima, with Likud a close second. So "Kadima" is the mode. The odd thing in this political situation was that the unity government of the Likud and Kadima parties made Benjamin Netanyahu of the Likud Party the prime minister and Shaul Mofaz of the Kadima Party the deputy prime minister—taking charge if Netanyahu is absent.

Table 3.9 Partisan Make-up of the Israeli Knesset, 2012

Party X	Seats f_x
0 Hadash	4
1 Haatzma`ut	5
2 Ichud Leumi	4
3 Israel Labor Party	8
4 Kadima	28
5 Likud	27
6 National Democratic Assembly	3
7 New Movement-Meretz	3
8 New National Religious Party	3
9 Ra`am-Ta'al	4
10 Shas	11
11 United Torah Judaism	5
12 Yisrael Beiteinu	15

Source: "Current Knesset Members." *The Knesset*. www.knesset.gov.il/mk/eng/mkindex_current_eng.asp?view=1.

The one care that you need to take when reporting the mode is that you need to report the category, not its frequency. If you look back to Table 3.8 to find the modal level of political competition, you can easily scan the frequency column and see that the biggest number there is fifty-one. But you wouldn't report fifty-one as the mode. To find the modal level of political participation, you would look to the left and find that the value of the category with a frequency of fifty-one is "Transitional." It is especially important to remember to look to the category label on the left when you want to report the mode for interval-level data because that is when you are most likely to make a mistake. Look back even farther now to Table 3.5 to find the mode of the number of executions for states from 2000 to 2009. When you look down the frequency column, the biggest number you see is six. But six is not the mode. To find the mode, you look to the left of the six to find the value of the category with the frequency of six. That value is one. The most frequent number of executions from 2000 to 2009 was one. Because more states had one execution than any other number of executions, the mode is one.

SUMMARIZING THE MATH: AVERAGES

When we look at a single variable, the first thing we usually want to know is: What is the average value? But the term "average" can have different meanings. Usually, we think of the arithmetic average, or mean. But the mean can be calculated only if the data are interval level—when they have both order and a scale. In addition, sometimes, if there are outliers in the data, the mean is skewed and so doesn't really reflect the central tendency of the data. A second possible way of measuring the "average" of an attribute is to calculate the median. This is the attribute held by the case in the center of the distribution. The median is reported either if the data are ordinal level (they have order, but the unit is not of a constant size) or if they are interval level but skewed. A final measure of central tendency is the mode. The mode is the category of your variable with the most cases. The mode is the only measure of central tendency that can be reported for nominal-level data (where the attributes can only be categorized, not ordered), but it can also be reported for interval- and ordinal-level data.

Calculate Averages from Raw Data

Follow these four steps to calculate averages from raw data:

1. Put the cases in order numerically from least to most.
2. To find the mode, find the value that occurs most often on your list.
3. To find the median, find the middle value in this list of ordered data. You can do this by moving your fingers from the outside to the middle. Or, if you have too many cases to do that, the middle case is the total number of cases plus one, divided by two. You would count from either the top or the bottom of your ordered list of value to find the middle case. If you have an even number of cases, you average the middle two cases.
4. To calculate the mean, add up all the values and divide by the number of cases.

A Political Example: U.S. Test Scores. Frequently, the quality of education in the United States is in the news. Many feel that the United States is falling behind the other nations of the world. If you look at math scores by high school students, how does the United States compare to other countries? Table 3.10 shows data collected by the Organization for Economic Cooperation and Development (OECD), an intergovernmental organization of industrialized countries for its member states. Is the United States actually below average? Calculate the mean, median, and mode. Which of these measures of central tendency does it make sense to use in reporting these data?

1. Put the cases in order numerically from least to most.
   419, 421, 445, 447, 466, 483, 483, 487, 487, 487, 489, 490, 492, 493, 494, 495, 496, 497, 497, 498, 501, 503, 507, 512, 513, 514, 515, 519, 526, 527, 529, 534, 541, 546

Table 3.10 Math Scores

Country	Math Literacy Scale	Country	Math Literacy Scale
Australia	514	Italy	483
Austria	496	Japan	529
Belgium	515	Luxembourg	489
Canada	527	Mexico	419
Chile	421	Netherlands	526
Czech Republic	493	New Zealand	519
Denmark	503	Norway	498
England	492	Poland	495
Estonia	512	Portugal	487
Finland	541	Slovak Republic	497
France	497	Slovenia	501
Germany	513	South Korea	546
Greece	466	Spain	483
Hungary	490	Sweden	494
Iceland	507	Switzerland	534
Ireland	487	Turkey	445
Israel	447	United States	487

Source: Howard L. Fleischman, Paul J. Hopstock, Marisa P. Pelczar, and Brooke E. Shelley, "Highlights From PISA 2009: Performance of U.S. 15-Year-Old Students in Reading, Mathematics, and Science Literacy in an International Context," *NCES* 2011004 December 2010. http://nces.ed.gov/pubsearch/pubsinfo.asp?pubid=2011004.

2. To find the mode, find the value that occurs most often on your list.

   By having the values in order, it is easy to see when there are multiple cases with the same value. Both 497 and 483 have two cases. But the value 487 has three cases, which makes it the mode. Because the United States is one of the cases with a math score of 487, by using the mode we would conclude that U.S. students have average math skills.

3. To find the median, find the middle value in this list of ordered data.

   There are too many cases in this list to use the two-finger approach. So I count the number of cases and find that there are thirty-four. The median case is (34 + 1)/2 = 17.5. The median value is halfway between the seventeenth and eighteenth cases. I start counting at the beginning of my ordered list of values and find that the seventeenth case has a value of 496 and the eighteenth case has a value of 497. The median is the average of 496 and 497: (496 + 497)/2 = 496.5. Using the median, the United States' score of 487 is below average.

4. To calculate the mean, add up all the values and divide by the number of cases. When I total the values, I get 16,853. To get the mean, I divide the total by the number of cases. Thus, the mean = 16,853/34 = 495.68. So the United States is below average if you use the mean. This is an interval-level variable and there are no outliers, which make the mean an appropriate measure of central tendency. Because the mean and median are very close, we know that the data are not skewed. We can conclude that the United States is below average on high school math skills.

Calculate Averages from Tables

Follow these three steps to calculate averages from tables:

1. Find the category that has the most cases. Look to the left to find the label connected to the category that has the most cases. This is the mode.

2. If your frequency table has a cumulative frequency column, find the median by seeing which category pushes the cumulative frequency over 50 percent. Look to the left to identify the category of the variable that is the median. If the frequency doesn't have a cumulative frequency, calculate the median case the same way as for raw data: $(N + 1)/2$. Count up the number of cases from the least up until you find the middle case. Look to the left to find the value of the median case.

3. To calculate the mean, add a column to your frequency table in which you multiply the value of your variable X by its frequency f_x. Add up the values in this column ($\Sigma X f_x$) and divide by the number of cases (N): $\Sigma X f_x / N$. Remember that the number of cases—N—in this table is not the same as the number of rows. To find the number of cases, add up the values in the frequency column: $N = \Sigma f_x$.

Table 3.11 Harvard Clerk Appointments by Supreme Court Justices, 2011, Raw Data

Supreme Court Justice	Number of Harvard Clerks
Roberts	2
Sotomayor	2
Breyer	1
Kagen	2
Kennedy	3
Ginsburg	1
Alito	0
Scalia	1
Thomas	0

Source: "List of Law Clerks of the Supreme Court of the United States." Wikipedia, The Free Encyclopedia, October 26, 2012. http://en.wikipedia.org/w/index.php?title=List_of_law_clerks_of_the_Supreme_Court_of_the_United_States&oldid=519974610.

A Political Example: The Best and the Brightest. When John F. Kennedy was elected president, he prided himself on appointing people who were the smartest in the country rather than political hacks. It ended up that many of his appointments, frequently called "the Best and the Brightest," were Harvard graduates. Do Harvard graduates still do well when it comes to governmental appointments? Look at Supreme Court clerks. Normally, each Supreme Court justice chooses four of the best law school graduates to clerk for them each year. How many of those appointments are graduates of Harvard Law School?

If we collect the number of Harvard clerks appointed by each Supreme Court justice during 2011, we find the data presented in Table 3.11. This is just the raw data, though. The frequency is found in Table 3.12.

Table 3.12 Harvard Clerk Appointments by Supreme Court Justices, 2011, Frequency

Number of Clerks X	Frequency f_x	Percent	Cumulative Percent	$X(f_x)$
0	2	22.2	22.2	$0 \times 2 = 0$
1	3	33.3	55.5	$1 \times 3 = 3$
2	3	33.3	88.8	$2 \times 3 = 6$
3	1	11.1	99.9	$3 \times 1 = 3$
Total	$N = \Sigma f_x = 9$	99.9		$\Sigma X(f_x) = 12$

1. Find the category that has the most cases.
   The frequency column shows two rows with three cases. This means that this variable is bimodal. Looking to the left, we can see that the values connected to these rows are one and two clerks from Harvard.

2. Find the median by finding the middle case. Look to the left to find the value of the median case.
   To find the median value, we can look at the Cumulative Percent column. This passes 50 percent on the second row. Looking to the left, we can see that the value connected with this row is 1. Alternatively, if I didn't have the Cumulative Percent column, I could have found the middle case by using the equation $(N + 1)/2$. Because there are nine justices, $N = 9$. The median case is $(9 + 1)/2$ or 5. I find the fifth case by adding up the frequency cells until I get to five. The first row has two cases and the second row has three. That means that the fifth case is in the second row. Because the value of that row is one, the median number of clerks from Harvard is one.

3. To calculate the mean, multiply each value of your variable (X) by its frequency (f_x). Add up the values in this column and divide by the number of cases.
   To find the mean from this table, I've added in a column labeled $X(f_x)$, which multiplies the value of X by its frequency. The total number of clerks is found at the bottom of this column. To get the mean number of Harvard clerks, we divide the total number of Harvard clerks (12) by the number of Supreme Court justices ($N = 9$). Mean = $12/9 = 1.33$. All three averages tell the same story: In 2011, Supreme Court justices averaged a little over one Harvard graduate out of their four clerks.

> **BOX 3.1 Numbers in the News**
>
> Although the OPEC oil embargo of the 1970s raised the temporary specter of the economic effects stemming from U.S. dependency on foreign oil, the attacks of September 11, 2001, raised these stakes. Because of fears of Middle East–based terrorism, dependency on foreign oil began to be seriously discussed as a national security issue. In the ensuing years, the United States and Canada have increased oil production, leaving both nations less dependent on Middle Eastern oil producers. The International Energy Agency tracks two key variables—oil production and oil consumption—in order to determine how dependent different countries are on Middle Eastern oil. In May 2013, it issued a report suggesting that because developing countries in Asia are more dependent on Middle Eastern oil, the power dynamic in the world has changed. The numbers are worth watching because they could have both geopolitical and economic consequences.[6]

USING SPSS TO ANSWER A QUESTION WITH AVERAGES

Although being able to do the math is important to understanding what the statistics mean, in the real world, you will have a statistical package like SPSS available to do the math for you. If you were just learning how to use SPSS and not learning the math, it would be tempting to report all the measures of central tendency, forgetting what they mean. Now that you've learned how to do the math, though, it will be more natural to remember the assumptions that each of the calculations made about the data. Although the mean is the most common measure of central tendency we use, it makes two assumptions. First, it assumes that the data have an interval level of measurement. Second, it assumes that there are no outliers skewing the mean. If there is an outlier, you would want to use the median as the measure of central tendency. Also, if the variable has an ordinal level of measurement, you would want to report the median. Finally, the mode is the only measure of central tendency you can use if you have a nominal level of measurement.

To use SPSS to find the measures of central tendency, you will simply add one step to the process you went through to get a frequency table in the prior chapter. This process is summarized in Box 3.2. Open your data—either by importing an Excel file, or by double-clicking on a previously created SPSS.sav file. Open the frequency command window under "Analyze," "Descriptive Statistics." One of the options in this window is "Statistics." If you click on that option, you will see a new window with many statistics you can request. On the right, you'll see a group of options under the heading "Central Tendency." Select "Mean," "Median," and "Mode." Click on "Continue" to bring yourself back into the frequency window and then click on "OK" to get your tables. In the output window, you'll see two boxes: one with the statistics and one with the frequency. If your variable is interval level, you'll notice two goofy things. First, there will probably be a footnote on the mode indicating that there are multiple modes so it is showing the smallest. This is because with truly interval-level data, there usually aren't very many cases with precisely the same value. If this is the case, the mode might not mean very much, although in the next chapter, we'll see one way that we can make it more meaningful. Second, you'll notice that the frequency table is huge because there are so many different values. Again, we'll learn a trick in the next chapter to make this more useful. But for now, ignore the frequency table and simply report the averages.

BOX 3.2   How to Get Measures of Central Tendency in SPSS

After opening your data in SPSS

>Analyze
>>Descriptive Statistics
>>>Frequencies
>>>>Click on the variable to highlight it.
>>>>→To bring the variable over to the "Variable" box
>>>>>Statistics
>>>>>>Mean
>>>>>>Median
>>>>>>Mode
>>>>>>Continue
>>>>OK

At this point, the frequency of the variable will show up in the "Output" window.

An SPSS Application: Income in Poor Places

You are interning at AmeriCorps' national office in Washington, D.C. You are in the Public Affairs Office, and Rosemary Shamieh, the publications specialist, approaches you about a pamphlet she is putting together. In it, she is comparing some of the American communities that are served by the program to less-developed countries. For example, one of the communities she is highlighting is Wheelwright, Kentucky. Wheelwright is a mining community whose claim to fame is being ranked as the one-hundredth poorest town in America, with a per capita income of $5,367. Because she wants to be accurate in her claim that it is like a less-developed nation, she asks you to find the average income in developing countries.

You first identify your cases by finding the list of developing countries on Wikipedia. You create a coding sheet listing all of those countries in the first column. In the second column, you want to enter the average income. You then remember that your International Relations professor used to love to show you data on a website called Gapminder.[7] He used to rave about how it contains all sorts of international data. After finding the website, you click on the "Data" option and search for income. You see a variable that looks like what you want, which is labeled "Income per Person." After entering the data from 2011 for all of the developing countries onto your coding sheet, you enter it into Excel and import it into SPSS.

1. Open your data in SPSS.
2. Clean your data using the "Define Variable Properties" command.
   I set the value of −9 to missing.
3. Get a frequency.
   I want the measures of central tendency, which are available in the frequency window. To get a frequency, I click on "Analyze" (in either of the windows), choose "Descriptive Statistics," and then choose "Frequencies." Once the frequency window is open, I move my variable into the variable box. Then I choose the "Statistics" option and mark "Mean," "Median," and "Mode" for inclusion. This is shown in Figure 3.1.

Figure 3.1 Get Measures of Central Tendency in SPSS

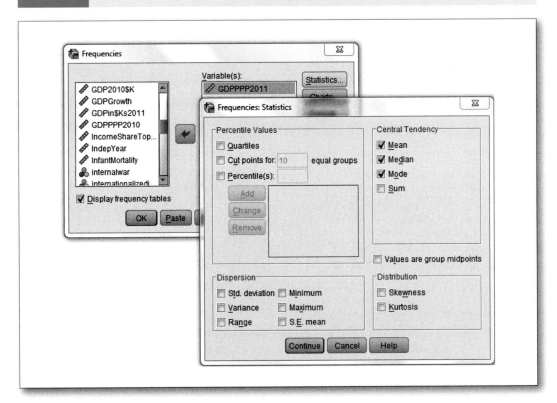

After I click on "Continue" and "OK," two tables appear in the Output window. Figure 3.2 shows a screen shot of these two tables.

Figure 3.2 Average Income for Developing Countries

Statistics

GDPPPP2011

N	Valid	148
	Missing	6
Mean		7762.625885
Median		4827.059320
Mode		387.2228[a]

a. Multiple modes exist. The smallest value is shown

GDPPPP2011

		Frequency	Percent	Valid Percent	Cumulative Percent
Valid	387.2228	1	.6	.7	.7
	425.3637	1	.6	.7	1.4
	484.1486	1	.6	.7	2.0
	511.2584	1	.6	.7	2.7
	591.4617	1	.6	.7	3.4
	608.1557	1	.6	.7	4.1
	668.0272	1	.6	.7	4.7
	694.7217	1	.6	.7	5.4
	798.5521	1	.6	.7	6.1
	865.5453	1	.6	.7	6.8
	900.3507	1	.6	.7	7.4

4. Choose the appropriate measure of central tendency for this variable.

The first table in Figure 3.2 shows the statistics: the mean income is $7762.63 and the median is $4827.06. The fact that the mean is so much higher than the median indicates that there are some developing countries with incomes much higher than is typical. Because these outliers have skewed the mean upward, the median is a more appropriate measure of central tendency for the variable. I notice that the mode of $387.22 has a note attached to it that says, "Multiple modes exist. The smallest value is shown." Looking down at the frequency, I see that each of the values has a frequency of one. Every country has a different average income, so the distribution is not just bimodal, it is 148-modal. It doesn't make sense to report a mode for these data.

5. Write a memo describing your results.

Finally, I write a memo summarizing my findings for my boss.

Memo

To: Rosemary Shamieh, Publications Specialist
From: T. Marchant-Shapiro, intern
Date: June 14, 2012
Subject: Incomes in Developing Countries

You asked whether the average income for Wheelwright, Kentucky, is similar to the average income of developing countries. From Gapminder, a reputable online compiler of international data, I collected the incomes of developing countries to find their average. Wheelwright's average income of $5367 is slightly above the median income of developing countries.

Wheelwright's average income is well below the mean income of developing countries, which is $7762.63. Unfortunately, that number is skewed upward because it includes oil-rich countries like Kuwait and Qatar, which are developing but have higher average incomes than the United States. Because of those outliers, the median is a better measure of the average income of developing countries. The median income is $4827.06.

Wheelwright's income is about $500 more than the median income of developing countries. This places it at the 58th percentile for developing countries. So 58 percent of the developing countries earn less than Wheelwright, and 42 percent earn more. If you place its average of $5367 in the context of the average U.S. income, $41,728.14, it is clear that Wheelwright is much more like a developing country than it is like the rest of the United States. It is not unreasonable for you to use Wheelwright as an example in your pamphlet.

Your Turn: Measures of Central Tendency

YT 3.1 Math is not the only thing we learn in school. Is it possible that U.S. students do better in comparison to other countries in reading? Table 3.13 shows the OECD data for reading.

Table 3.13 Reading Scores

Country	Reading Literacy Scale	Country	Reading Literacy Scale
Australia	515	Czech Republic	478
Austria	470	Denmark	495
Belgium	506	England	494
Canada	524	Estonia	501
Chile	449	Finland	536

Country	Reading Literacy Scale	Country	Reading Literacy Scale
France	496	New Zealand	521
Germany	497	Norway	503
Greece	483	Poland	500
Hungary	494	Portugal	489
Iceland	500	Slovak Republic	477
Ireland	496	Slovenia	483
Israel	474	South Korea	539
Italy	486	Spain	481
Japan	520	Sweden	497
Luxembourg	472	Switzerland	501
Mexico	425	Turkey	464
Netherlands	508	United States	500

Source: Howard L. Fleischman, Paul J. Hopstock, Marisa P. Pelczar, and Brooke E. Shelley, "Highlights From PISA 2009: Performance of U.S. 15-Year-Old Students in Reading, Mathematics, and Science Literacy in an International Context," *NCES* 2011004 December 2010. http://nces.ed.gov/pubsearch/pubsinfo.asp?pubid=2011004.

1. Calculate the mean, median, and mode. Be sure to show all your work.
2. Which of these measures of central tendency does it make sense to use in reporting these data?
3. Is the United States below average for reading?

YT 3.2   The data in Table 3.14 are the number of executions performed between 1990 and 1999 in the states in which capital punishment occurred. What was the average number of executions per state during this decade?

1. Calculate the mode, median, and mean from these tabular data. Be sure to show all your work.
2. Which is the most appropriate measure of central tendency?

Table 3.14   State Death Penalty Executions, 1990–1999

Number of Executions X	Number of States f_x	Number of Executions X	Number of States f_x
1	4	3	5
2	3	4	1

(Continued)

Table 3.14 (Continued)

Number of Executions X	Number of States f_x	Number of Executions X	Number of States f_x
5	1	21	1
7	2	22	1
9	1	23	1
10	1	40	1
12	3	65	1
19	2	166	1

Source: Halperin, Rick, "Death Penalty News and Updates." January 27, 2012. http://people.smu.edu/rhalperi/.

Apply It Yourself: Calculate the Percent of Income Earned

The year is 2011. Occupy Wall Street is just heating up. Many of the demonstrators are carrying signs proclaiming themselves as one of the 99 percent. Their argument is that the richest 1 percent controls proportionally too much of the wealth of the United States. You are interning in the office of Michael R. Bloomberg, mayor of New York City. The demonstrators have been occupying Zuccotti Park for so long that the mayor is considering shutting it down to protect public safety. You are observing during a meeting that Bloomberg calls to discuss the issue with top-level staff. Bloomberg, himself worth about $20 billion, asks one of his aides whether the United States is atypical in its distribution of wealth. The aide promises Bloomberg an answer by the end of the day. As soon as the meeting is over, the aide motions you over and assigns you to collect data on the percent of income earned by the wealthy of each country to see what is normal, warning you that the question requires immediate attention; Mayor Bloomberg expects a memo by the end of the day. You run to your office and search for the ever handy Gapminder website. In it, you find a variable that includes data collected by the World Bank[8] showing what percent of income was earned by the richest 10 percent of each country. The data are not perfect—it is the top 10 percent, not 1 percent, and the data available for each country come from different years. But considering the time constraints, you put together a dataset including the value for the most recent year, from 1997 to 2007, for which data are available for each country. In the WorldData dataset, use the variable IncomeShareTop10% to answer the mayor's question.

1. Open "WorldData." Because this is already an SPSS data file, you can double-click on it and (assuming SPSS is installed on your computer) your computer will automatically use SPSS to open it. Otherwise, you can always open SPSS first and then click on "File" and "Open" to find and open the file.

CHAPTER 3 Measures of Central Tendency: That's Some Mean Baseball   **75**

2. Clean the data first by making sure that all missing values are set to missing. Second, because all of the values are supposed to be percentages, they should range from 0 to 100—make sure that is the case.

   >Data
   >>Define Variable Properties
   >>>"IncomeShareTop10%"→Variable to Scan
   >>>Continue
   >>>>–9=Missing
   >>>>OK

3. Get a frequency for the variable, including the statistics for mean, median, and mode.

   >Analyze
   >>Descriptive Statistics
   >>>Frequencies
   >>>>IncomeShareTop10%→Variable
   >>>>>Statistics
   >>>>>>Mean
   >>>>>>Median
   >>>>>>Mode
   >>>>>>Continue
   >>>>OK

4. Choose the appropriate measure of central tendency for this variable. Compare this value to the percent of income earned by the top 10 percent in the United States—you can find this by looking in the data window.

5. Write a memo comparing those two values. For this memo, you do not need to include a table, but do report the numbers and attach an SPSS printout for grading purposes.

Key Terms

Mean (p. 53)
Median (p. 58)
Measure of central
   tendency (p. 53)

Mode (p. 62)
Outlier (p. 58)

CHAPTER 4

Measures of Dispersion

Missing the Mark

In baseball, one of the advantages some teams have over others is the amount they are able to spend on players' salaries. Presumably, the teams that spend more on salaries should be able to recruit better players. At least, this is the underlying premise of the movie *Moneyball*.[1] The movie begins at the end of the 2001 season when the Oakland A's were able to make it to the playoffs, but then lost to the Yankees. Team manager Billy Beane has high hopes for the next season because of the quality of his players. But then his best players get recruited by other teams (especially the Yankees) who are able to pay higher salaries. As a result, Beane has to put another team together from scratch, knowing he can't afford the salaries paid to superstar players. Table 4.1 shows how total payroll varied among Major League Baseball teams in 2002, the season that the movie describes.

It's easy enough to eyeball the data. Although the teams all have about the same number of players, there is a huge disparity in the salaries paid to those teams. You can see that the median payroll is $60 million, but that doesn't communicate how the payrolls for the different teams are distributed around that average. By looking at all of the data, you can see that about half the teams have payrolls between $45 and $80 million. That's a fairly wide spread among the middle spenders, but the extremes show an even greater disparity. Where the Yankees have $126 million to divide between twenty-eight players, the A's only have $40 million to divide between twenty-seven. This disparity suggests that it is not enough to report the average for a set of data. We also want to know how well that average describes the distribution of all the cases around the middle. So in addition to wanting a measure of the center of the data, we also want a measure of its dispersion.

The importance of analyzing dispersion as well as averages came home to the U.S. military during the first Gulf War. The military was very proud of its new precision-guided missiles. During the war, the spokesman proudly released to the media film footage of how these missiles were able to hit targeted buildings with pinpoint accuracy. This was of particular importance because Baghdad was filled with civilians—whatever our beef with Saddam Hussein for invading Kuwait, no one wanted innocent bystanders to pay the price. But as Hussein allowed reporters into Baghdad, it became increasingly clear that the footage

Table 4.1 Payroll for Major League Baseball Teams in 2002

Team Case	Total Payroll (in $Millions) X
1. New York Yankees	$ 125,928,583
2. Boston Red Sox	$ 108,366,060
3. Texas Rangers	$ 105,726,122
4. Arizona Diamondbacks	$ 102,819,999
5. Los Angeles Dodgers	$ 94,850,953
6. New York Mets	$ 94,633,593
7. Atlanta Braves	$ 93,470,367
8. Seattle Mariners	$ 80,282,668
9. Cleveland Indians	$ 78,909,449
10. San Francisco Giants	$ 78,299,835
11. Toronto Blue Jays	$ 76,864,333
12. Chicago Cubs	$ 75,690,833
13. St. Louis Cardinals	$ 74,660,875
14. Houston Astros	$ 63,448,417
15. Los Angeles Angels	$ 61,721,667
16. Baltimore Orioles	$ 60,493,487
17. Philadelphia Phillies	$ 57,954,999
18. Chicago White Sox	$ 57,052,833
19. Colorado Rockies	$ 56,851,043
20. Detroit Tigers	$ 55,048,000
21. Milwaukee Brewers	$ 50,287,833
22. Kansas City Royals	$ 47,257,000
23. Cincinnati Reds	$ 45,050,390
24. Pittsburgh Pirates	$ 42,323,599
25. Miami Marlins	$ 41,979,917
26. San Diego Padres	$ 41,425,000
27. Minnesota Twins	$ 40,225,000
28. Oakland Athletics	$ 40,004,167
29. Washington Nationals	$ 38,670,500
30. Tampa Bay Rays	$ 34,380,000

Source: "Major League Baseball Salaries," *USA Today Salaries Database*, 2012. http://content.usatoday.com/sportsdata/baseball/mlb/salaries/team.

being released by the military did not reflect what was actually happening on the ground. The guided missiles represented only a small proportion of the attacks—most were conducted with older, less accurate weapons. And the guided missiles weren't nearly as accurate as the footage suggested—they would frequently miss their intended targets. The inaccuracy meant that there was a distribution of landings around where the missiles were aimed, and as a result, collateral damage was not as minimal as the military led the media (and us) to believe. The death toll pointed to the importance of knowing not only that, on average, the military hit its mark, but also how the misses were distributed around the target.

In this chapter, we will look at distributions in order to find measures of dispersion. The narrower those measures, the better the average describes the attribute we have measured. But for a variable with a high level of dispersion, the average fails to reflect the distribution. Usually, measures of dispersion take two general forms: ranges within which the population falls and numerical measures of how far the population is (on average) from the mean.

RANGES

Leaving the violence of the military behind, let us move on to the cutthroat process of applying to law school. One aspect to consider is the cost of tuition. A recent *New York Times* article[2] suggests that with the scarcity of jobs, the debt incurred by a law student might not be a financially sound decision. Indeed, if you look at the average cost of tuition at the top twenty law schools, you might be overwhelmed. These numbers are given in Table 4.2.

Table 4.2 Tuition at the Top Twenty Law Schools in 2010

School	Tuition	School	Tuition
Yale	$48,340	Northwestern	$47,472
Harvard	$45,026	Cornell	$49,020
Stanford	$44,121	Georgetown	$43,750
Columbia	$48,004	UCLA	$45,967
University of Chicago	$44,757	University of Texas, Austin	$42,814
NYU	$46,196	Vanderbilt	$44,074
UC Berkeley	$48,152	USC	$46,264
University of Pennsylvania	$46,514	Washington University	$42,330
University of Michigan	$46,250	George Washington University	$42,205
University of Virginia	$43,800	N = 20	
Duke	$45,271		

Source: "Best Law Schools," *U.S. News & World Report,* 2011. http://grad-schools.usnews.rankingsandreviews.com/best-graduate-schools/top-law-schools/law-rankings.

These data give a mean tuition of $45,387.75 and a median of $45,045.50. Both of these averages reflect the same high cost of going to a top-tier law school.

But not all schools charge the same amount of tuition, so before rejecting law school out of hand, it is worth looking at the distribution of tuition among all law schools, not just the top twenty. We get the range by finding both the lowest and the highest tuition charged by any U.S. law school. Frequently, people will discuss the range in terms of those two endpoints. But technically, the range is the single number you get by subtracting the former from the latter.

$$\text{Range} = \text{Highest Value} - \text{Lowest Value}$$

Among U.S. law schools, Cornell charges the most tuition at $49,020 per year, whereas Southern University in Louisiana charges the least at $12,580. So, plugging those two numbers into the equation, we get

$$\text{Range of Law School Tuition} = \$49,950 - \$12,580$$

$$= \$36,440$$

A range of $36,440 is fairly wide. So if you are conducting a cost-benefit analysis of whether to attend law school, you should keep in mind that the cost of tuition is not a constant—where you choose to go can have a big impact on your calculations.

In addition to considering the cost when you are applying to law school, your adviser will tell you to categorize a variety of law schools into the likelihood of you being admitted: Reach, Likely, Safe. The reach schools are the ones you would love to attend, but aren't sure you can get accepted; the likely schools are the ones that normally accept students with your qualifications; and the safe schools are the ones to which you are sure to be admitted—you might even be good enough to make law review. Your advisor will recommend applying to at least one or two in each of these categories. This strategy of applying to schools of varying quality reflects the law schools' practice of accepting a wide range of students. The way you determine which schools are reach, likely, and safe is by looking at the distribution of students that the different schools admit.

Normally, you will look at the student body's distribution on two attributes: GPA and LSAT scores. The students admitted to Yale, for example, have a median GPA of 3.90 (on a 4.0 scale) and a median LSAT score of 173 (out of a possible 180). The law schools do not normally give the range in terms of absolute best and absolute worst GPAs and LSATs for the students they admit. Instead, they give what is known as the interquartile range, or the range within which the middle two-fourths of the population lie. In the same way that they found the median by identifying the score of the student who tested right in the middle of all of the scores, they can identify the middle scores within the top and bottom halves of the students. To find the lower value, you find the value for which 25 percent of the cases fall below. This is called the first quartile. To find the higher value, you find the value for which 75 percent of the cases fall below. This is called the third quartile. Technically, the interquartile range is the value connected to the 75th percentile minus the value for the 25th percentile:

$$\text{Interquartile Range} = \text{Value (75th percentile)} - \text{Value (25th percentile)}$$

But, as with the range, instead of giving the single number from that subtraction, you'll usually give the two endpoints of the interquartile range.

So for Yale, the interquartile range for GPA is given as 3.82–3.96. By definition, half of the students admitted score in the interquartile range. If you know the median and the interquartile range, you can divide the student body into fourths. At Yale's law school, a quarter of the admitted students have GPAs below 3.82; a quarter, between 3.82 and the median of 3.90; a quarter, between 3.90 and 3.96; and a quarter have GPAs above 3.96. Similarly, because Yale's interquartile range for LSAT scores is 170–176, we know that half of their students score in that range. If you are in that interquartile range for both GPA and LSAT score, you can feel comfortable placing that school on your "likely" list. If you score above the 75th percentile, you can put the school on your "safe" list. If you score below the 25th percentile, the school is a reach. How much of a reach it is depends on how far below the 25th percentile you are. At most schools, a quarter of the students fall in a range of three points on the LSAT. This suggests that applying to a school where your LSAT score is more than three points below their 25th percentile is a very unrealistic reach.

Businesses combine the range and the interquartile range in what they call a "five-point summary." The five points are the minimum and maximum (from the range), the first and third quartile (from the interquartile range), and the median. If you think about it, if you know those five points of the distribution of a variable, you know quite a lot about what the distribution looks like. Sometimes, statisticians make a visual presentation of those five points in what they call a **box-and-whiskers plot**. The box is around the interquartile range, with a line for the median contained within the box. Then two lines come out of the box, extending whiskers to the minimum and maximum data points.

Table 4.3 Law School Tuition, 2010*

Tuition X	Frequency f_x	Percent	Cumulative Percent
$15,000	5	2.6	2.6
$20,000	9	4.6	7.2
$25,000	20	10.2	17.4
$30,000	46	23.6	41.0
$35,000	49	25.2	66.2
$40,000	38	29.4	85.6
$45,000	24	12.3	97.9
$50,000	4	2.1	100.0
Total	195	100.0	

Source: "Best Law Schools," *U.S. News & World Report*, 2011. http://grad-schools.usnews.rankingsandreviews.com/best-graduate-schools/top-law-schools/law-rankings (February 7, 2011).

*Rounded to the nearest $5,000.

Let's return to the law school tuition data to see how this works. Table 4.3 shows a frequency for tuitions—I've rounded it to the nearest $5,000 just to make the table a manageable size. To find the five points, you need to have a cumulative percent column. You'll recall from the previous chapter that the cumulative percent is the percent of cases included in the current and all previous rows. So for the first row, we begin with 2.6 percent because there are no previous rows. In the second row, we add its 4.6 percent to the previous cumulative percent of 2.6 percent to get 7.2 percent. In the third row, we add its 10.2 percent to the previous 7.2 percent to get a cumulative 17.4 percent, and so on.

With the table set up this way, it is easy to find the five key data points. The minimum is the first value of $15,000 and the maximum is the last value of $50,000. We get the other three points by looking at the cumulative percent column. The first quartile point is found in the category that just exceeds 25 percent. We can see that the third row only includes up to the 17.4th percentile, so that does not contain the 25th percentile. The next row includes every value more than the 17.4th percentile up to 41 percent. Because 25 percent is between those two values, we know that the 25th percentile falls in the fourth row. Looking over to the left, we find that the tuition connected to this row is $30,000. We can conclude that the cutoff between the first and second quartile is $30,000. As in the previous chapter, the median is the category that includes the 50th percentile. Looking back at the cumulative percent column, we see that happening in the fifth row. Looking to the left, we see that the median tuition is $35,000. To find the 75th percentile, we find the row where the cumulative percent exceeds 75 percent. This happens in the sixth row, which is connected to a tuition of $40,000.

Normally, the five-point summary is set in brackets like $\{Q_0, Q_1, Q_2, Q_3, Q_4\}$, where Q_0 and Q_4 are the minimum and maximum, Q_1 and Q_3 designate the interquartile range, and Q_2 is the median. So for law school tuition, we would give the five-point summary as $\{\$15,000, \$30,000, \$35,000, \$40,000, \$50,000\}$. You can eyeball the summary and see that there are some law schools that charge as low as $15,000 and as high as $50,000 for tuition, but half of the law schools charge in the range between $30,000 and $40,000.

A box-and-whiskers plot can help you visualize the distribution of the data even better. You begin to create a box-and-whiskers plot by drawing a line with a scale, making sure it goes high enough to include your maximum value. Second, above the scale, draw vertical lines for each of the five key points. Third, draw lines connecting the vertical lines for the first and third quartiles at both the top and the bottom. This will create a box that contains the median. Fourth, make the whiskers by drawing two lines: the first out from the middle of the first quartile line to the minimum and the second from the third quartile to the maximum. If you follow this procedure for the tuition, you should get a plot like Figure 4.1. (Notice when you present data in pictorial form, you label it a "Figure." This is in contrast with the tables you have prepared previously, which present data in columns.) Although the figure does not add any information to the numbers contained in Table 4.3, looking at the box and whiskers adds some intuition regarding the meaning of the numbers. Notice that the left whisker is longer than the right whisker. Because they each contain one quartile, both whiskers contain the same number of schools. The difference is that starting at the minimum, the first quartile had to extend much farther to collect enough schools to contain a quartile. In contrast, the most expensive schools are clustered very tightly together. Because the left whisker is longer than the right whisker, you can conclude that although there are some low-cost law schools, there aren't as many of them as there are of the high-cost ones. In general, the distribution is fairly symmetrical. It's fairly clear that, in general, law schools charge between $30,000 and $40,000. If you look carefully, you might be able to find law schools that charge less. And if you're thinking about a law school that charges between $40,000 and $50,000, you might want to be careful that you are getting some added value for the extra cost.

Figure 4.1 Box-and-Whiskers Plot for Law School Tuition

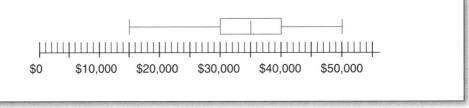

Table 4.4 Unemployment Rate for Swing States, Raw Data

State	Unemployment Rate (%)
Case	X
Colorado	8.1
Florida	8.6
Iowa	5.1
Minnesota	5.6
Nevada	11.6
New Hampshire	5.0
New Mexico	6.7
North Carolina	9.4
Ohio	7.3
Pennsylvania	7.4
Virginia	5.6
Wisconsin	6.8

Source: "Local Area Unemployment Statistics: Unemployment Rates for States," *Bureau of Labor Statistics,* 2012. www.bls.gov/web/laus/laumstrk.htm.

DISTANCE FROM MEAN

Although the range is useful for understanding the extremes of the distribution, it is not very helpful in understanding how tightly around the average the data are organized. It would be nice to have a measure that actually reflects the values of all the cases, not just the extremes. Normally, the measure we use to do that is the standard deviation and its cousin, the variance. These two measures of dispersion look at how far, on average, each case is from the mean. Just as we used slightly different procedures in calculating the mean if we had raw data than if we had tabular data, we will once again separate the procedures for those two kinds of data when calculating the standard deviation.

Calculate the Standard Deviation from Raw Data

In the 2012 presidential election, the economy was, as it usually is, a key issue. In the summer of 2012, Republican super PACs ran an ad campaign among twelve swing states focusing on the issue of unemployment.[3] The success of the ad campaign depended on the importance of unemployment for those particular states. What was the distribution of unemployment among them? The raw data are found in Table 4.4.

The standard deviation compares the value of each case with the mean. Making this comparison is a multistep process, so we are going to set up a work table that includes a separate column for each step in the process. The first step is to calculate the mean. Just like we did in the previous chapter, we add up all the values of our variable and divide by the number of cases. Summing the values of X, we get 87.2. Counting the number of rows, we find that $N = 12$. By dividing the sum of values of our unemployment variable by the number of cases, we find that the average unemployment rate for the swing states is 7.27 percent. (Remember that the lower case Greek letter μ—pronounced "myu"—is the symbol for the mean.)

Table 4.5 Unemployment Rate for Swing States, Frequency

State Case	Unemployment Rate (%) X	Distance from Mean $X - \mu$	Squared Distance $(X - \mu)^2$
Colorado	8.1	0.83	0.6889
Florida	8.6	1.33	1.7689
Iowa	5.1	−2.17	4.7089
Minnesota	5.6	−1.67	2.7889
Nevada	11.6	4.33	18.7489
New Hampshire	5.0	−2.27	5.1529
New Mexico	6.7	−0.57	0.3249
North Carolina	9.4	2.13	4.5369
Ohio	7.3	0.03	0.0009
Pennsylvania	7.4	0.13	0.0169
Virginia	5.6	−1.67	2.7889
Wisconsin	6.8	−0.47	0.2209
$N = 12$	$\Sigma X = 87.2$ $\mu = 87.2/12$ $= 7.27\%$	Sum of Squares = $\Sigma(X - \mu)^2 = 41.7468$ Variance = SS/N = 3.4789 $\sigma = \sqrt{\text{Variance}} = \sqrt{3.4789}$ Standard Deviation = $\sigma = 1.87\%$	

Source: "Local Area Unemployment Statistics: Unemployment Rates for States," *Bureau of Labor Statistics*, 2012. www.bls.gov/web/laus/laumstrk.htm.

The next step is to find how far each state is from the mean. Table 4.5 adds two columns to the basic frequency in order to do the work for calculating the standard deviation. In the third column, we subtract the mean from the value for that case. That gives us how far each case is from the mean. At some point, you might be tempted to just add up the differences in column 3. Unfortunately, if you ever do this, you'll find that the negative values cancel out the positive values, so the third column totals to zero. When we think about distances, we actually want positive numbers. It would be nice if we could take the absolute value of the difference between the mean and each value because then we would have all positive distances. But mathematically, absolute values are hard to compute. The tradition in statistics is to square the difference in order to make the distance positive. The fourth column squares the difference between the value of X and the mean. Because in column 4, the squared distance from the mean is always positive, we can add up this column. This is called the sum of squares (SS). We want the average distance, so the next step after summing the column is to divide by N. At this point, we have calculated the statistic known as the **variance**, or the average squared distance from the mean. But in the process of calculating the variance we squared the distance, so the distance has been exaggerated. Thus, the final step is to take the square root, so that the magnitude is more in line with the notion with which we began of how far, on average, each case is from the mean. The square root of the variance is called the **standard deviation**. Once again, there is a Greek letter that statisticians use: The standard deviation is given by the lower-case Greek letter sigma, which looks like a lower case "b" that has fallen over: σ. After going through these steps, we end up with a standard deviation of 1.87 percent. If you look at the data, the mean and standard deviation make sense. The mean of 7.27 percent is just about in the middle. If you sandwich the mean with a standard deviation on each side (7.27% ± 1.87%), you will normally include about two-thirds of the cases. In this case, the sandwich (5.40% < X < 9.14%) includes eight cases, which is precisely two-thirds of our N of 12.

Table 4.6 State Unemployment Rates, May 2012, Frequency Distribution

Unemployment Rate (%) X	Number of States f_x
3	1
4	2
5	5
6	7
7	14
8	9
9	9
10	2
11	1
Total	50

Source: "Local Area Unemployment Statistics: Unemployment Rates for States," *Bureau of Labor Statistics*, 2012. www.bls.gov/web/laus/laumstrk.htm.

Calculate the Standard Deviation from Tabular Data

As with calculating the mean, when you get many cases, it is easier to calculate the standard deviation from a table than from raw data. We don't know whether the average unemployment rate of 7.27 percent for the swing states is above or below the average for the country as a whole. Let's calculate the mean and standard deviation for all the states using tabular data. Table 4.6 shows this frequency distribution, although for ease of calculations, I've rounded the unemployment rate to the nearest percent.

We now set up a work table (in Table 4.7) to make the necessary series of calculations straightforward. Remember that to calculate the mean for tabular data, we couldn't just add up the values of X like we did with the table of raw data because for some values of X, there is more than one state with that level of unemployment. So we create a new column in which we multiply the level of unemployment X by the number of states (f_x) in that row. We then sum that column. Finally, we divide that total by N to get the mean unemployment rate of 7.20 percent.

After calculating the mean in the third column, we add a fourth column in which we subtract the mean from each value of X: $X - \mu$. In the fifth column, we square that difference $(X - \mu)^2$ for each row. In computing the standard deviation from tabular data, we need a sixth column to account for the fact that each row contains multiple cases. In it, we multiply the squared distance from the mean by the number of cases in each row f_x: $(X - \mu)^2 f_x$. At the bottom, we add up all the values in that last column: $\Sigma(X - \mu)^2 f_x$. We call this our sum of squares. Next, we divide that sum by N—this gives us the variance. Finally, we take the

Table 4.7 Work Table: State Unemployment Rate, May 2012

Unemployment Rate X	Number of States f_x	Xf_x	$X - \mu$	$(X - \mu)^2$	$(X - \mu)^2 f_x$
3	1	3	−4.2	17.64	17.64
4	2	8	−3.2	10.24	20.48
5	5	25	−2.2	4.84	24.20
6	7	42	−1.2	1.44	10.08
7	14	98	−0.2	0.04	0.56
8	9	72	0.8	0.64	5.76
9	9	81	1.8	3.24	29.16
10	2	20	2.8	7.84	15.68
11	1	11	3.8	14.44	14.44
	$\Sigma f_x = 50$ $N = 50$	$\Sigma Xf_x = 360$ $\mu = 360/50$ $\mu = 7.20\%$	Sum of Squares = SS = $\Sigma(X - \mu)^2 f_x = 138.00$ Variance = SS/N = 2.76 $\sigma = \sqrt{\text{Variance}} = \sqrt{2.76}$ Standard Deviation = $\sigma = 1.66\%$		

Source: "Local Area Unemployment Statistics: Unemployment Rates for States," *Bureau of Labor Statistics*, 2012. www.bls.gov/web/laus/laumstrk.htm.

square root of the variance to get the standard deviation of 1.66 percent. Remember that if you go above and below the mean by a distance of one standard deviation, the resulting range should include about two-thirds of the cases. In this case, 60 percent of the states have unemployment rates of 7.20 percent ± 1.66 percent (or 5.54% < X < 8.86%). Recall from the previous section that the swing states had a mean unemployment rate of 7.27 percent with a standard deviation of 1.87 percent. It looks like the swing states of the previous section are fairly typical of the country in terms of their unemployment rates. We can expect that the swing states would respond to the ad campaign in the same way as the rest of the country.

Proportions as a Special Case. You might have noticed that all of the variables we've used in this chapter have had an interval/ratio level of measurement. The notion of dispersion assumes at a minimum that the values have order. And because the standard deviation requires the calculation of the mean, you need at least interval-level data for that. You could potentially find a five-point summary for ordinal-level variables, but, depending on the data, that may or may not have much meaning.

The one exception is with dichotomous variables—variables that may be nominal but have only two categories. In that case, you actually can calculate a mean and standard deviation, although the mean would have a specific interpretation. Take, for example, race. Race is a nominal-level variable, and so technically, you can't calculate a mean. But suppose you have a variable X for which you were to code all minorities as "1" and all non-Hispanic Whites as "0." Table 4.8 shows the distribution of minorities in the United States.

If we set this up in the standard worksheet for tabular data, we can calculate the mean and standard deviation. Notice in Table 4.9 that the mean is 0.36. What does that mean? Think about the values of your variable. If non-Hispanic whites are coded 0 and minorities are coded 1, what the mean is telling you is that 36 percent of Americans are minorities. We actually already saw that in the percentage column of Table 4.8. We could restate that percent as a proportion (given by a lower case "p") so that p = 0.36. So, as long as you code a dichotomous variable with one value as 0 and the other as 1, the mean is telling you what proportion of the cases are coded 1.

In the work table, I've calculated the math for the standard deviation. The interesting thing about this is that the population size ends up being irrelevant when calculating the

Table 4.8 Minority Population in the United States, 2010 Census: Distribution

Minority? X	Frequency (in millions) f_x	Percent
0 White, non-Hispanic	196.8	63.7
1 Minority	111.9	36.2
Total	308.7	99.9

Source: "Overview of Race and Hispanic Origin: 2010," *United States Census Bureau*, 2011. www.census.gov/prod/cen2010/briefs/c2010br-02.pdf.

Table 4.9 Minority Population in the United States, 2010 Census: Calculated

Minority? X	Population in Millions f_x	Xf_x	$X - \mu$	$(X - \mu)^2$	$(X - \mu)^2 f_x$
0	196.8	0	−0.36	0.13	25.51
1	111.9	111.9	0.64	0.41	45.83
	$\Sigma f_x = 308$ $N = 308$	$\Sigma Xf_x = 111.9$ $\mu = 111.9/308$ $\mu = 0.36$		$SS = \Sigma (X - \mu)^2 f_x = 71.34$ Variance $= SS/N = 0.23$ $\sigma = \sqrt{\text{Variance}} = \sqrt{.23}$ Standard Deviation $= \sigma = 0.48$	

Source: "Overview of Race and Hispanic Origin: 2010," *United States Census Bureau*, 2011. www.census.gov/prod/cen2010/briefs/c2010br-02.pdf.

standard deviation for a proportion. I've used N in Table 4.9 in the equations as appropriate, but I could just as easily have used the following equation, which uses only the mean or proportion (p) and doesn't use N at all:

$$\sigma = \sqrt{p(1-p)}$$
$$= \sqrt{0.36(1-0.36)}$$
$$= \sqrt{0.36(0.64)}$$
$$= \sqrt{0.23}$$
$$= 0.48$$

This finding isn't magic—it's hidden in the algebra that I used in calculating the standard deviation in Table 4.9. Simply by the nature of using proportions, the Ns end up cancelling out. So all you need to know is what proportion of cases are in the two categories ($p = \mu$) and you can calculate the standard deviation.

Demographers have been making predictions about how, if current trends continue, the United States will become a majority minority country in the not-so-distant future. The 2010 census documented the first time that white births were exceeded by minority births within the United States. The Census Bureau now predicts that whites will be in the minority by 2040.[4] We can see that trend by comparing the current population distribution of the 2010 census with the prior distribution of the 2000 census. In the 2000 census, non-Hispanic whites made up 69.1 percent of Americans. This means that 30.9 percent of Americans were minorities. In this case, the proportion of minorities is 0.309. To get the standard deviation for 2000, we can simply plug that into the equation:

$$\sigma = \sqrt{p(1-p)}$$
$$= \sqrt{0.309(1-0.309)}$$
$$= \sqrt{0.309(0.691)}$$
$$= \sqrt{0.21}$$
$$= 0.46$$

Notice that the standard deviation was less in 2000 ($\sigma = 0.46$) than in 2010 ($\sigma = 0.48$). The higher standard deviation reflects the fact that the United States has become more diverse. The closer the proportions of the two groups, the higher the standard deviation. Table 4.10 shows the standard deviations associated with various proportions. You can see that the standard deviation keeps increasing until $p = 0.5$. At that point, the standard deviation begins to decrease again.

The more equal the two proportions, the higher the standard deviation. Look at Table 4.11, which contains the distributions for Montana and Mississippi. If I asked you which state has the higher variance, you are going to be tempted to answer Montana because the difference between 87.8 and 12.2 is much greater than the difference between 58.0 and 42. But in reality, a community is most diverse when the proportions are equal. If you live in Montana, which is almost 90 percent white, you experience a

Table 4.10 Standard Deviations Associated with Various Proportions

p	1 − p	p(1 − p)	$\sigma = \sqrt{p(1-p)}$
0.1	0.9	0.09	0.30
0.2	0.8	0.16	0.40
0.3	0.7	0.21	0.46
0.4	0.6	0.24	0.49
0.5	0.5	0.25	0.50
0.6	0.4	0.24	0.49
0.7	0.3	0.21	0.46
0.8	0.2	0.16	0.40
0.9	0.1	0.09	0.30

low ethnic and racial variation. Because the state is so homogeneous, there is little variability. In contrast, Mississippi is heterogeneous—it is only 58.0 percent non-Hispanic white. If you live in Mississippi, you experience a high level of variation. Plugging the two proportions into the equation for the standard deviation, you get the following two standard deviations:

$$\sigma_{\text{Montana}} = \sqrt{p(1-p)}$$
$$= \sqrt{.122(.878)}$$
$$= .327$$

$$\sigma_{\text{Mississippi}} = \sqrt{p(1-p)}$$
$$= \sqrt{.42(.58)}$$
$$= .494$$

Because Mississippi is more racially diverse, it has a higher standard deviation than Montana.

The knowledge that equal proportions translate into a high standard deviation gives you a language to discuss dispersion for nominal and ordinal data. In some instances, you can use the interquartile range to describe ordinal data, but there is no number that you can report to summarize the dispersion of a nominal-level variable. With your understanding of calculating the standard deviation for dichotomous variables, you can, however, make a qualitative evaluation of whether there is a high or low level of dispersion. If you see that the categories are of roughly equal proportions, you can say that there is a high level of variation and so the average (whether median or mode) does not tell the whole story. If one category dominates, you can say that there is a low level of variation and so the average does a good job summarizing the variable.

Table 4.11 Minority Population in Montana and Mississippi

Minority? X	Montana Percent	Mississippi Percent
0 White, non-Hispanic	87.8	58.0
1 Minority	12.2	42.0
Total	100.0	100.0

Source: "State and County QuickFacts," *U.S. Census Bureau*, 2012. http://quickfacts.census.gov/qfd/states/30000.html.

SUMMARIZING THE MATH: DISPERSION

There are two major approaches to describing how dispersed a variable is. One approach is to give ranges connected to specific points of the distribution. To find these key points, we need only add a cumulative percent column to our frequency. The other approach looks at how far each case is from the mean by calculating the standard deviation. In calculating the standard deviation, we set up our work table differently depending on whether we are working with raw or tabular data.

Calculate the Range, Interquartile Range, and Five-Point Summary

Normally, we use the different kinds of ranges to describe data that are already in a table.

- In a frequency table, the data should already be in order from smallest to largest. You find the range by subtracting the smallest value of X (or Q_0) from the largest value of X (or Q_4) to get a single number that corresponds to how widely the values are distributed. Frequently, though, people will report the range in terms of the two end points.

- To get the interquartile range, add a column to your frequency table in which you calculate the cumulative percent. The value of X that is associated with the row where the cumulative percent crosses over 25 percent is the lower end of the interquartile range. The value of X where it crosses over 75 percent is the upper end of the interquartile range. Technically, you subtract these two values to get a single number for the interquartile range. But usually, it is reported as being between these two points. You can conclude that the middle half of the cases lie in this range.

- The five-point summary is usually given in brackets as $\{Q_0, Q_1, Q_2, Q_3, Q_4\}$. The minimum ($Q_0$) and maximum ($Q_4$) are the two endpoints from the range, Q_1 and Q_3 define the interquartile range, and the median (Q_2) is the value of the variable for the 50th percentile. In addition to reporting these points numerically in brackets, you can present them visually in a box-and-whiskers plot, where you place a box around the interquartile range containing a line for the median and draw lines out from the ends of the box to the minimum and maximum values.

A Political Example. Take the unemployment data from Table 4.6 and turn them into a proper frequency table with columns for percent and cumulative percent. This is found in Table 4.12.

1. Find the range by subtracting the smallest value of X (Q_0) from the largest value of X (Q_4).
   In the first column, I see that the highest percentage of unemployment experienced by a state is 11 percent. The lowest is 3 percent. The range is 11 − 3 = 8 percent.

Table 4.12 State Unemployment Rates, May 2012

Unemployment Rate in Percent X	Number of States f_x	Percent	Cumulative Percent
3	1	2.0	2.0
4	2	4.0	6.0
5	5	10.0	16.0
6	7	14.0	30.0
7	14	28.0	58.0
8	9	18.0	76.0
9	9	18.0	94.0
10	2	4.0	98.0
11	1	2.0	100.0
Total	50	100.0	

Source: "Local Area Unemployment Statistics: Unemployment Rates for States," *Bureau of Labor Statistics*, 2012. www.bls.gov/web/laus/laumstrk.htm.

2. The value of X, which is associated with the row where the cumulative percent crosses over 25 percent, is the lower end of the interquartile range (Q_1). The value of X where it crosses over 75 percent is the upper end of the interquartile range (Q_3).

   In the fourth column, I see that the cumulative percent crosses 25 percent in the fourth row. This row has a value of 6 percent. The cumulative percent crosses 75 percent in the sixth row. It has a value of 8. The interquartile range is 8 − 6 = 2 percent.

3. The five-point summary is usually given in brackets as $\{Q_0, Q_1, Q_2, Q_3, Q_4\}$. You can present these points visually on a box-and-whiskers plot: Above a scale, you place a box around the interquartile range containing a line for the median. You then draw lines out from the ends of the box to the minimum and maximum values.

   In addition to the four points I've already found, I need the median (Q_2), which is found where the cumulative percent exceeds 50 percent. This is in the fifth row, which corresponds to 7 percent unemployment. Thus, the five-point summary is {3, 6, 7, 8, 11}. If I make a box-and-whiskers plot, it looks like Figure 4.2. From this, I can conclude that half of the states have unemployment rates between 6 and 8 percent, but a few have as low as 3 percent or as high as 11 percent.

Figure 4.2 Box-and-Whiskers Plot for State-Level Unemployment

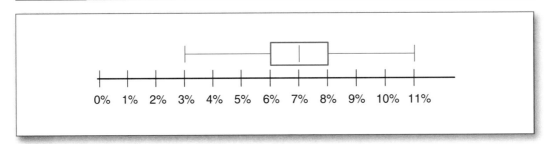

Calculate the Standard Deviation from Raw Data

To calculate the standard deviation from raw data, you begin with the same work table you used to calculate mean and then add two more columns.

1. Calculate the mean (μ) value of your variable by adding up all the values (X) and dividing by the number of cases (N): μ = ΣX/N.

2. Add a column to your table in which you subtract the mean from the value: X − μ.

3. Add another column in which you square the difference from the mean: $(X − μ)^2$.

4. At the bottom of that column, add up all the values of the squared difference: $Σ(X − μ)^2$.

5. Get the average squared difference, or the variance, by dividing by the number of cases: Variance = $Σ(X − μ)^2/N$.

6. Get the standard deviation by taking the square root of the variance: Standard Deviation = σ = $\sqrt{\text{Variance}}$.

A Political Example. In June 2012, the College Board, the organization responsible for the SATs and Advanced Placement exams, set up 857 desks on the national Mall in Washington, D.C. Each desk represented one of the 857 students who drop out of high school every hour. Their goal was to raise education as an issue in the presidential election. The 857 is a nationwide average. How do the different states compare in dropout rates? In this section, we'll find the average dropout rate (given as the percent of students who dropped out of high school that year) for the twelve swing states addressed by the unemployment campaign. The raw data are found in Table 4.13.

1. Calculate the mean (μ) value of your variable by adding up all the values (X) and dividing by the number of cases (N): μ = ΣX/N.

I set up a work table shown in Table 4.14, where the first column is the case, the second column is the dropout rate. At the bottom of the second column, I sum the dropout rates and get a total of 45.8. I have twelve rows of data, so $N = 12$. The mean is $45.8/12 = 3.82\%$.

2. Add a column to your table in which you subtract the mean from the value: $X - \mu$.
   I add a third column where I take the difference between the dropout rate for the state and the mean dropout rate.

3. Add another column in which you square the difference from the mean: $(X - \mu)^2$.
   I add a fourth column to the work table that squares the difference from the third column.

4. At the bottom of that column, add up all the values of the squared difference: $\Sigma(X - \mu)^2$.
   At the bottom of the fourth column, I sum the squares to get 20.22.

Table 4.13 Work Table: Dropout Rate for Swing States

State Case	Dropout Rate X
Colorado	6.4%
Florida	3.3%
Iowa	2.9%
Minnesota	2.8%
Nevada	5.1%
New Hampshire	3.0%
New Mexico	5.2%
North Carolina	5.2%
Ohio	4.3%
Pennsylvania	2.6%
Virginia	2.7%
Wisconsin	2.3%

Source: "Digest of Education Statistics," *National Center for Education Statistics*, 2011. http://nces.ed.gov/programs/digest/d10/tables/dt10_113.asp.

Table 4.14 Dropout Rate for Swing States, Calculated

State Case	Dropout Rate X	Distance from Mean $X - \mu$	Squared Distance $(X - \mu)^2$
Colorado	6.4%	2.58	6.66
Florida	3.3%	−0.52	0.27
Iowa	2.9%	−0.92	0.85
Minnesota	2.8%	−1.02	1.04
Nevada	5.1%	1.28	1.64
New Hampshire	3.0%	−0.82	0.67
New Mexico	5.2%	1.38	1.90

(Continued)

Table 4.14 (Continued)

State Case	Dropout Rate X	Distance from Mean X − μ	Squared Distance (X − μ)²
North Carolina	5.2%	1.38	1.90
Ohio	4.3%	0.48	0.23
Pennsylvania	2.6%	−1.22	1.49
Virginia	2.7%	−1.12	1.25
Wisconsin	2.3%	−1.52	2.31
N = 12	ΣX = 45.8 μ = 45.8/12 = 3.82%	Sum of Squares = SS = Σ(X − μ)² = 20.22 Variance = SS/N = 1.68 σ = √Variance = √1.68 Standard Deviation = σ = 1.30%	

Source: "Digest of Education Statistics," *National Center for Education Statistics*, 2011. http://nces.ed.gov/programs/digest/d10/tables/dt10_113.asp.

5. Get the average squared difference, or the variance, by dividing the sum of squares by the number of cases: Variance = Σ(X − μ)²/N.
   To get the variance, I divide the sum of squares by the number of cases: 20.22/12 = 1.68.

6. Get the standard deviation by taking the square root of the variance: Standard Deviation = σ = √Variance.
   The standard deviation is the square root of the variance: σ = √Variance = √1.68 = 1.30%.

BOX 4.1 Numbers in the News

For years, the United States education system has been compared negatively to other industrial countries on the basis of average test scores. A recent study by the Economic Policy Institute[5] suggests that comparing averages may not be the best way to evaluate the quality of education received. Test scores are highly correlated with the social class of the students. The authors suggest that the United States has had historically low average scores because the class distribution of its students has a higher variation: The United States has around the same percentage of students from the highest social class, but the proportion from the lowest class is about double that from, for example, Canada. A more accurate evaluation would compare scores for students within a given social class.

Calculate the Standard Deviation from Tabular Data

To calculate the standard deviation from tabular data, you begin with the work table you used to find the mean for tabular data and add three columns.

1. Begin by calculating the mean. Remember that to get the number of cases, you add up the frequency column: $N = \Sigma f_x$. And remember that you need to add a third column, which multiplies the value of X by the frequency of that value: Xf_x. To get the total values of X, you sum that column: ΣXf_x. To get the mean, you divide that sum by the number of cases: $\mu = \Sigma Xf_x/N$.

2. Once you've calculated the mean, you add a fourth column in which you subtract the mean from the value of X: $X - \mu$.

3. In the fifth column, you square that difference to make all the values positive: $(X - \mu)^2$.

4. In the last column, you multiply the squared difference by the number of cases in that row: $(X - \mu)^2 f_x$.

5. At the bottom of that column, add up all the values of the squared difference to get the sum of squares: $\Sigma(X - \mu)^2 f_x$.

6. You divide that sum by the number of cases to get the variance: Variance = $\Sigma(X - \mu)^2 f_x/N$.

7. Finally, you take the square root of the variance to get the standard deviation: Standard Deviation = $\sigma = \sqrt{\text{Variance}}$.

A Political Example. What is the distribution of the dropout rate for the entire country? Table 4.15 shows a frequency for all the states with the dropout rate rounded to the nearest percent.

1. Begin by calculating the mean. To get the mean, you first multiply each value times its frequency; second, sum those products; third, divide that sum by the number of cases: $\mu = \Sigma Xf_x/N$.

I set up a work table shown in Table 4.16, with four columns in addition

Table 4.15 State Dropout Rates, Frequency

Dropout Rate in Percent X	Number of States f_x
2	7
3	12
4	11
5	12
6	4
7	3
Total	49*

Source: "Local Area Unemployment Statistics: Unemployment Rates for States," *Bureau of Labor Statistics*, 2012. www.bls.gov/web/laus/laumstrk.htm.

*The dropout rate is not available for Vermont.

Table 4.16 State Dropout Rate, Calculated

Dropout Rate X	Number of States f_x	Xf_x	$X - \mu$	$(X - \mu)^2$	$(X - \mu)^2 f_x$
2	7	14	−2.06	4.24	29.71
3	12	36	−1.06	1.12	13.48
4	11	44	−0.06	0.00	0.04
5	12	60	0.94	0.88	10.60
6	4	24	1.94	3.76	15.05
7	3	21	2.94	8.64	25.93
10	2	20	2.71	7.3441	14.6882
11	1	11	3.71	13.7641	13.7641
	$\Sigma f_x = 49$ $N = 49$	$\Sigma Xf_x = 199$ $\mu = 199/49$ $\mu = 4.06\%$	Sum of Squares = SS = $\Sigma(X - \mu)^2 f_x = 94.82$ Variance = SS/N = 1.94 $\sigma = \sqrt{Variance} = \sqrt{1.94}$ Standard Deviation = $\sigma = 1.39\%$		

Source: "Digest of Education Statistics," *National Center for Education Statistics*, 2011. http://nces.ed.gov/programs/digest/d10/tables/dt10_113.asp.

to the two found in Table 4.15. At the bottom of the frequency, I get the number of cases by summing the frequencies for each row. In the third column, I multiply the value of X (from column 1) by its frequency (in column 2). At the bottom of that third column, I sum all its values and divide by the number of cases to get the mean dropout rate for all the states: $\mu = 199/49 = 4.06\%$.

2. Once you've calculated the mean, add a column in which you subtract the mean from the value of X: $X - \mu$.
   This is in the fourth column.

3. In the next column, you square that difference to make all the values positive: $(X - \mu)^2$.
   In the fifth column, I square the value from the fourth column.

4. In the last column, you multiply the squared difference by the number of cases in that row: $(X - \mu)^2 f_x$.
   In the sixth column, I multiply the value in the fifth column by the number of cases found in the second column.

5. At the bottom of that column, add up all the values of the squared difference: $\Sigma(X - \mu)^2 f_x$.
   The sum of squares is 94.82.

6. You divide that total by the number of cases to get the variance: Variance = $\Sigma(X - \mu)^2 f_x / N$.
   Variance = 94.82/49 = 1.94.

7. Finally, you take the square root of the variance to get the standard deviation:
   Standard Deviation = $\sqrt{\text{Variance}}$.

σ = *Standard Deviation* = $\sqrt{1.94}$ = 1.39%.

Calculate the Standard Deviation for a Dichotomous Variable

Although normally you can calculate a standard deviation only for interval-level data, you can also calculate it for dichotomous nominal- or ordinal-level variables. For the dichotomous variable, you will want to assign a value of 0 to one of the categories and a value of 1 to the other category. You could calculate the standard deviation using either of the prior two techniques (with raw or tabular data). But the math ends up simplifying for dichotomous variables, so that all you need to know is the proportion of cases in the category you gave a value of 1. In this case, the standard deviation is given by

$$\sigma = \sqrt{p(1-p)}$$

where p is the proportion of cases in the category.

A Political Example. Czechoslovakia formed as an independent nation after World War I, during the breakup of the Austro-Hungarian Empire. It was an ethnically diverse country, which led to a lot of tension between the various ethnic groups. At the time of the breakup of the Soviet bloc in 1991, Czechs made up 62.8 percent of the country. What was the standard deviation for the ethnicity of Czechoslovakia?

$$\sigma = \sqrt{p(1-p)}$$

$$= \sqrt{0.628(1-0.628)}$$

$$= \sqrt{0.628(0.372)}$$

$$= \sqrt{0.234}$$

$$= 0.483$$

USE SPSS TO ANSWER A QUESTION WITH MEASURES OF DISPERSION

In SPSS, you request the measures of dispersion in the same window in which you have already requested measures of central tendency. The first step in any data analysis is always to make sure that the data are clean. Using the "Define Variable Properties" in SPSS, scan the values, cleaning any that look wrong and making sure that any missing values are actually being treated as missing by SPSS. Once you know the data are clean, you will, as usual, begin your analysis in the "Analyze" window. We are still learning univariate descriptive statistics, so under "Analyze," you find "Descriptive Statistics" and "Frequency." After selecting the variable you are analyzing, you open the "Statistics" window and request any measures of dispersion and measures of central tendency you might want. Box 4.2 summarizes this process.

BOX 4.2   How to Get Measures of Dispersion in SPSS

After opening your data in SPSS and cleaning them:

>Analyze
>>Descriptive Statistics
>>>Frequencies
>>>>Click on the variable to highlight it
>>>>→To bring the variable over to the "Variable" box
>>>>Statistics
>>>>>Mean
>>>>>Median
>>>>>Mode
>>>>>Range
>>>>>Standard Deviation
>>>>>Percentiles "25" "75"
>>>>>Continue
>>>OK

At this point, the frequency of the variable will show up in the "Output" window.

The more complicated part of univariate statistics is in presenting the data in an understandable way. Translating an SPSS frequency into a professional-looking table is fairly straightforward if you are dealing with nominal- or ordinal-level data because they normally do not have too many categories. Unfortunately, as we saw in the previous chapter, interval-level data frequently have so many categories that you wouldn't want to include them all in a frequency table. The point of tables is to summarize data in a way that is easy to process mentally. But the human brain cannot deal with twenty categories. In order to avoid brain overload, we will normally collapse interval-level data into fewer categories for the tables. Previously in this chapter, I simplified the math by rounding the data. That is not my goal here. Because the computer is going to be doing the math, there is no reason to lose data by rounding. So you will still report the statistics (measures of central tendency and measures of dispersion) for the uncollapsed data. My goal in collapsing the data is only to clarify the presentation of the data in my table. So keep in mind that the subsequent collapsing process applies only to the presentation of the information in the table.

The most important rule of collapsing data is to do it in such a way that the data are easy to understand. First, because our brains have a hard time keeping track of more than about five things at once, you probably want to collapse the values into three to five categories. Second, keep the categories of equal size (in terms of the original scale rather than in numbers of cases) so that the variable doesn't lose its meaning as an interval-level variable. Third, you want to avoid having categories with no cases in them. Fourth, each case must fall into exactly one category.

In order to make sure that each case falls into exactly one category, you need to pay particular attention to the precise values of the original variable. Make sure that you have included both the smallest and largest values. Pay attention to the decimal values for your cutoff points between categories—the upper end of a category needs to include its largest value, but not any larger values. Frequently, we will label categories in even amounts using overlapping endpoints. When we do that, it means that the endpoint is actually included in the lower end of the range, but the range extends only up to (but not including) the upper end of the range. For example, if I were to collapse the unemployment rates for the countries of the world, I would want to collapse them into 10 percent ranges. My labels would be "0–10 percent," "10–20 percent," "20–30 percent," "30–40 percent," "40–50 percent," and so forth. The first range would include any countries with no unemployment all the way up to 9.99 percent (using however many decimals were actually included in the data). The second category would begin with those countries with precisely 10.00 percent all the way up to 19.99 percent, and so forth.

It is easiest to begin the collapsing process by getting (and printing) a frequency for the original variable. Notice the maximum and minimum values—you need to be sure to include them in your categories. Keeping in mind that you want at least three categories, but probably not more than five, decide where some logical cutoff points would be. Remember that the goal is not to have the same number of cases in each category. Rather, you want the width of the categories on the original scale to be of equal size. You might want to divide the range by five to get a starting point for the width of the categories. Then look at the frequency you printed up to see if that width makes sense. The highest priority here is that the ranges make sense and are of equal size—you can adjust the number of categories to assure this. Once you've chosen the ranges for the categories, draw lines between them on your frequency so that you know precisely what values are going to be in each category.

BOX 4.3 How to Collapse Variables in SPSS

After opening your data in SPSS:

>Transform
>>Recode into Different Variables
>>>"*Variable*" → Input Window
>>>Variable Name = "*collapsed variable name*" (only 8 characters)
>>>Variable label = "*collapsed variable label*" (longer and you can use punctuation)
>>>\>Change
>>>\>Old and New Values
>>>>\>Range, Lowest through Value
>>>>>"*upper limit of lowest range*"
>>>>>Value = "1"
>>>>>\>Add
>>>>\>Range
>>>>>"*lower limit of second range*"
>>>>>"*upper limit of second range*"
>>>>>Value = "2"
>>>>>\>Add
>>>>\>Range, Value through Highest
>>>>>"*lower limit of highest range*"
>>>>>Value = "3"
>>>>>\>Add
>>>>\>Continue
>>>\>OK

At this point, it is a good idea to make sure that the new variable is properly collapsed. Using the "Define Variable Properties" command, pull up the new *variable: Make sure that the recoded categories are all there and include the correct number of cases. You will also want to enter the ranges into the value label box for each category.*

Once you've decided on the precise ranges of your collapsed categories, follow the process summarized in Box 4.3. (For future reference, remember that this table is found in Appendix 2, along with all of the other "How to Use SPSS" boxes.) To actually collapse the variable, you will use the "Transform" command. You have several options here. I recommend that you "Recode into Different Variables." If you make a mistake (and you inevitably will), this will guarantee that the original data are still available so that you can redo your work correctly. Move the variable over to the "Input Variable" box. Give the variable a name that has eight or fewer characters. When you click on "Change," the name of your new variable will appear in the "Input Variable" next to the name of the original variable. Click on "Old and New Values" to open the window where you can define the ranges of the categories in the way you want the variable collapsed. One of the options in the window is "Range, Lowest through Value." If you use this option, you need only enter the upper end of the bottom category, and SPSS will be sure to include all values less than this. Similarly, there is an option "Range, Value through Highest" where you need to include only the smallest value of the top category and SPSS will include all larger values in this category. For the middle categories, you need to use the "Range" command. This is where you have to be careful in defining the endpoint of each of your categories. The lower limit will be the whole number at the bottom of the range; the high end will be the decimal value of the value just under the beginning of the next category. Remember that each case must be in one and only one category.

After you complete the command for creating the collapsed variable, use the "Define Variable Properties" command to pull up the *new* variable. Look over the categories listed to make sure that the recoded categories are all there. Also compare the number of cases in each category to make sure that they line up. Next, go to the right-hand side to the boxes for value labels—enter the ranges into the appropriate box for each category. Finally, get a frequency of the new variable to translate into a professional-looking frequency table for use with your memo. (Remember that instructions for creating a frequency table and writing a memo are in Appendix 1.)

An SPSS Application: Law School Tuition

It is January 2011. You are interning at the *New York Times*. Last week, the *Times* ran an article attacking the financial soundness of attending law school. In response, the newspaper has received several replies from law schools stating that the article misstated the statistics on employment rates, cost of tuition, and salary rates. There has already been a correction made to the article regarding one of the individuals used as a source in the story. But the editor would like you to research the statistics and answer the following question: What do most law students pay each year in tuition? In your memo, include the appropriate measures of central tendency and measures of dispersion.

1. Open and clean the data in SPSS.
   Using "Define Variable Properties," I see that there are no missing data and the numbers all look right. I take notes on the range and decide on how I will collapse it into logical subranges. I decide to collapse into the following categories: $10,000–19,999, $20,000–29,999, $30,000–39,999, and $40,000–49,999.

2. Get a frequency for your variable, including the appropriate measures of central tendency and dispersion.
   I get a frequency for Tuition, as shown in Figure 4.3. I check the output window (shown in Figure 4.4) to make sure that the data in the frequency look fine. I also copy the Descriptives table into a Word document because I am going to need it for my memo.

Figure 4.3 Frequency for Tuition

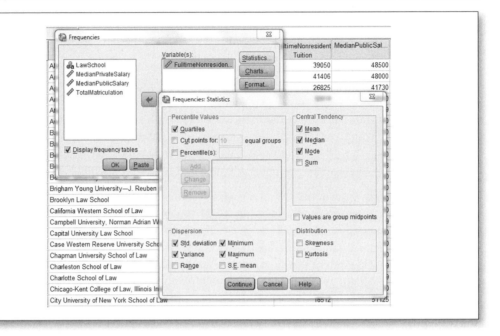

Figure 4.4 Output from Frequency

→ **Frequencies**

[DataSet1] C:\Users\marchantsht1\Documents\Quantitative Analysi

Statistics

Full time Nonresident Tuition

N	Valid	195
	Missing	0
Mean		34150.39
Median		34220.00
Mode		30850[a]
Std. Deviation		7342.764
Variance		53916187.26
Minimum		12580
Maximum		49020
Percentiles	25	29553.00
	50	34220.00
	75	39130.00

a. Multiple modes exist. The smallest value is shown

Full time Nonresident Tuition

		Frequency	Percent	Valid Percent	Cumulative Percent
Valid	12580	1	.5	.5	.5
	15330	1	.5	.5	1.0
	16085	1	.5	.5	1.5

3. In SPSS, recode your original variable into a new variable in order to collapse the values into categories.

   I use the "Transform," "Recode into Different Variable" command to create a collapsed variable with the $10,000 intervals I had decided upon. Figure 4.5 shows where the commands are. Figure 4.6 shows how to designate the old and new variable names.

Figure 4.5 Recode into Different Variable Command

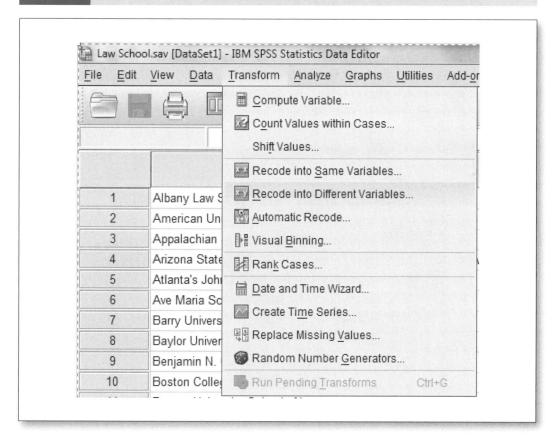

Figure 4.6 Giving a Name and Label to a New Variable

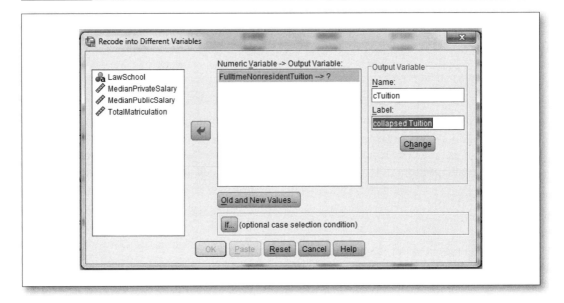

After clicking "Change" to designate this name and label for the output variable, I click on "Old and New Variables" to code the changes. These commands are shown in Figure 4.7. After clicking "Add" one last time, and then "Continue," SPSS returns me to the original recode page and I click "OK" to process the full request.

Figure 4.7 Entering Recode Commands

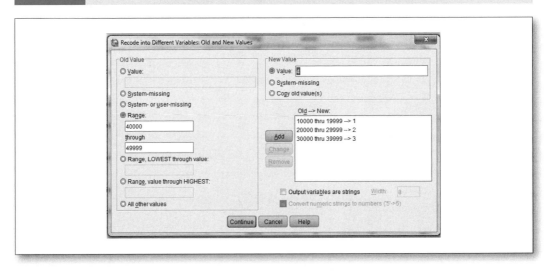

4. Use "Define Variable Properties" to attach labels to your values.
   For the new collapsed variable, I identify the intervals included in each of the categories.

5. Get the frequency for your collapsed variable.
   I do this in the same way as before except that I request the collapsed variable, and I do not need the statistics because I will be reporting those from the uncollapsed variable. This frequency is shown in Figure 4.8.

Figure 4.8 Frequency of the Collapsed Variable with Correct Value Labels

collapsed Tuition

		Frequency	Percent	Valid Percent	Cumulative Percent
Valid	$10-20,000	6	3.1	3.1	3.1
	$20-30,000	45	23.1	23.1	26.2
	$30-40,000	103	52.8	52.8	79.0
	$40-50,000	41	21.0	21.0	100.0
	Total	195	100.0	100.0	

6. Use the frequency of the collapsed variable to create a frequency table and include it with your memo according to the instructions in Chapter 2.
   Instead of searching through Chapter 2 for the instructions, I remember that Appendix 1 has all the instructions on professional writing. I look through there for the page titled "Create a Frequency Table in Word." Using those instructions, I take the frequency that SPSS gave me as output and transform it into a professional-looking table.

Memo

To:	Editor, *New York Times*
From:	T. Marchant-Shapiro, intern
Date:	January 11, 2011
Subject:	Law School Tuition

You asked me to follow through on the article about the financial implications of attending law school by analyzing whether the article's description of the cost of tuition was fair and accurate. In order to do that, I found the cost of tuition for all U.S. law schools as reported by *U.S. News & World Report*. Although the article accurately represented the cost of tuition at top-tier schools, many schools have lower tuition than the $43,000 reported in the article.

The attached table shows the tuition rates for U.S. law schools. These have a range of $36,440—from a low of $12,580 at Southern University Law Center in Baton Rouge, Louisiana, to a high of $49,020 at Cornell. The median tuition is $34,220, which is very close to the mean of $34,150. Half of all law schools charge $30,000–$40,000 a year for tuition.

The $43,000 tuition referenced in the article does not reflect what most law schools charge. In reality, three out of four law schools charge less than $40,000 per year. The average of $34,000 is still a lot of money. Perhaps the author of the article could have still made his argument using that number. In the future, he would be well advised to gather more data.

Tuition for U.S. Law Schools in 2010

Annual Tuition*	Frequency	Percent
$10–20,000	6	3.1
$20–30,000	45	23.1
$30–40,000	103	52.8
$40–50,000	41	21.0
Total	195	100.0

Source: "Best Law Schools," U.S. News & World Report, 2011. http://grad-schools.usnews.rankingsandreviews.com/best-graduate-schools/top-law-schools/law-rankings (February 7, 2011).

*Based on non-resident tuition.

Your Turn: Measures of Dispersion

YT 4.1 In 1976, Jimmy Carter challenged Gerald Ford for the U.S. presidency. In the lead up to the debates, Carter's staff were worried that Ford might have an advantage simply because he would look more presidential. Although Ford had not won a national election, he was still the sitting president. In addition, they were concerned that he stood three inches taller than Carter. The lore of the presidency talks about George Washington and Abraham Lincoln as natural leaders because their height commanded respect. So as part of the negotiations about the 1976 debates, Carter's staff insisted that he have a small riser to stand on behind the podium so that neither of the two competitors had a height advantage. A similar negotiation has occurred during all subsequent debates. Although the assumption of these negotiations is that height can enhance leadership abilities, the question is, "Do presidents tend to be taller than average?" Table 4.17 shows a frequency for the heights of all the presidents in inches. Find the five-point summary for the data.

Table 4.17 Presidential Height in Inches

Height X	Number of Presidents f_x	Percent	Cumulative Percent
64	1	2.3	2.3
66	2	4.7	7
67	2	4.7	11.7
68	5	11.6	23.3
69	3	7.0	30.3
70	5	11.6	41.9
71	3	7.0	48.9
72	12	27.9	76.8
73	3	7.0	83.8
74	4	9.3	93.1
75	1	2.3	95.4
76	2	4.7	100.1
Total	43	100.1	

Source: "Heights of Presidents and Presidential Candidates of the United States," *Wikipedia, The Free Encyclopedia*, May 31, 2013. http://en.wikipedia.org/wiki/Heights_of_presidents_and_presidential_candidates_of_the_United_States.

YT 4.2 — Is it possible that the focus on height is a TV thing? Using the raw data in Table 4.18, calculate the mean and standard deviation for the height of presidents during the TV era.

Table 4.18 Height of Presidents during the TV Era

President	Height	President	Height
Dwight D. Eisenhower	70.5	Ronald Reagan	73
John F. Kennedy	72	George H. W. Bush	74
Lyndon Baines Johnson	76	Bill Clinton	74
Richard Nixon	71.5	George W. Bush	71.5
Gerald Ford	72	Barack Obama	73
Jimmy Carter	69.5	$N = 11$	

Source: "Heights of Presidents and Presidential Candidates of the United States," *Wikipedia, The Free Encyclopedia*, May 31, 2013. http://en.wikipedia.org/wiki/Heights_of_presidents_and_presidential_candidates_of_the_United_States.

YT 4.3 If height really is a component in leadership ability (as opposed to just making candidates more attractive for television), we would expect that all of the presidents would be above average, not just the recent ones. Table 4.19 repeats the first two columns from Table 4.17. Set up a work table and calculate the mean and standard deviation from the data.

Table 4.19 Height of All U.S. Presidents

Height X	Number of Presidents f_x	Height X	Number of Presidents f_x
64	1	71	3
66	2	72	12
67	2	73	3
68	5	74	4
69	3	75	1
70	5	76	2

YT 4.4 After the breakup of the Soviet bloc, Czechoslovakia faced increased ethnic tensions and so voted in 1992 to peacefully dissolve into two countries: the Czech Republic and the Slovak Republic. Although the boundary was drawn so that most Czechs ended up in the Czech Republic and most Slovaks ended up in the Slovak Republic, there is still some ethnic diversity in each of the two countries. Calculate the standard deviation for each country.

A. Czech Republic: 90.4 percent Czech
B. Slovak Republic: 85.75 percent Slovak

Apply It Yourself: Evaluate Graduates' Salaries

Now it is your turn to be an intern at the *New York Times*. The editor wants you to evaluate the law school article's claims about the salaries earned by law school graduates and write him a memo with your findings. What do most law students earn after graduation? Read the article "Is Law School a Losing Game?" by David Segal, available at www.nytimes.com/2011/01/09/business/09law.html?_r=1&scp=2&sq=law%20school&st=cse.

Use the "Law School" dataset to analyze the median salaries earned by graduates from the various U.S. law schools.[6] Find the appropriate measures of central tendency and measures of dispersion for graduates going into private practice. Do the same for those going into public practice. Be sure to check your frequencies: a value of −9 needs to be set to missing. Be sure to collapse the variables so that you can include intelligible frequencies for both variables in your memo.

1. After opening "Law School" dataset in SPSS, make sure that the variables "MedianPrivateSalary" and "MedianPublicSalary" are clean using "Define Variable Properties." In particular, check to make sure that "−9" is set to missing. Take notes on where you think it would be logical to divide the ranges into intervals for the collapsed variables. (For standardization purposes, your professor might prefer you to use ranges of $10,000 for collapsing public salary and ranges of $30,000 for collapsing private salary.)

2. Get a frequency for MedianPrivateSalary, including the appropriate measures of central tendency and dispersion.

    >Analyze
    >>Descriptive Statistics
    >>>Frequency
    >>>>MedianPrivateSalary → Variable Box
    >>>>Statistics
    >>>>>Mean
    >>>>>Median
    >>>>>Quartiles
    >>>>>Range
    >>>>>Standard deviation
    >>>>>Minimum
    >>>>>Maximum
    >>>>OK

3. In SPSS, recode your original variable into a new variable in order to collapse the values into categories.

    >Transform
    >>Recode into Different Variables
    >>>Variable Name=*"cPrivateSalary"*
    >>>Variable Label=*"collapsed Private Salary"*
    >>>Old and New Variables
    >>>>Range, Lowest through Value
    >>>>>*"high end of lowest catgory"*
    >>>>>Value=*"1"*
    >>>>Add
    >>>>Range

>*"low value of second category"*
>*"high value of second category"*
>Value="2"
>>Add
>>Range
>>>*Same process for other categories*
>>Range, Value through Highest
>>>*"low end of highest category"*
>>>Value="*number of categories*"
>>>Add
>>Continue
>Change
>OK

4. Use "Define Variable Properties" to attach labels to the values of your collapsed variable.

5. Get the frequency for your collapsed variable.

>Analyze
>>Descriptive Statistics
>>>Frequency
>>>>"*cPrivateSalary*"→Variable Box
>>>OK

6. Repeat this process for MedianPublicSalary. (For standardization purposes, your professor might prefer you to use four ranges of $10,000 in your collapsed variable, with the last range extending up to the outlier.) When you begin to recode this variable, click "Reset" to remove your prior commands so that you can begin with a clean slate.

7. Use the frequencies of the collapsed variables to create frequency tables according to the instructions in Chapter 2 and include them with your memo. Be sure to include the SPSS output with your memo, including the original statistics table and the two frequencies of your collapsed variables.

Key Terms

Box-and-whiskers plot (p. 80)
Dichotomous variable (p. 86)
Interquartile range (p. 79)

Range (p. 79)
Standard deviation (p. 84)
Variance (p. 84)

CHAPTER 5

Continuous Probability

So What's Normal Anyway?

In games of chance, each possible outcome has a single objective probability. Casinos value their ability to control the odds on such games because they can assure themselves a profit. Sometimes players will win, but the casino can be assured that over the long haul, they will make over a 15 percent profit. Casinos do not, however, hold players who change those odds in high esteem. The movie *21*[1] depicts five students at MIT being trained to count cards by their math professor. Their goal is to use information to change the odds in their favor. If you choose one card from a new deck, you know in advance that the probability of getting an ace is 4/52. The probability of getting a card worth ten points is higher because not only are there four 10s worth ten points, but also four kings, queens, and jacks, giving you a probability of 16/52. But as cards get dealt, the probability changes as those sixteen cards are either dealt or not dealt. Card counters keep track of those changing odds in order to increase the probability of winning. Needless to say, casinos dislike card counters, and *21* depicts the brutality of establishments faced with losses in their profit margin.

Similarly, the outcomes of sporting events do not have a single objective probability connected to them. Odds makers rely on information about various factors that affect the outcome of a game to determine the payout of sporting events. For example, in football they will collect information about individual players' height, weight, speed, and strength to quantify a power rating. They then aggregate all of the players' power ratings into a rating of each team's defense and offense. For any given game, the odds makers will compare the two teams' ratings to determine a point spread. Their goal is to draw a line where half of the bets will be on each side, with a payout that guarantees for them the same kind of profit margin that casinos want. In general, they are able to do that because most gamblers are neophytes in comparison to the odds makers because they are less informed. Some gamblers, however, turn the sport into a business, gathering as much if not more information than the odds maker. Their goal is to find the instances when the line drawn by the odds maker is off. When they find this kind of discrepancy, they bet on their understanding that they have an increased probability of winning. *60 Minutes*[2] did a piece on Billy Walters, whose métier is sports gambling. He has a slew of consultants on his payroll collecting information about factors that could affect the outcome of games, from the weather to sports injuries. Many of those consultants are mathematicians who analyze the data to come up with a more nuanced prediction of the outcome of the game than a single point spread, one that describes a continuous range of probabilities connected to a range of possible point spreads. If the book maker has misplaced the line, Walters can take advantage of it. Walters does not always win, but his information allows him to estimate the probabilities well enough to win millions in any given year.

Similarly, political events frequently will be connected to a continuum of probabilities. For example, one of the fears since the end of the Cold War has been of state collapse. As a result, the CIA funded "The Political Instability Task Force" to develop models identifying risk factors in order to quantify the probability of state failure for any country at any given point in time. As economies worsen, as governments become less accountable to their citizenry, and as infrastructure deteriorates, the likelihood of state failure increases.

Statisticians describe such continuous probabilities using many different shaped curves. But the curve that they use the most often is the normal curve. Odds makers choose a spread that they believe is most likely to occur in the game. On either side of that line, the probability decreases. If you were to graph the probabilities, you would get a curve. At the left edge, the probability would be very low. As you headed toward the center, the probability would increase slowly at first and then more rapidly. Once you approach the middle, the increase would start to level off until it peaked. The probability would then begin to fall, slowly at first and then more rapidly. Further to the right, the probability would begin to level off and approach zero. The shape of this curve looks a little like the Liberty Bell—without the crack.

A bell curve can actually take many shapes—some skinny and tall, some flat and spread out. But the specific shape we normally use in statistics is called, quite appropriately, the normal curve. This curve repeats itself in nature. The velocity of gas molecules is normally distributed. The size of any given plant or animal is normally distributed. Doctors may say that the normal temperature for a human is 98.6°F, but in reality, we each have our own normal temperature that (if we aggregate over the whole human family) is normally distributed around the average of 98.6°. In this chapter, we will look at how our knowledge that many events are normally distributed can help us understand the probability of their occurrence.

THE NORMAL CURVE

Just as the normal curve can be found in nature, it can also be found in human endeavors. Suppose you were watching a batting competition between Major League batters. If you were to graph how far they hit the ball, you would also probably come up with a normal curve distribution. Babe Ruth would undoubtedly be way to the right of the curve—maybe over the fence. A few players would hit bunts. But most of the players would bat somewhere in the outfield. Similarly, in any given class I teach, I will have a few outstanding students; a few who are just taking up space; and many who, depending on how much work they put in, perform somewhere in the middle.

Although some normal curves are natural, some are created by design. For example, the administrators of the LSAT score the exams in such a way as to artificially create a normal curve. By definition, the highest score is assigned "180" and the lowest, "120." The average score is set at "150." For any given administration of the LSAT, the College Board calculates the standard deviation in the same way we did in Chapter 4. It then scales the scores so that one standard deviation equals 10 points on the final score. Students are assigned scores on the exam based on where they place in relationship to everyone else who took the test that day.

Figure 5.1 Normal Curve

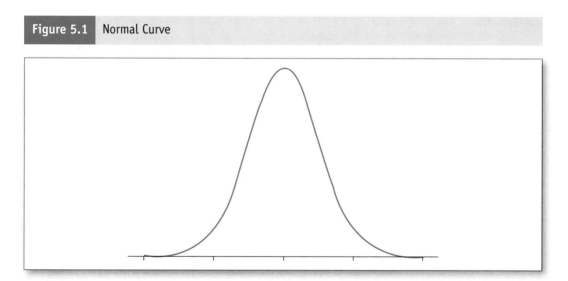

The normal curve has some interesting characteristics that make it useful for statisticians. Look at the normal curve in Figure 5.1. Because it is symmetrical and unimodal, the mean, median, and mode for a normal curve are all at the same point. Think about it. Because the curve is **symmetrical**, or shaped like a bell curve, there aren't any outliers to skew the mean away from the median, either up or down. Adding in the fact that the curve is **unimodal**—meaning it has a single mode or peak—we know that the peak of the curve has to be in the center of the distribution, which places the mode in the same location as the mean and median. Because all bell-shaped curves are symmetric and unimodal, all bell curves share the characteristic that the mean, the median, and the mode are identical.

The normal curve differs from other bell curves in the specific distribution that it follows. About two-thirds of the cases fall within the range of one standard deviation on each side of the mean. If you were to increase that range to two standard deviations on each side of the mean, you would find that 95 percent of the cases fall in that range. For example, another of the manmade normal distributions is IQ. The test is set up so that the average score is 100. The results are scaled so that one standard deviation equals 15 points. Remember in Chapter 4 when I told you that you can check your math for calculating the standard deviation by verifying that about two-thirds of the cases fall within a sandwich of one standard deviation on each side of the mean? That rule comes from our understanding about normal curves. If the variable is normally distributed, two-thirds of the cases lie within one standard deviation of the mean. This means that two-thirds of us have IQs between 85 and 115 ($\mu \pm 1$ standard deviation = 100 ± 15) and 95 percent have IQs between 70 and 130 ($\mu \pm 2$ standard deviations = 100 ± 30). Figure 5.2 shows the distribution for IQ. For any value, we can find how many standard deviations away from the mean it is. This distance from the mean in terms of standard deviations is called a z-score.

Figure 5.2 Distribution for IQ

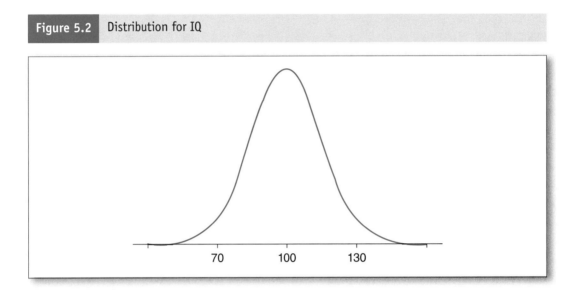

Z-SCORES

You probably don't want to see the mathematical equation for the normal curve—it's not very pretty. Calculating it by hand is a royal pain. Luckily, today, with modern computers, calculating the probability from that equation is not a very big deal—you can easily do an Internet search and find a program to calculate it for you. But without computers, it was much more difficult. So statisticians very carefully mapped out the probabilities associated with being at various standard deviations from the mean. These probabilities are found in "z tables" that appear as appendices in the back of most statistics textbooks, including this one. Even with the ease of computer software, as you are working in a class, it is still easier to use a z table than a computer. Find the z table on page 443 in Appendix 3 at the end of this book. I strongly recommend that you put a sticky on that page. You will be returning to it repeatedly and so you might just as well have it handy. Usually, at the top of a z table you will see a picture of a normal curve that looks like Figure 5.3. It has a line marking the mean/median/mode. To the right of that is a shaded area that ends at another line that is marked "z." That picture reminds you what probability is contained in the z table. Specifically, it indicates that the table gives you the probability that any case falls between the mean and the given z-score. This will become clearer as we walk through examples of reading the table.

Figure 5.3 z Table Distribution

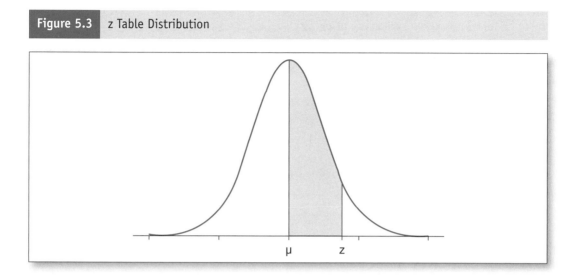

Using a z Table

The most confusing thing about reading a z table is that it looks like a contingency table, but isn't. The cell entry indicates the probability that a case would fall between the mean and this z-score. (Look back up to the picture at the top of the table: Remember that the shaded area is the probability given in the cells of the table.) The labels for both the rows and the columns each indicate a different part of the z-score. In the table, the z-scores have three place values—ones, tenths, and hundredths. The values to the left of the rows identify the ones and tenths. Above the columns is the hundredths place. In Figure 5.4, I show you how to find the probabilities of two different z-scores. First, suppose you want to find the probability of being one standard deviation above the mean. This is a z-score of 1.00. In the z table, come down to the row labeled "1.0" and then over to the first column (labeled "0.00") in order to find the probability associated with a z-score of 1.00. The value in this cell is 0.3413. You can see this in Figure 5.4. So $P(z = 1.00) = 0.3413$. Now look back up at the picture of the normal curve and pay attention to the shading. You interpret $P(z = 1.00) = 0.3413$ by saying, "The probability that a case lies between the mean and a z-score of 1.00 is 0.3413." Or you could say "34.13 percent of the cases lie between the mean and one standard deviation above the mean."

Try another example. What proportion of cases lie between the mean and 1.46 standard deviations above the mean? Look up $z = 1.46$. Begin with both your pointer fingers in the upper left-hand corner of the table on the letter "z." Bring your left finger down until it touches the row labeled "1.4." Bring your right hand to the right until it touches the label "0.06." Now bring your right finger down until it comes to the same row as your left finger. In that cell, you see 0.4279. This is the probability connected to a z-score of 1.46. We would write $P(z = 1.46) = 0.4279$. Now look at the picture and remember the interpretation: "0.4279 is the probability of being between the mean and a z-score of 1.46." This process is also depicted in Figure 5.4.

Figure 5.4 Finding the Probability of z = 1.00 and z = 1.46

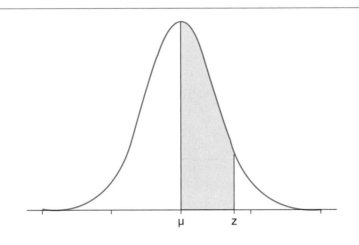

z	0.00	0.01	0.02	0.03	0.04	0.05	0.06
0.0	0.0000	0.0040	0.0080	0.0120	0.0160	0.0199	0.0239
0.1	0.0398	0.0438	0.0478	0.0517	0.0557	0.0596	0.0636
0.2	0.0793	0.0832	0.0871	0.0910	0.0948	0.0987	0.1026
0.3	0.1179	0.1217	0.1255	0.1293	0.1331	0.1368	0.1406
0.4	0.1554	0.1591	0.1628	0.1664	0.1700	0.1736	0.1772
0.5	0.1915	0.1950	0.1985	0.2019	0.2054	0.2088	0.2123
0.6	0.2257	0.2291	0.2324	0.2357	0.2389	0.2422	0.2454
0.7	0.2580	0.2611	0.2642	0.2673	0.2704	0.2734	0.2764
0.8	0.2881	0.2910	0.2939	0.2967	0.2995	0.3023	0.3051
0.9	0.3159	0.3186	0.3212	0.3238	0.3264	0.3289	0.3315
1.0	0.3413	0.3438	0.3461	0.3485	0.3508	0.3531	0.3554
1.1	0.3643	0.3665	0.3686	0.3708	0.3729	0.3749	0.3770
1.2	0.3849	0.3869	0.3888	0.3907	0.3925	0.3944	0.3962
1.3	0.4032	0.4049	0.4066	0.4082	0.4099	0.4115	0.4131
1.4	0.4192	0.4207	0.4222	0.4236	0.4251	0.4265	0.4279

Finding the Probability of Negative z-Scores

I've made you look at the picture at the top of the z table often enough that you are probably wondering, "What do I do for a value less than the mean?" You might think you need a table that pictures the shaded area to the left of the mean instead of the right, as in Figure 5.5. The figure above the z table doesn't have any negative z-scores. Does that mean that statisticians are all from Lake Wobegon (where all the children are above average)? No, it just means that we don't like to waste paper. (And that we like to remind our students of first principles.) Remember, the first characteristic we noticed about the normal curve is that it is symmetrical. You can see in Figure 5.5 that the left half of the curve is a mirror image of the right. So the probability of being within a z-score below the mean is identical to being within a z-score above the mean. Ignoring the negative sign, you would know that $P(z = -1.00) = P(z = 1.00) = 0.3413$ and that $P(z = -1.46) = P(z = 1.46) = 0.4279$. Just keep in mind that probabilities are always between zero and one—they cannot be negative. So negative z-scores are always associated with positive probabilities.

Figure 5.5 A Negative z-Score

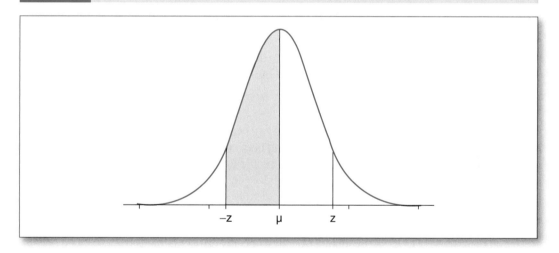

One more example: Try calculating the probability of falling between the mean and a z-score of –0.75. You would ignore the negative sign and look to the intersection between the row marked "0.7" and the column marked "0.05." In the cell, you see 0.2734. So $P(z = -0.75) = P(z = 0.75) = 0.2734$. You conclude that the probability of scoring between the mean and a z-score of –0.75 is 0.2734. You should be starting to feel comfortable reading a z table now.

Find the Probability of Different Ranges: One-Tailed

The properties of the normal curve allow us to use the probability of one part of the curve in order to calculate the probabilities of other parts. Remember that probabilities always range between 0 and 1. Because we know that the normal curve is symmetrical, we can conclude that the probability of being anywhere above the mean is 0.5000. (Look at the z table—do you see how, as the z-score gets larger, the probability in the cells approaches 0.5000?) Conversely, the probability of being below the mean is also 0.5000. That knowledge allows us to calculate probabilities for ranges other than the one depicted at the top of the z table.

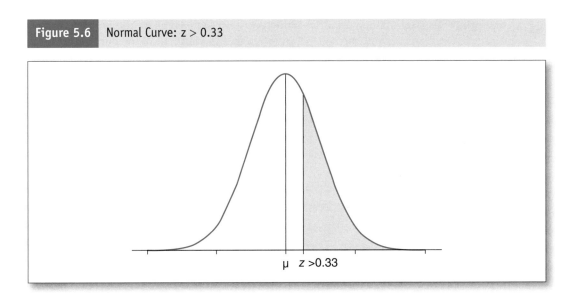

Figure 5.6 Normal Curve: z > 0.33

One concept that we will refer to frequently later in this book is a one-tailed test. Rather than asking what the probability is of being between the mean and the z-score, one-tailed probabilities ask the question, What is the probability of being in a tail? Suppose you know that you scored 0.33 standard deviations above the mean on a test. What proportion of students did better than you? Draw the picture of what you want—it should look like Figure 5.6. Remember again that the probability of being in half of the curve is 0.5000. You know from the z table that P(z = 0.33) = 0.1293. But that is the probability of being between the mean and a z-score of 0.33. In the picture you drew for this problem, that is the unshaded portion of the right half of the curve. You know that the probability of being in the entire right-hand side of the curve equals 0.5. This means that within the right half of the curve, the probability of being in the unshaded section plus the probability of being in the shaded section totals 0.5. If you know the probability of the unshaded section, you can subtract that from 0.5 to get the probability of the shaded portion. In this case, the probability of being in the tail equals 0.5000 – 0.1293 = 0.3707. Thirty-seven percent of the people did better than you.

Conversely, you can also calculate the probability of being in the lower tail (as in Figure 5.7) using the same logic. You just have to flip your picture and ignore the negative sign. Suppose you did below average on a test: z = –0.48. You want to comfort yourself that someone had to have done worse. So you ask, what is the probability that someone got a z-score of lower than –0.48? Draw a normal curve with a line at z = –0.48. You want to know the probability of doing worse, so you shade to the left of your z-score. The resulting drawing should look like Figure 5.7. You check the z table—remembering that P(z = –0.48) = P(z = 0.48)—and find P(z = 0.48) = 0.1844. The picture at the top of the z table reminds you that 18.44 percent of the class scored between you and the mean. So you subtract that from 0.5000 to get the proportion that scored worse than you: 0.5000 – 0.1844 = 0.3156. You did better than 31.56 percent of the class.

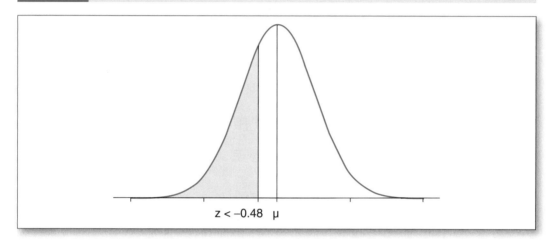

Figure 5.7 Normal Curve: z < –0.48

In addition to being able to calculate the probability of being in the tails, you can also calculate the probability of being below a certain point, as shown in Figure 5.8. Suppose

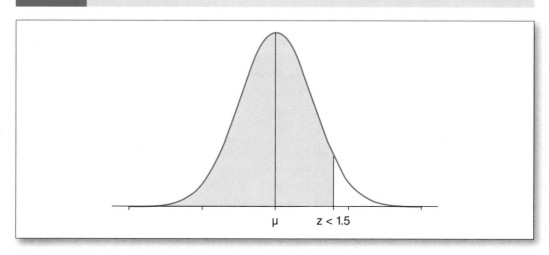

Figure 5.8 Normal Curve: z < 1.50

you know that you scored 1.50 standard deviations above the mean on a test. What percentage of the students did you outscore? Draw a picture of a normal curve. Your z-score is above the mean. You want to know about everyone who did worse than you, so on your curve, you will shade all the way to the left, as in Figure 5.8. You can look up a z-score of 1.50 and find that $P(z = 1.50) = 0.4332$. But out of habit, you look at the picture at the top of the z table and remember that 0.4332 is only the proportion of students who scored between you and the mean. You also want to include all those students who scored below the mean. You know that is 50 percent of the students, so you add 0.5000 to 0.4332 and get .9332. You scored better than 93.32 percent of the students!

BOX 5.1 Numbers in the News

In 2010, WikiLeaks, a nonprofit organization that publishes online documents made available by whistleblowers, began publishing classified diplomatic cables sent from U.S. embassies to the State Department. The revelations reached a head on November 28 when WikiLeaks released 220 inflammatory documents for publication in various well-respected international newspapers. Two days later, Interpol issued a red notice (something like an arrest warrant) for WikiLeaks founder Julian Assange in connection to a sex crime investigation in Sweden. Assange cried foul, accusing Interpol of issuing the warrant for political reasons because of the released documents.[3] On the surface, the timing appears to be too perfect to be coincidental. But to determine whether or not it was due to chance, you would need to compare the timing for this red notice with a distribution of the timing for similar cases. How long does it usually take between the filing of a complaint and the issuing of a red notice? It is unlikely to be issued immediately. But over time, as Interpol collects more evidence, the probability would slowly increase. At some point, the probability would reach its apogee and then begin to decline. If the Assange red notice was issued in that span of time when the probability was high, it is unlikely to be politically motivated. It is more likely that Interpol was politically motivated if it issued the red notice at an unusual time.

Find the Probability of Different Ranges: Two-Tailed

Thus far, we have been calculating one-tailed probabilities. But sometimes we want to calculate two-tailed probabilities. In this case, we are looking at a range that crosses the mean but does not include the two tails on either side of it. If a university had normally distributed grades, the average GPA would be 2.0 and a standard deviation would be 1.0. What is the probability that any given student will have a GPA between a D and a B? Begin by drawing a picture like Figure 5.9. Think of the shaded area as having two parts—below and above the mean. You will calculate the probability of being in each shaded section and then add the two probabilities together. Here, the two sides are mirror images of each other, so $P(-1.0 \leq z \leq 1.0) = 2 \times P(z = 1.0) = 2 \times 0.3413 = 0.6826$.

But the range doesn't have to be symmetrical. If it isn't, divide the range at the mean. Then just add probabilities connected to the shaded sections on each side of the mean

Figure 5.9 Normal Curve: $-1.0 \leq z \leq 1.0$

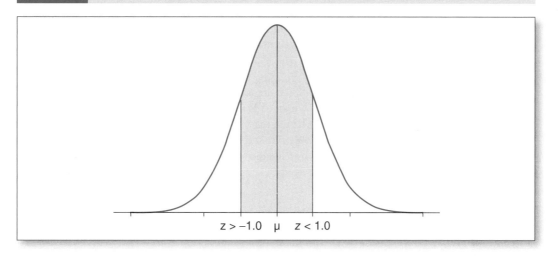

together. If you wanted to know how many students have GPAs between a C– ($z = -.3$) and A– ($z = 1.7$), calculate each side separately. $P(z = -.3) = .1179$ and $P(z = 1.7) = 0.4554$. So $P(-0.3 \leq z \leq 1.7) = .1179 + 0.4554 = 0.5733$. By adding the probabilities of the two sides together, you know that 57.33 percent of the students have GPAs between a C– and an A–.

What proportion of students would have GPAs outside the middle range of 1.0 and 3.0? Always start by drawing a picture. For this question, it would look like Figure 5.10. Again, think of the curve as having two sections—one below and one above the mean. This time,

Figure 5.10 Normal Curve: $z \leq -1.0$ or $1.0 \leq z$

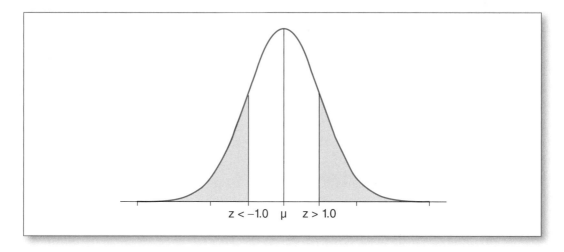

though, the two sections are in the tails. If you wanted the probability of being one *or* the other tail, you find the probability of each and add them together. We already saw that the probability of being between the mean and one standard deviation is $P(z = 1.00) = 0.3413$. Because we're interested in the tail instead of the probability of being between the mean and one standard deviation, we want to subtract the probability of that z-score from 0.5000. One tail is $0.5000 - 0.3413 = 0.1587$. The other tail is the same, so $P(z \leq -1.0$ or $1.0 \leq z) = 2 \times P(z = 1.0) = 2 \times 0.1587 = 0.3174$.

You could also figure the probability of being in the tails if the tails were different sizes. How would you figure the probability of being outside the range between $z = -.75$ and $z = 1.80$? Just calculate each independently and add them together. $P(z \leq -0.75) = 0.5000 - 0.2734 = 0.2266$. $P(1.80 \geq z) = 0.5000 - 0.4641 = 0.0359$. Adding those together, we get $P(z \leq -0.75$ or $1.80 \leq z) = 0.2266 + 0.0359 = 0.2625$. For a two-tailed distribution, just add the two tails together.

FINDING A Z-SCORE

In the previous section, you learned how to find the probabilities connected to various z-scores. Unfortunately, people usually report raw scores rather than z-scores. That means that before you can actually use your understanding of the normal curve to determine a probability, you need to translate the raw score into a z-score. I've already indicated that a z-score is the same as how many standard deviations a case is from the mean. An individual with a z-score of 1.00 is one standard deviation above the mean. Conversely, an individual with a z-score of −1.00 is one standard deviation below the mean.

To calculate a z-score, you need the mean and standard deviation for a variable in addition to the raw score you want to analyze. Once we learned that for IQ scores the mean was 100 and the standard deviation was 15, it was easy enough to eyeball the fact that an IQ of 85 was one standard deviation below the mean of 100. It isn't always so easy, so you should use the following equation to calculate the z-score from a particular value (X):

$$z = (X - \mu)/\sigma$$

where μ is the mean and σ is the standard deviation.

Suppose you find out that your IQ is 120. To calculate your z-score, you would substitute 120 for X, 100 for the mean, and 15 for the standard deviation.

$$z = (120 - 100)/15$$
$$z = 20/15$$
$$z = 1.33$$

An IQ of 120 has a z-score of 1.33 because it is 1 and 1/3 standard deviations above the mean.

Alternatively, suppose you take the LSAT and get 145. Remember that the mean LSAT score is 150 and one standard deviation is 10 points.

$$z = (145 - 150)/10$$

$$z = -5/10$$

$$z = -0.50$$

So a score of 145 on the LSAT is equivalent to a z-score of −0.5. In this case, a negative z-score indicates that it's to the left of the mean. So a z-score of −0.5 is half a standard deviation below the mean. As long as you know the mean and standard deviation, you can translate any value of your variable into a z-score. Once you've determined the z-score connected to a particular value, you can find the probability you want by using the techniques of the previous section.

USE PROBABILITIES TO CALCULATE Z SCORES

Occasionally, you'll be asked to do the opposite of what we've done so far in this chapter. Instead of finding a probability for a particular value, you will be given a probability and asked to find the value connected to it. In order to do this, you still begin by drawing a normal curve and shading in the appropriate region. This time, though, instead of writing the value below the bottom scale, you write the probability that you were given at the beginning. Keep in mind that the z table will only give you the probability between the mean and the z-score. If you've shaded a different region than that, you'll need to make the same kind of adjustments we did above (adding or subtracting from 0.5 for each side of the curve) in order to identify the probability of being between the mean and the z-score. Once you know that probability, you find that value in the cells of the z table. After finding the cell with the probability closest to the probability determined by your problem, you look at the headings for the row and column. Adding those two values together gives you the z-score connected to the probability in question. Once you've got the z-score, if you know the mean and standard deviation, you use the equation for the z-score to calculate the value of X connected to it.

For example, most schools consider the top 10 percent of students for their gifted and talented program. What would be the minimum IQ for a student being admitted to the gifted and talented program? The curve that we are thinking about looks like Figure 5.11 with 10 percent in the shaded portion above the point we're looking for. We know that 50 percent of the students will be below the mean and 50 percent will be above it. If we are looking for the z-score that includes 10 percent above it, we know that 40 percent will be between that z-score and the mean. So we want a z-score that has a probability of $0.5000 - 0.1000 = 0.4000$. All you need to do is scan the z table to find the entry that is closest to 0.4000. The closest probability to .4000 is 0.3997. Keep your finger on that cell. If you look to the left, the label for that row tells you that the first two digits of the z-score are "1.2." Look up, and the label for that column tells you that the third digit is "0.08." Putting those together, you know that a probability of 0.3997 is associated with $z = 1.28$.

Figure 5.11   Normal Curve Where Upper Tail Includes 10 Percent

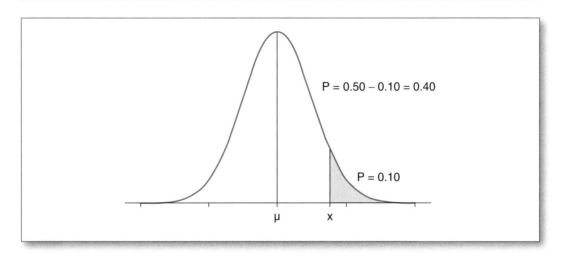

But we aren't as interested in the z-score as we are in the IQ of the student with a z-score of 1.28. Knowing the z-score, we can work backwards to find what value of X corresponds with it. Remember that

$$z = (X - \mu)/\sigma$$

We just found that $z = 1.28$. Earlier in the chapter, we learned that the mean score for IQ is 100 with a standard deviation of 15. Plugging those numbers into the equation, we get

$$1.28 = (X - 100)/15$$

We know that we can multiply both sides of an equation by the same number, so we multiply both sides by 15 to cancel out the 15 in the denominator:

$$15 \times 1.28 = 15 \times (X - 100)/15$$
$$19.2 = X - 100$$

And we know that we can add the same number to both sides of equation, so we add 100 to both sides so that X is left alone:

$$19.2 + 100 = X - 100 + 100$$
$$119.2 = X$$

We are left with $X = 119.2$. So gifted and talented programs are geared toward students with IQs greater than 119.

Figure 5.12 Normal Curve Where Middle Range Includes 50 Percent

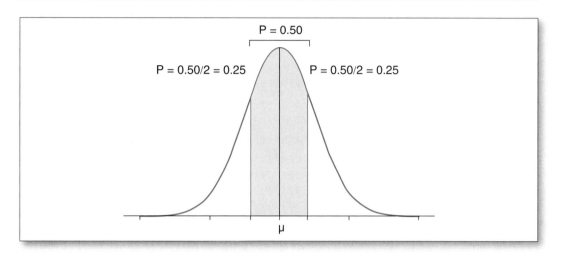

Similarly, we can find the end points of a range if we are given the probability of falling within it. Suppose you are a member of the school board and have proposed a new reading program that you want to pilot. Knowing that the best students and the worst students learn differently, you want to pilot the program among average students. You want to identify the range of IQs for the middle half of students. Begin by drawing a normal curve like Figure 5.12. Your range will extend on either side of the mean. You are going to deal with the shaded portions on each side of the mean separately. Because you want the full shaded region to include half, the probability of being on each side of the mean is 0.5/2 = 0.25. To get the z-score for the upper end of the range, you find the cell that contains a probability closest to 0.2500. The closest probability to 0.2500 is 0.2486, which has a z-score of 0.67. Because the range is symmetrical, you know that the bottom half of the shaded portion has a z-score of −0.67. Finally, you translate the z-scores into IQs using the equation:

$$z = (X - \mu)/\sigma$$

You will be calculating two values of X, one for each end of the range. Begin with the lower end of the range where $z = -0.67$. Remember that the mean is 100 and standard deviation is 15 for IQs.

$$-0.67 = (X_1 - 100)/15$$
$$15(-0.67) = 15(X_1 - 100)/15$$
$$-10.05 = X_1 - 100$$
$$100 - 10.05 = X_1 - 100 + 100$$
$$89.95 = X_1$$

For the upper end, you do the same thing, except use a z-score of 0.67.

$$0.67 = (X_2 - 100)/15$$
$$15(0.67) = 15(X_2 - 100)/15$$
$$10.05 = X_2 - 100$$
$$100 + 10.05 = X_2 - 100 + 100$$
$$110.05 = X_2$$

For your pilot program, you want to target students with IQs between 90 and 110.

Regardless of what your range is, you follow the same procedure. First, draw a picture of a normal curve, shading in the range in question. Second, draw a line in for the mean. You will analyze each half of the curve separately, remembering that each half has a probability of 0.5. Identify the probability of falling between the mean and the value you want to find. Look in the cells of the z table in Appendix 3 to find the cell closest to that probability. Identify the z-score for that cell. From the z-score, use the values of the mean and standard deviations in order to calculate the value of X.

SUMMARIZING THE MATH: PROBABILITIES OF NORMALLY DISTRIBUTED EVENTS

An interval-level variable will usually have a continuous probability distribution. These distributions can take on many different shapes, but the most common shape is a normal curve. Like all bell-shaped curves, the normal curve is symmetrical and unimodal. As a result, the mean, median, and mode are identical, and half the cases lie above and half below the average. The normal curve differs from other bell curves in precisely how the probability is distributed along the curve. We can find that probability by using a z table, which standardizes the position on the distribution into a distance measured in units the size of a standard deviation for the variable. By knowing the mean and standard deviation of a variable, we can calculate the z-score for a particular value of the variable. We can then look up the z-score on a z table to find the probability of a case falling between the mean and that z-score.

Calculate z-Scores

A z-score is how many standard deviations a value is away from the mean. If it is above the mean, it has a positive z-score; below, a negative z-score. To calculate the z-score, you find the distance of the value from the mean and divide by the size of the standard deviation:

$$z = (X - \mu)/\sigma$$

A Political Example: Voter Choice in the 2008 Election. In 2000, we all watched the election results on November 7. That night, and on into the next day, we saw a big map of the United States that slowly filled the center of the country with red and the edges with blue. One of

CHAPTER 5 Continuous Probability: So What's Normal Anyway? **127**

the legacies of that election is our understanding of what it means to be a red state or a blue state. That distinction, a legacy of the winner-take-all implementation of the electoral college, has encouraged us to think of the states as being on polar extremes. But is that characterization accurate? If you look at the states more closely, is it possible that support for the parties approximates a normal distribution rather than the bimodal distribution that the red state/blue state dichotomy encourages?

We can answer this question by looking at the percent of vote for President Obama in 2008 by congressional district. Table 5.1 shows this frequency. The mean vote by congressional district for Barack Obama in 2008 was 53.76 percent. But, of course, support for him was not uniformly distributed. He did better in some districts than others. The fact that Obama outran John McCain results in a distribution that is skewed slightly up (the mean is greater than the median of 52.00 percent). But on balance, the distribution is fairly symmetrical and it is certainly unimodal. The standard deviation around the mean is 14.77. (Note that although I collapsed the data for Table 5.1, I calculated the mean and standard deviation from the uncollapsed data.)

Table 5.1 Frequency of Congressional District Vote for President Obama, 2008

Percent Vote for Obama	Frequency	Percent
20–40	70	16.1
40–60	230	52.9
60–80	108	24.8
80–100	27	6.2
Total	435	100.0

Source: Congressional Quarterly, "House Races in 2010," 2010. http://innovation.cqpolitics.com/atlas/house2010_rr?referrer=rightrail.

Note: President Obama did not receive less than 20 percent of the vote in any district.

Calculate the z-score for three congressional districts in which I have lived: Connecticut CD3, X = 62; Utah CD3, X = 29; and Wisconsin CD3, X = 58.

1. Connecticut District 3 is the solidly Democratic district where I currently live. To translate its 62 percent into a z-score, I use the equation

$$z = (X - \mu)/\sigma$$

X is the support of Obama in this district (62.00), μ is the average support (53.76), and σ is the standard deviation for the distribution (14.77). Substituting those values into the equation, we get

$$z = (62.00 - 53.76)/14.77$$
$$= 8.24/14.77$$
$$= .5579$$

So the z-score for the Connecticut District 3 is 0.5579. This district is about half a standard deviation above the mean.

2. Utah District 3 is the strongly Republican district that now includes the city in which I grew up. To find its z-score, I use the same equation as before, except X = 29.

$$z = (X - \mu)/\sigma$$
$$z = (29.00 - 53.76)/14.77$$
$$= -24.76/14.77$$
$$= -1.68$$

The negative value indicates that this district is below the average. So we would say, Utah District 3 is 1.68 standard deviations below the mean.

3. Wisconsin District 3 (where I lived before moving to Connecticut) is part of a swing state. (Are you seeing a pattern here? What are the odds that I will keep moving into the 3rd district?) Again using the same equation, but with X = 58, I get

$$z = (X - \mu)/\sigma$$
$$z = (58.00 - 53.76)/14.77$$
$$= 4.24/14.77$$
$$= .2871$$

That district is 0.2871 standard deviations above the mean.

Find the Probability of a z-Score

To find the probability connected to a particular z-score, you look it up in a z table. This looks like a contingency table, but the cells contain the probability of being between the mean and the z-score indicated by the headings to the row and column containing the cell.

1. Looking up a z-score on a z table will give you the probability that a value will be between the mean and that value. To use a z table, you find the row that is labeled with the correct value of the ones and tenths. You then move over to the column that is labeled with the correct hundredths value for the z-score. Within that cell, you will find the probability of a case being between the mean and that z-score.

2. You should always draw a picture of a normal curve and shade in the portion about which the question is asking you. If your picture looks different from the one at the top of the z table, you need to be sure to adjust the probability you get from the z table to get the probability you actually want.

3. Because the normal curve is symmetrical, the probability associated with a negative z-score is the same as the probability of a positive z-score.

4. Because each half of the normal curve has a probability of 0.5, if you want to know the probability of being beyond the z-score (instead of between it and the mean), you can just subtract the probability in the table from 0.5000.

5. If you want to know the probability for a range that extends across the mean, you can cut the range into two parts: the part above the mean and the part below the mean. You would then calculate the probability of being below the mean and add it to the probability of being above the mean.

A Political Example: Voter Choice in the 2008 Election. We can use the z-scores from the previous examples to find the probabilities of being in various ranges. (1) What is the probability that a congressional district supported Obama more than the mean of 53.76 percent, but less than Wisconsin CD3's vote of 58 percent? (2) What is the probability that a district would support Obama more than Connecticut CD3? (3) What is the probability that a congressional district would support Obama less than Utah CD3? (4) What is the probability that a congressional district would support Obama less than Utah CD3 or more than Connecticut CD3?

1. What is the probability that a congressional district supported Obama more than the mean of 53.76 percent, but less than Wisconsin CD3's vote of 58 percent?

    a. Find the z-score on the z table.

    We know from the previous section that for Wisconsin CD3, z = 0.2871. If we look that up in the z table, we find that there is a probability of 0.1141 connected to a z-score of 0.29. Thus, there is a probability of 0.1141 that in 2008, a congressional district would support Obama between the mean and Wisconsin CD3.

    b. Draw a picture of what you are asking, and compare it to the figure at the top of the z table.

    The picture I drew looks like the one at the top of the z table. So I know that the probability given in the cell is the answer to my question: There is a probability of 0.1141 that in 2008, a congressional district would support Obama between the mean and Wisconsin CD3.

2. What is the probability that a district would support Obama more than Connecticut CD3?

    a. Find the z-score on the z table.

    We know from the previous section that for Connecticut CD3, z = 0.5579. If we look that up in the z table, we find that there is a probability of 0.2123 connected to a

Figure 5.13 Normal Curve for Wisconsin CD3, Shaded Curve

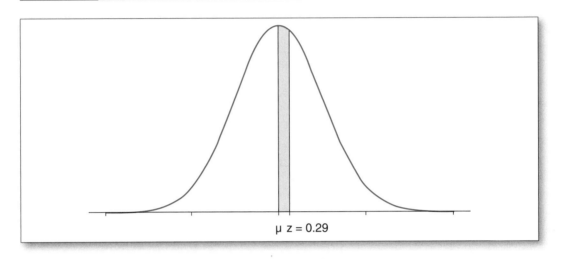

z-score of 0.56. Thus, there is a probability of 0.2123 that in 2008, a congressional district would support Obama between the mean and Connecticut CD3.

b. Draw a picture of what you are asking and compare it to the figure at the top of the z table.

I have shaded the tail of the curve rather than the space between the mean and my z-score. Because I know that half the cases are in the right half of the curve, I know that I can subtract the probability I got in the z table from 0.5000 to get the

Figure 5.14 Normal Curve for Connecticut CD3, Shaded Tail

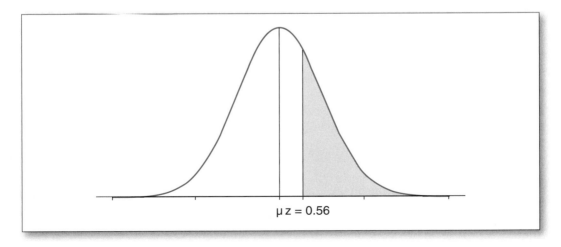

probability of being in the tail. The probability of a congressional district giving more support to Obama than Connecticut CD3 is 0.5000 − 0.2123 = 0.2877.

3. What is the probability that a congressional district would support Obama less than Utah CD3?

   a. Find the z-score on the z table.

   We know from the previous section that for Utah CD3, z = −1.6764. There are no negative z-scores on the z table, but because the normal curve is symmetrical, we know that the probability of a negative z-score is the same as if it were positive. If we look up 1.6764 in the z table, we find that there is a probability of 0.4535 connected to a z-score of 1.68. Thus, there is a probability of 0.4535 that in 2008, a congressional district would support Obama between the mean and Utah CD3.

   b. Draw a picture of what you are asking and compare it to the figure at the top of the z table. (See Figure 5.15.)

Figure 5.15 Normal Curve for Utah CD3, Shaded Mean and Value

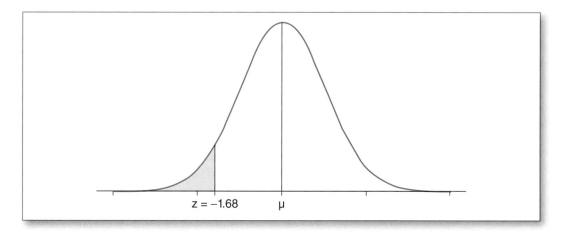

The z table has the area between the mean and the shaded value. It is fine to flip it to deal with the negative z-score. But that would give me the probability of the unshaded section of the left side of the curve. Because I know there is a probability of 0.5 of being in the left half, I can subtract the probability of the unshaded half from 0.5 to get the shaded half. The probability of giving less support than Utah CD3 is 0.5000 − 0.4535 = 0.0465.

4. What is the probability that a congressional district would support Obama less than Utah CD3 or more than Connecticut CD3?

   a. Find the z-score on the z table.

This time, I have two z-scores so I have two probabilities. We already found that Connecticut CD3's z = 0.5579 yields a probability of 0.2123 that in 2008, a congressional district would support Obama between the mean and Connecticut CD3. We also found that Utah CD3's z = −1.6764 gives a probability of 0.4535 that in 2008, a congressional district would support Obama between the mean and Utah CD3.

b. Draw a picture of what you are asking and compare it to the figure at the top of the z table. (See Figure 5.16.)

Figure 5.16 Normal Curve for Area between Utah CD3 and Connecticut CD3

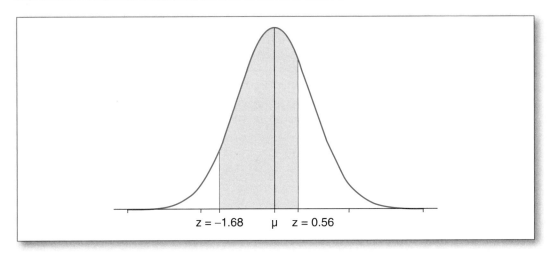

Because I have shaded portions on both sides of the mean, I am going to cut this curve into two pieces at the mean and find the probabilities connected to the shaded portions of each side. I know that the z table gives me the probability of being between the mean and the z-score, so I know that I can add those probabilities together to find the probability of being in the shaded portions. Because the Utah section has a probability of 0.4535 and the Connecticut section has a probability of 0.2123, the probability of being between these two points is 0.4535 + 0.2123 = 0.6658.

Use a Probability to Calculate a Value

Sometimes, instead of being asked to find the probability of a distribution connected to particular values of your variable, you are given the probability of a distribution and asked to find the values connected with it.

1. Begin by drawing a picture of the normal curve, shading the area for which you know the percentage.

CHAPTER 5 Continuous Probability: So What's Normal Anyway?   **133**

2. Keeping in mind that each half of the curve has a probability of 0.5000, find the probability of being between the mean and the value you are trying to find.

3. Find the cell in the z table that is closest to that probability.

4. Look to the headings for the row and column that contain that cell in order to identify the z-score.

5. Using the mean and standard deviation, calculate the value of X for that z-score:

$$z = (X - \mu)/\sigma$$

$$\sigma z = \sigma(X - \mu)/\sigma$$

$$\sigma z = X - \mu$$

$$\sigma z + \mu = X - \mu + \mu$$

$$\sigma z + \mu = X$$

A Political Example: Voter Choice in the 2008 Election. One last question: I actually began this section with the question of what proportion of districts are in the extremes as opposed to the middle. Specifically, how far on either side of the mean do I have to go to include 70 percent of the congressional districts?

1. Begin by drawing a picture of the normal curve, shading the area of which you know the percentage. (See Figure 5.17.)

Figure 5.17 Normal Curve Where Middle Range Includes 70 Percent

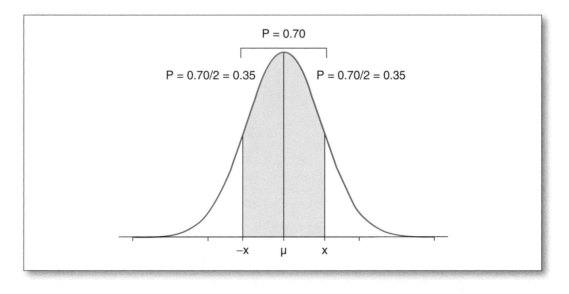

2. **What is the probability of being between the mean and the value you are trying to find?**
   The curve crosses onto both sides of the mean, so I cut it in half and find the probability of being on each side of the mean. Because the distribution is symmetrical around the mean, to include 70 percent of the districts, I know that the shaded portion on each side of the mean will contain 35 percent of the districts. So I am going to find the z-score that corresponds to a probability of 0.3500.

3. **Find the cell in the z table that is closest to that probability.**
   I scan to contents of the cells in the z table until I find two numbers of each side of 0.3500. I see 0.3485 and 0.3508. The value 0.3508 is slightly closer to 0.3500 so I choose it.

4. **Look to the headings for the row and column that contain that cell in order to identify the z-score.**
   A probability of 0.3508 corresponds to a z-score of 1.04. So I know that the range that sandwiches 1.04 standard deviations on each side of the mean contains 70 percent of the districts.

5. **Using the mean and standard deviation, calculate the value of X for that z-score.**
   I need to translate the z-score of 1.04 into actual support for Obama in 2008. What values of X correspond with the z-scores 1.04 and −1.04? For the upper limit, I do the following algebra, substituting 1.04 for z and the mean and standard deviation we've been using all along:

$$z = (X - \mu)/\sigma$$

$$1.04 = (X - 53.76)/14.77$$

$$14.77\,(1.04) = 14.77(X - 53.76)/14.77$$

$$15.36 = X - 53.76$$

$$15.36 + 53.76 = X - 53.76 + 53.76$$

$$69.12 = X$$

For the lower limit, I do the following algebra, this time substituting −1.04 for z:

$$z = (X - \mu)/\sigma$$

$$-1.04 = (X - 53.76)/14.77$$

$$14.77(-1.04) = 14.77(X - 53.76)/14.77$$

$$-15.36 = X - 53.76$$

$$-15.36 + 53.76 = X - 53.76 + 53.76$$

$$38.40 = X$$

CHAPTER 5 Continuous Probability: So What's Normal Anyway? **135**

We can conclude that 70 percent of the congressional districts gave Obama a vote of between 38.40 percent and 69.12 percent.

USE SPSS TO ANSWER A QUESTION WITH CONTINUOUS PROBABILITY

Assuming that a variable is normally distributed, knowing the z-score of the cases allows you to find the outliers—those cases that are either above or below average on a characteristic. The general rule of thumb is that 95 percent of cases are within two standard deviations of the mean. It can be helpful to find those cases that are outside that range. Perhaps, with further analysis, you could determine what makes them unusual.

In SPSS, you can find the z-scores for a variable and save them as a new variable. Once you have your data in SPSS and have cleaned it, you go into "Analyze," "Descriptive Statistics," and "Descriptives." Choose the variable for which you want the z-scores and bring it over to the "Variable" box. You will then click in the box to request SPSS to "Save Standardized Values as Variables." (Sometimes, statisticians will call z-scores standardized values.) Once you click on "OK," SPSS will calculate the z-score for each of your cases and enter them all as a new variable. Box 5.2 summarizes this process. Open your data window and you'll see this new variable in the column on the right.

BOX 5.2   How to Use SPSS to Calculate Z-Scores

After opening your data in SPSS and cleaning them:

>Analyze
>>Descriptive Statistics
>>>Descriptives
>>>>"*Variable*"→Variable
>>>>Save Standardized Values as Variables
>>>>OK

At this point, the variable will show up in the last column of the data window with a "Z" in front of the name.

You next need to find which cases exceed two standard deviations from the mean. First, you want to identify the cases. Under "Data," choose "Select Cases." Click on "If condition is satisfied" and then the "If" button. You can now use mathematical operations to choose cases. For this chapter's purposes, you want to find those cases that have z-scores greater than 2. Choose your standardized variable (the one with a "Z" in front of it) and bring it over to the Variable box. To get those cases greater than 2, you would enter ">2" after the variable name. Click on "Continue" and "OK." This process is summarized in Box 5.3. Look in the data window and see that there are slashes in the case numbers of most of your cases. That means that SPSS will not use those cases in the following analysis. If you come down to one of the cases without a slash and then look at your new standardized variable, you'll see that its value exceeds 2.0.

BOX 5.3   How to Use SPSS to Select Cases

After opening your data in SPSS and cleaning them:

>Data
>>Select Cases
>>>If condition is satisfied
>>>If
>>>>"Variable"→"mathematical operators for condition"
>>>>>For example: "ZMurder > 2"
>>>>Continue
>>>OK

At this point, if you enter the data window, you will see slashes through the case numbers for all cases not meeting your criteria.

Finally, you need to identify the cases you found. Go to "Analyze" and "Reports." One of the options is "Case Summaries." Choose this option and then choose any variables you want displayed. You will probably want the case identifier, the original variable, and the standardized variable. After you "OK" the command, the Output window will produce a table that lists the values for the variables you requested for each of the cases you selected. This process is summarized in Box 5.4. (Notice that this can be a helpful process when you are cleaning data: If you find a value that is unexpected, you can select for that value and then identify the case with the "Case Summary" command.)

BOX 5.4   How to Use SPSS to Identify Selected Cases

After opening your data in SPSS, cleaning it, and selecting for particular values:

>Analyze
>>Reports

CHAPTER 5   Continuous Probability: So What's Normal Anyway?   **137**

> Case Summaries
>   *"Case Identifier" "Variable"*→Variables
> OK

This will give a table that gives the values of each variable requested for all of the selected cases.

An SPSS Application: Crime Data

It is May 2010. You are interning at the Justice Department and in charge of opening mail for the day. When you open a letter addressed to Attorney General Eric Holder, you see it is from Mitch Landrieu, the Mayor of New Orleans. In it, Mayor Landrieu indicates that the murder rate of New Orleans has been skyrocketing since Hurricane Katrina and requests DOJ intervention. Realizing the importance of the letter, you take it to your immediate supervisor. She smiles, thinking that this is a chance to make your experience as an intern truly memorable. "You're right," she says, "Attorney General Holder is going to want you to deliver this letter to him in person. However, he won't be back in the office until this afternoon. Why don't you take the next couple of hours to gather data about whether the murder rate in New Orleans really is any different from other big cities in the country? You can deliver your memo along with Mayor Landrieu's letter as soon as the AG gets back this afternoon."

Knowing that the FBI is tasked with collecting crime data for the country in the Uniform Crime Report, you check out their website and download the dataset that enumerates crimes by city.[4] Because you want only the large cities, you sort the Excel file for population and select out those cities with populations over 250,000. At this point, you realize that the crimes are given in raw counts, so you standardize those to rates per 100,000 population. You then import the data into SPSS and begin your analysis. You remember your statistics professor telling you that a case that is more than two standard deviations above the mean can definitely be considered higher than normal, so you decide to translate the murder rates into z-scores to find out if New Orleans really does have an exceptionally high murder rate.

1. After opening your data in SPSS, make sure that it is clean using "Define Variable Properties."
   There are no unusual values and no data missing.

2. Get z-scores for the variable while you request its descriptive statistics.
   Figure 5.18 shows what it looks like as you request z-scores in this way. Notice that there is a check in front of "Save standardized values as variables." This will ensure that a new variable will be created that gives the z-score for each case on this variable.
   In the Descriptive Statistics table (shown in Figure 5.19), I notice that the mean murder rate is 10.8 with a standard deviation of 9.4. I can conclude that most metropolitan areas have murder rates between 0 and 20.

Figure 5.18 How to Find z-Scores with SPSS

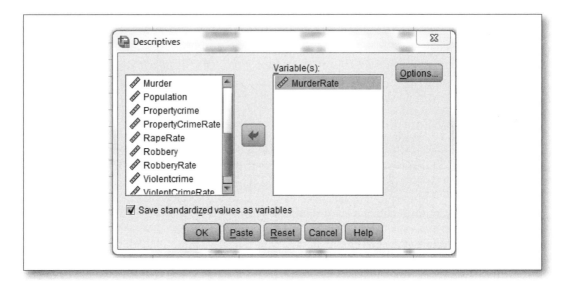

Figure 5.19 Descriptive Statistics for Murder Rate of Cities with Population over 250,000

Descriptive Statistics

	N	Minimum	Maximum	Mean	Std. Deviation
Murder Rate	74	.7702	49.1136	10.844295	9.3636864
Valid N (listwise)	74				

3. **Select those cases that are greater than 2 standard deviations above the mean.**
Rather than hunting for those cases that are outliers, I use the "Select" command to identify those cases that are at least two standard deviations above the mean. This process is shown in Figure 5.20. Once I've clicked "Continue" and "OK" to process the request, I go into the Output window and see slashes over most of the case identifiers. At this point, I could look for the cases without slashes (just as I could have looked for the cases with z-scores over 2. Figure 5.21 shows a screen shot in which Detroit and Baltimore (both cities with high murder rates) do not have the slashes.

CHAPTER 5 Continuous Probability: So What's Normal Anyway? **139**

Figure 5.20 Selecting z-Scores over 2

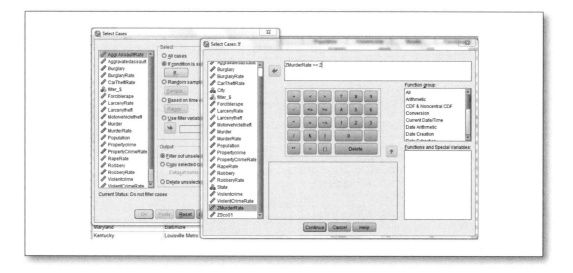

Figure 5.21 Data Selected for High Murder Rate

11	California	San Jose
12	Hawaii	Honolulu
13	Michigan	Detroit
14	Florida	Jacksonville
15	California	San Francisco
16	North Carolina	Charlotte-Mecklenburg
17	Texas	Austin
18	Ohio	Columbus
19	Texas	Fort Worth
20	Tennessee	Memphis
21	Massachusetts	Boston
22	Maryland	Baltimore
23	Kentucky	Louisville Metro

4. **Identify the cases you have selected.**
   But rather than hunting for the cases and risking missing some, I have SPSS tell me what they are by requesting a "Report," which includes "Case Summaries." Figure 5.22 shows where these commands are located. Once the "Summarize Cases" window is open, I indicate that I want the report to include "City," "State," "MurderRate," and "ZMurderRate," as shown in Figure 5.23.

Figure 5.22	Finding the Commands for Case Summaries

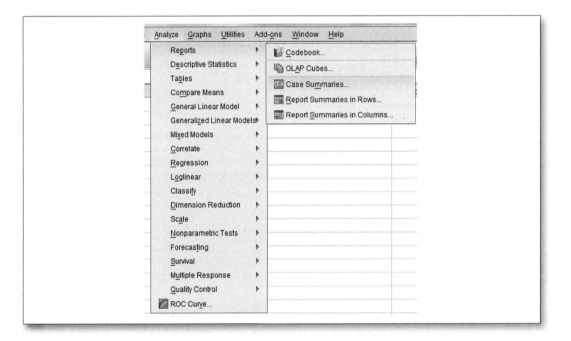

5. **Using the summary information, write your memo.**
   Figure 5.24 shows the summary. Mayor Landrieu is correct. New Orleans has a murder rate four standard deviations above the mean. St. Louis has the second highest murder rate, with a z-score of 3. Detroit and Baltimore have z-scores of 2.5, and Newark has a z-score of 2.2.

Memo

To:	Attorney General Eric Holder
From:	T. Marchant-Shapiro, intern
Date:	May 22, 2010
Subject:	New Orleans murder rate

Figure 5.23 Get Identifying Information about the Requested Cases

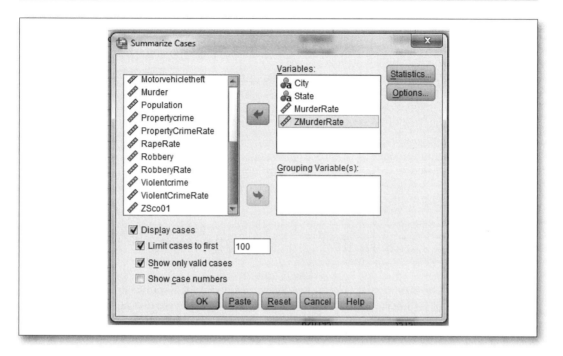

Figure 5.24 Case Summary for High Murder Rate

Case Summaries[a]

	City	State	Murder Rate	Zscore: Murder Rate
1	Detroit	Michigan	34.4656	2.52265
2	Baltimore	Maryland	34.8476	2.56345
3	New Orleans	Louisiana	49.1136	4.08699
4	St. Louis	Missouri	40.5461	3.17202
5	Newark	New Jersey	32.0994	2.26995
Total N	5	5	5	5

a. Limited to first 100 cases.

This memo comes in regard to a request made by Mayor Mitch Landrieu. The mayor has requested that the Department of Justice intervene because of the high murder rate in New Orleans. I was asked to see whether the murder rate in New Orleans was substantially higher than other metropolitan areas. Using the FBI's Uniform Crime Report, I found that Mayor Landrieu is correct: New Orleans has a substantially higher murder rate than any other city in the United States.

The average murder rate for metropolitan areas with populations over 250,000 is 10.8 murders per 100,000 population. Most cities have murder rates of fewer than 22. Five cities stand out as exceptions: New Orleans, St. Louis, Detroit, Baltimore, and Newark. The latter three (Detroit, Baltimore, and Newark) have from 32 to 34 murders per 100,000 population each year. St. Louis has 40. New Orleans has 49.

Mayor Landrieu does have cause to be concerned about the murder rate in his city. It is more than twice as high as most cities. I would also advise you to keep tabs on St. Louis. Although its murder rate is not as high as New Orleans, it is well over the typical range and so is also cause for concern.

Your Turn: Continuous Probability

YT 5.1

The 2010 midterm election was a landslide for the Republican Party in the House of Representatives. After the Republicans gained sixty seats in the House, the media began talking about the election as an indication that the United States no longer supported Democratic policies. But the single district winner-take-all system obscures the fact that the Republican wins were not unanimous. Table 5.2 shows the proportion of votes the Republican candidate received in each of the congressional districts. The mean percentage was 50.7 with a standard deviation of 16.87.

Table 5.2 Frequency of Congressional District Two-Party Vote for Republican Candidate, 2010

Percent Vote for Republican	Frequency	Percent
0–20	18	4.4
20–40	92	22.7
40–60	156	38.4
60–80	134	33.0
80–100	6	1.5
Total	406*	100.0

Source: Congressional Quarterly, "House Races in 2010," 2010. http://innovation.cqpolitics.com/atlas/house2010_rr?referrer=rightrail.

*Notice that this totals less than the total of 435 congressional districts because some representatives were unchallenged and so were not included in the analysis.

For each of the following representatives, calculate their z-score:

1. John Boehner (68.6 percent Republican)
2. Nancy Pelosi (16.0 percent Republican)
3. Nita Lowey (38.0 percent Republican)
4. Michele Bachmann (56.9 percent Republican)
5. Barney Frank (44.6 percent Republican)
6. Don Young (69.2 percent Republican)

YT 5.2 Using the same data as in Your Turn 6.1, calculate the probability that a district will lie in the following ranges. Remember that the mean percentage was 50.7 percent with a standard deviation of 16.87. Be sure to draw a picture of the normal curve and shade the ranges identified in each question.

1. What is the probability that a district will have a Republican percentage between Michele Bachmann (X = 56.9%) and the mean?
2. What is the probability that a district will have a Republican percentage between Nita Lowey (X = 38.0%) and the mean?
3. What is the probability that a district will have a Republican percentage between Michele Bachmann (X = 56.9%) and Barney Frank (X = 44.6%)?
4. What is the probability that a district will have a Republican majority (X > 50.0%)?
5. What is the probability that a district will have a Republican vote of either more than John Boehner (X = 68.6%) or less than Nancy Pelosi (X = 16.0%)?
6. What is the range around the mean that contains 80 percent of the districts?

Apply It Yourself: Evaluate the Murder Rate

It is July 2012. You are interning in the mayor's office in Aurora, Colorado. All summer you've been looking forward to the release of *The Dark Knight Rises*. There is a midnight showing on July 20, the day it opens, but you have to work that Friday, so you decide to go Friday night instead. Friday morning you get to work and the office is in uproar: At the midnight showing, James Holmes, a former graduate student at the University of Colorado, Denver, allegedly opened fire on the audience. Suddenly, Aurora is in the news. This event combines with memories of the Columbine shooting to cause national comments about the violence of Colorado. Mayor Steve Hogan wants some data to respond to the accusations. You are tasked with evaluating whether Aurora has a higher than average murder rate for cities its size. Using the FBI's Uniform Crime Report, you find the murder rates for U.S. cities with populations between 200,000 and 350,000.

1. After opening "Crime, cities 200-350k" in SPSS, make sure that it is clean using "Define Variable Properties."

2. Get z-scores for the murder rate variable.

   >Analyze

   >Descriptive Statistics

   >Descriptives

   "MurderRate"→Variable

   >Save Standardized Values as Variables

   >OK

3. Select those cases that are greater than 2 standard deviations above the mean.

   >Data

   >Select Cases

   >If condition is satisfied

   >If

   "ZMurderRate"→"ZMurderRate > 2"

   >Continue

   >OK

4. Identify the cases you have selected.

   >Analyze

   >Reports

   >Case Summaries

   "City" "State" "MurderRate" "ZMurderRate"→Variables

   >OK

5. In the data window, find Aurora and identify its murder rate and z-score.

6. Write your memo. (Please include a copy of the case summary for grading purposes.)

Key Terms

One-tailed (p. 118)
Standardized value (p. 135)
Symmetrical (p. 113)

Two-tailed (p. 120)
Unimodal (p. 113)
z-score (p. 114)

CHAPTER 6

Means Testing

Sampling a Population

In the 2012 Summer Olympics, Chinese swimmer Ye Shiwen set a world record of 58.68 seconds for the freestyle component of the 400-meter medley. Instead of being impressed by the young woman's ability to best even the time of the men's gold medal winner, John Leonard, executive director of the World Swimming Coaches Association, called her performance "disturbing." Contrasting Ye's performance to the normal pattern for elite swimmers, Leonard said, "Every time we see something . . . this 'unbelievable,' history shows us that . . . there was doping involved."[1] Although the unusual is possible, it will frequently lead to suspicions of a sinister cause.

Like Olympic coaches, epidemiologists watch to find patterns out of the ordinary. They know that diseases happen. But they also know the normal frequency of different diseases. They are on the lookout for "hotspots" where an unusually large incidence of a particular disease appears focused. Most communities will not have precisely the average incidence. Some will have a higher incidence, some a lower one. Epidemiologists want to find the communities where the incidence is so high it is unlikely to be due to chance. The movie *A Civil Action* depicts such a community.[2] It is based on true events set in motion when a doctor in Boston noticed that an unusually high number of his young leukemia patients were denizens of Woburn, Massachusetts.[3] His patients got together to find out why so many of their neighbors suffered from similar problems and concluded that water pollution from local businesses was the likely culprit. The movie tracks the families' lawyers as they gather physical evidence to supplement the statistical evidence in the lawsuit.

Similarly, the case that identified smoking as a cause of lung cancer was based on statistical evidence. Because the incidence of lung cancer is so much higher among smokers than non-smokers, scientists testified that the difference couldn't be due to chance. They concluded that there must be something about smoking that incites normal lung cells to grow in uncontrolled ways. The tobacco industry replied that scientists have never "proven" that smoking causes cancer. In other words, scientists have not observed smoke inciting lung cells into cancer. The industry is correct: statistics never "proves" anything. Statistics can only indicate correlation, not causation.

As a result, as statisticians we are very careful to set a high standard of correlational evidence. Because we know that our findings might be the result of chance, we try to minimize that possibility. Normally, our standard is that we draw conclusions only when there is less than a 5 percent chance of our observation being due to random fluctuations. When our findings meet that standard, we say that they are statistically significant. In order to draw that conclusion, we compare how the population is distributed with the distribution of our subsample. By using our knowledge of the normal curve, we can find the probability of observing the difference that we see if, in reality, our subpopulation is not different from the population. In this chapter, we will first describe the criteria we use in defining what is normal, and second, describe a statistical technique called means testing in order to make the comparison between a population mean and a sample distribution.

TYPE I AND TYPE II ERRORS

An unusual event is not necessarily a guarantee of an external cause. As you look at a probability distribution, the height describes the probability of getting that result randomly. As that height shrinks, you know that it is becoming less likely to be the result of chance. Conversely, it is becoming more likely to be the result of some independent variable. But the fact that it is more likely to have a cause other than chance does not rule out the possibility of it being a random event. As you move to the extreme edge of the probability curve, its probability never quite reaches zero. Where John Leonard looked at Ye Shiwen's five-second improvement and saw doping as the cause, former Olympian Ian Thorpe saw it as the natural improvement that is possible in talented young athletes.[4] How do we draw the line between what is unusual and what is unbelievable when we know that whereever we draw the line, we might make a mistake? Ye Shiwen's performance might have nothing to do with doping, whereas other competitors might well have used performance-enhancing drugs but not raised any eyebrows because their performances did not improve enough to place them in the tail of the normal distribution. In medical terms, we would say that incorrectly accusing Ye Shiwen of doping is a false positive finding, and missing dopers who don't perform extraordinarily is a false negative finding. In statistical terms, we call a false positive finding a Type I error (that's the Roman numeral one) and a false negative is a Type II error (you say "type two").

It would be nice if there was a simple test that concluded with certainty what is random and what is not, but we are dealing here with probabilities. As a result, as we move the cutoff line further to the right to prevent false positives, we are simultaneously increasing the probability of a false negative. Conversely, as we move the line further to the left to prevent false negatives, we increase the probability of a false positive. Because both kinds of errors have potential costs, finding the balance between them is fundamentally a judgment call.

In 2009, the U.S. Preventative Services Task Force (USPSTF) issued new recommendations regarding routine mammograms. Although they recommended a continuation of routine mammograms for women older than fifty, they discontinued their prior recommendation for women in their forties. They concluded that universal mammograms for that age group increase the probability of false positives by 70 percent. In contrast, relying exclusively on self- and doctor exams to initially identify breast cancer would increase the probability of false negatives by only 0.005 percent.[5] The USPSTF concluded that saving one life by the more inclusive testing was not worth the anxiety suffered by the 1,330 women who would be diagnosed incorrectly, half of whom would end up having unnecessary biopsies. In contrast, the American Cancer Society refused to change its recommendation, concluding that "the lifesaving benefits of screening outweigh any potential harms."[6] In the end, although both organizations are free to make public policy recommendations, individual women need to decide whether or not to get a mammogram. They are in a better position to make that decision if they understand that the accuracy of the mammogram has a probabilistic distribution—a positive diagnosis does not guarantee breast cancer.

Because lives are not usually on the line when it comes to social science research, we do not usually draw the line as far to the left as either the American Cancer Society or the U.S. Preventative Services Task Force. Both organizations are willing to tolerate a very high rate of false positive diagnoses in order to catch and treat as many cases of breast cancer as possible. In contrast, in our attempts to understand how the world works, we prefer to assume that the events we see are random unless we are persuaded beyond reasonable doubt that they are not. This is very similar to Blackstone's eighteenth-century formulation regarding criminal law: "It is better that ten guilty persons escape than that one innocent suffer." The ten guilty people who escape would be Type II errors, the false positives, made by the judicial system, whereas the conviction of an innocent person would be a Type I error, a false negative.

As with the judicial system, as social scientists, we normally try to minimize the risk of a Type I error. We understand that this comes at the cost of increasing the probability that we will mistakenly conclude that real events are only random. We do not actually even measure the probability of Type II errors. Instead, we focus on measuring the probability of Type I errors. Before we report any results, we want to be very confident that what we observe is not due to chance. So as scientists, we set fairly high standards that we expect our data to exceed. Normally, we will report only results that have a 5 percent chance or less of being due to random error. In Chapter 5, I referred to a "one-tailed test" in connection to a picture that was shaded in one of its tails. This is reproduced in Figure 6.1. We say that our results are statistically significant at the 0.05 level (which we pronounce as "point oh five") if the shaded portion contains only 5 percent of the curve. Or, there is only a 5 percent chance of observing these results if, in reality, there is actually nothing going on. Sometimes, we talk about data being statistically significant at the 0.01 level (we say "point oh one level"), and very occasionally, we talk about being statistically significant at the 0.10 level ("point one oh level"). These levels of significance correspond respectively with having a 1 percent probability or 10 percent probability of being in the tail. Statistical significance is the probability of a Type I error: the probability of seeing an extreme sample mean when, in reality, the subpopulation is no different from the population.

Figure 6.1 Normal Curve: One-Tailed

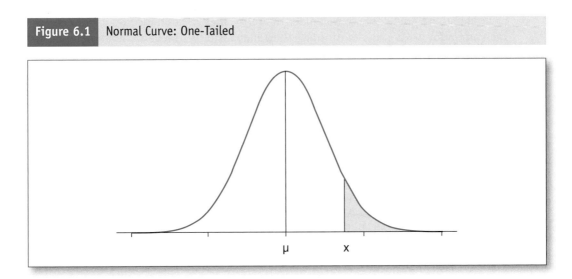

MEANS TESTING

Thus far in this book, we have been discussing descriptive univariate statistics. We call them descriptive statistics because we are dealing with measurements of all of the cases in a population. We call them univariate statistics because we are simply describing the distribution of a single variable across that population. Our understanding of how that distribution ought to look if it is randomly distributed allows us to begin to identify subpopulations that are not randomly distributed. If a group has a distribution that is substantially different from the population, then we can conclude that the difference is probably not the result of chance. Or, that some characteristic of the group has led its members to have different attributes of the original variable. This allows us to begin to ask the "why" questions that are at the heart of political science research. We no longer need to limit our research to describing how a characteristic varies across a population. We can now ask: Why is there variation in this characteristic? In this chapter, we will focus on means testing. This is a technique that compares the distribution of an attribute for a population with the distribution for a subsample. In this comparison, we ask the question: What is the probability of observing a difference simply due to chance?

Compare Populations with Samples

Remember that with means testing, we are comparing the distribution of a variable for an entire population with the distribution of the same variable for a subset of cases. When we see that the mean for the subset is different from that for the population, we wonder whether something distinguishes that group from the population in such a way as to cause that difference. Many political scientists have noticed that African countries are different from the rest of the world on a wide range of economic, social, and political characteristics.

It is quite understandable that we would ask, Are those differences due to chance, or are they indicative of basic underlying differences?

In order to answer that question, we can use a means test. But before we do, we want to make sure that the only difference between our population and our sample is whether or not they are in Africa. One thing that differentiates African countries is that most were still under colonial domination into the twentieth century. If we want to address the impact of colonization on Africa, we should limit our analysis to a population of countries of the world that were colonies in the twentieth century. In addition, the colonial experience of some colonizers may have been more conducive than others in the development of economic and political structures. Perhaps those countries colonized by Great Britain developed differently from those colonized by France or any other European power. Because different imperial regimes treated their colonies in different ways, let's limit our analysis to the population of nations that, as colonies, were ruled by Great Britain in the twentieth century. Does the subset of African countries look different politically and economically from the population of all former British colonies?

A common measure of economic strength is GDP per capita, which is obtained by taking the gross domestic product of a nation and dividing by the size of the population. Do the former African colonies have a different level of GDP per capita from the former colonies as a whole? In all the data we have used up to this point, we have actually been able to measure the variables for the full population. The statistics we base on those measures are called parameters. Table 6.1 shows the frequency for GDP per capita for all of the twentieth-century British colonies. Relying on the same techniques we have used up to this point, I can use these data to calculate the population mean for British colonies. In 2011, these former colonies averaged a per capita GDP of $13,307.87.

Table 6.1 GDP per Capita for All Twentieth-Century Former British Colonies, 2011

GDP per Capita X	Number of Countries f_x	Percent
0–$10,000	31	57.4
$10–20,000	10	18.5
$20–30,000	5	9.3
$30,000+	8	14.8
$\mu = \$13,307.87$	$N = 54$	

Source: Compiled from Gapminder, "Income Per Person," www.gapminder.org/data/.

Our question is whether African countries look different from other countries. If we take the subset of African nations, do they have a GDP that is substantially different from the full population of British colonies? Our general standard is that we want to be 95 percent certain that they are different. Or, conversely, we want there to be only a 5 percent chance of a Type I error. In order to draw that kind of conclusion, we need to find not only that the mean of the subgroup is different from the population, but also that its distribution is narrow enough to exclude the difference being caused by chance. In order to draw those kinds of conclusions, I need to factor into my statistics the uncertainty that arises because I am measuring a sample rather than a population.

One consequence of leaving parametric statistics behind is that mathematicians use different symbols for the statistical measures of samples and parameters. Up until now, we have generally used Greek letters to stand for statistical measures. But for samples, mathematicians use Latin-based letters. So where we once used the Greek μ (mu) for the mean of the population, we will now use the Latin-based $\bar{X}$ (pronounced X-bar) for the mean of a sample. For standard deviation, the Greek σ (sigma) we used for the population becomes a simple "s" for a sample. And the number of cases changes from an upper-case "N" for the population to a lower-case "n" for a sample. Table 6.2 summarizes the difference between the symbols used for population and sample statistics.

Table 6.2 Symbols for Population and Sample Statistics

	Mean	Standard Deviation	Number of Cases
Population Symbol	μ	σ	N
Sample Symbol	$\bar{X}$	s	n

But the symbols are not the only things that can change with samples; the equations can change as well. The equation of the mean remains unchanged on the assumption that the sample is an unbiased estimate of the population. (Notice the importance of that assumption: If a sample is not representative of a relevant section of the population, the results will be biased.) Thus far, we have given the equation for the population mean as

$$\mu = (\Sigma X)/N$$

For the sample mean, we modify the equation simply by substituting the appropriate symbols:

$$\bar{X} = (\Sigma X)/n$$

Mathematically, the only real difference between calculating a population mean and a sample mean is that the symbols change.

But although an unbiased sample ought to be centered on the same spot, its standard deviation will not be the same. In Figure 6.2, I've graphed all of the values of X in a frequency distribution. You can see how this distribution would have a certain spread around

Figure 6.2 A Possible Distribution

```
                XXXXXXXX
            XXXXXXXXXXXXXXXX
          XXXXXXXXXXXXXXXXXXXX
         XXXXXXXXXXXXXXXXXXXXXXX
        XXXXXXXXXXXXXXXXXXXXXXXXX
       XXXXXXXXXXXXXXXXXXXXXXXXXXXX
     XXXXXXXXXXXXXXXXXXXXXXXXXXXXXXXXX
   XXXXXXXXXXXXXXXXXXXXXXXXXXXXXXXXXXXXXXX
```

the mean. I've then taken a sample of those Xs and put them in color. The mean of the sample should be pretty close to the mean of the population. But you can see how the spread around the mean would necessarily shrink. So a population standard deviation will always be bigger than a sample standard deviation. To compensate for underestimating the standard deviation of a sample, we divide by a smaller number, n − 1 instead of N, in order to push our estimate of the standard deviation up just a bit. So we begin with the equation for the population standard deviation:

$$\sigma = \sqrt{(\sum(X-\mu)^2)/N}$$

But we not only substitute the sample symbols, we also divide by n − 1 instead of N:

$$s = \sqrt{(\sum(X-\bar{X})^2)/(n-1)}$$

In calculating the standard deviation of a sample, we will use a work table similar to the one we used to calculate the standard deviation of a population. Table 6.3 begins with the GDP for the former British colonies in Africa. The first column identifies the case; the second, its GDP per capita. At the bottom of the second column, we total the values of GDP. We then divide that total by the number of cases to get the sample mean. In this case, $\bar{X}$ = $4218.48. In the third column, we find the difference between the value of each case and the mean. In the fourth, we square the mean. At the bottom of the fourth column, we find the sum of squares. (You will find that the sum of squares is used repeatedly among the statistical measures of interval-level variables.) Unlike the work table for a population, now that you are calculating the standard deviation for a sample instead of a population, you divide the sum of squares by n − 1 instead of N to get the variance. Finally, you take the square root of the variance to get the standard deviation. In this case, s = $4945.91.

Table 6.3 GDP per Capita for Select Former British Colonies

Country	GDP per Capita X	$X - \bar{X}$	$(X - \bar{X})^2$
Botswana	13625.12	9406.64	88484789.55
Cameroon	2033.23	−2185.25	4775304.89
Egypt	6116.61	1898.13	3602911.54
Gambia	798.55	−3419.93	11695906.84
Ghana	1641.30	−2577.18	6641851.08
Kenya	1496.47	−2722.01	7409353.14
Lesotho	1874.85	−2343.63	5492615.64
Malawi	865.55	−3352.93	11242171.10
Mauritius	12771.76	8553.28	73158518.36
Nigeria	2396.62	−1821.86	3319174.95
Seychelles	16875.13	12656.65	160190695.56
Sierra Leone	924.41	−3294.07	10850908.36
Somalia	943.04	−3275.44	10728536.67
South Africa	9482.09	5263.61	27705597.60
Sudan	3181.93	−1036.55	1074444.40
Swaziland	4728.59	510.11	260216.09
Tanzania	1348.37	−2870.12	8237560.11
Uganda	1277.81	−2940.67	8647565.93
Zambia	1476.93	−2741.55	7516070.08
Zimbabwe	511.26	−3707.22	13743491.99
n = 20	ΣX = $84,369.59		SS = $\Sigma(X - \bar{X})^2$ = 464,777,683.90
	$\bar{X}$ = (ΣX)/20 = $4218.48		SS/(n − 1) = 464,777,683.90/(20 − 1)
			Variance = 24,461,983.36
			s = $\sqrt{24,461,983.36}$ = $4945.91

Source: Compiled from Gapminder, "Income Per Person," www.gapminder.org/data/.

Standard Error and t-Tests

Now that we know the population GDP for British colonies and the sample mean and standard deviation for the per capita GDP of former colonies in Africa, we are finally in a position to compare those means to see if the difference could be due to random error. In order to do that, we need to understand the probability distribution for sample means. We know that, on average, the sample mean will be an unbiased estimate of the population mean. But even though this is our best estimate of the mean, because of random error in our sampling, any specific sample mean we calculate will only approximate the population mean. If we kept taking samples from the population, we would expect that the estimate of the mean will bounce around the population mean. If we were to graph the distribution of these estimates, we would find that the estimates would center on the population mean, but would form their own standard deviation around that mean. We call that distribution the standard error of the mean.

Nate Silver (originally known as a baseball sabermetrician) has applied his understanding about statistics to the political world. He collects all the surveys that have been done in a particular election and models them all as different estimates of a single underlying voter preference. He knows that these polls aggregate into the distribution of the standard error. From that knowledge, he's able to run simulations based on the underlying probabilities and do a remarkably good job in predicting the outcome of races. His column, "Five Thirty Eight," reports his predictions for the *New York Times*. His skill in predicting the outcome of the 2012 presidential election (at the state level!) led to universal adulation, including at least one column[7] that applied a series of Chuck Norris jokes to Nate Silver. What I am trying to say is that you, too, can be as powerful as Chuck Norris if you understand the standard error of the mean.

The standard error is dependent on the standard deviation of the values of X and the size of the sample according to the following formula:

$$\text{Standard Error} = \sigma/\sqrt{n}$$

We divide the standard deviation by the square root of the sample size because we would expect that the bigger the sample, the tighter our estimates will be around the population mean. In Chapter 5, when we used the z table, we were calculating the probability that any given case would be at any given point along the range of possible values. But now that we are taking a sample of multiple points from the population, we are narrowing down the probable range considerably. And the larger the sample size, the more likely our sample mean will be close to the population mean. As a result, the standard error is much tighter around the mean than is the standard deviation. Frequently, though, the reason we are taking a sample is that we do not actually know the standard deviation for the population. Because of that, we will substitute our sample standard deviation (s) for our population standard deviation (σ). For our African sample, we substitute our sample standard deviation (s = $4945.91) for the population standard deviation σ:

$$\text{Standard Error} = s/\sqrt{n}$$

$$\text{s.e.} = \$4945.91/\sqrt{20}$$

$$= \$1105.94$$

Remember that the standard error of the mean is the same thing as the standard deviation of estimates of the mean around the actual mean. If our samples were big enough, this would be normally distributed around the mean. But as samples get smaller, the distribution gets mushed down into a flatter, wider distribution because small samples don't do as good of a job of estimating—there will be more error. With a small sample size, the distribution is called a "Student's t" distribution. This is not because only students use it. Rather, the statistician who developed it was working for Guinness Brewery at the time and, because of a previous bad experience, they prohibited their employees from publishing for fear of leaking trade secrets. William Sealy Gosset thought his discovery of the t-distribution was more of an important mathematical finding than a trade secret, so he decided to publish his finding anonymously so he wouldn't get into trouble. As a result, he signed this article (and others) "Student."[8] The t-distribution is usually published in the appendix of statistics textbooks right after the z table. Find it at the end of this book now. (See Appendix 4, page 445.)

BOX 6.1 Numbers in the News

In the 2012 presidential election, many polls predicted that Mitt Romney would receive more votes than he actually did. Because Gallup pollsters were concerned that their prediction was so far off, they worked with political scientists at the University of Michigan to find the cause.[9] Their report identified various differences between the poll's sample and the population of voters, each of which overrepresented Romney voters. For example, in sampling the Midwest, the poll overrepresented Midwesterners from the central time zone (who were more supportive of Romney) and underrepresented those from the eastern time zone (who were more supportive of Obama). By identifying those variables where the poll sample differed from the population of voters, Gallup was able to modify its research design to provide more accurate election predictions.

The t-score is very similar to the z-score. With the z-score, you compared a particular value of X with the mean to discover how many standard deviations X is from the mean. With a t-score, you compare a sample mean ($\bar{X}$) with a population mean (μ) to see how many standard errors away from the mean the sample is. We gave the z-score as

$$z = (X - \mu)/\sigma$$

The t-score looks very similar:

$$t = (\bar{X} - \mu)/s.e.$$

With the t-score, you can identify the probability of observing the sample mean if, in reality, the population mean is accurate.

Because the standard error is dependent on the sample size, the t-distribution is also dependent on the sample size. There is a different distribution for each possible sample size. If you wanted as much information as a z table contains, you would need to have a different t table for each possible sample size. Rather than have a whole series of different tables, we limit what goes into the table to the t-scores that are connected to the important levels of statistical significance. Because we are interested in data that are statistically significant at the 0.10, 0.05, and 0.01 levels, t tables will, at a minimum, include those three probabilities in the columns. Each row is connected to a different sample size. For that sample size, the entry in each cell tells you the minimum t-score necessary to be statistically significant at the level given in the heading for the column.

The most confusing thing about the t table is that it doesn't actually identify the various sample sizes. Instead, the rows are labeled as "d.f." or "degrees of freedom." The number of degrees of freedom is one less than n, the sample size. Do you remember in algebra how you could solve for variables if you had multiple equations containing those variables? The rule your algebra teacher gave you was that you need to have the same number of equations as unknowns for which you wanted to solve. The logic for calculating sample statistics is that you can solve for as many as you have data points. You've already used one data point to solve for the mean, so you can only solve for n − 1 after that—so that's how many degrees of freedom you have left when you solve for the standard deviation. The trick to remembering this is that with the sample mean, you divide by n, and with the sample standard deviation, you divide by n − 1.

Let's apply all that we've learned in order to answer the question we asked originally about the economic strength of African countries. Our question was, "Are African countries really different from other countries?" Our hypothesis was that if we look at twentieth-century British colonies, those in Africa would not be as wealthy. We found already that, on average, former British colonies have a mean GDP per capita of $13,307.87. For the sample of African colonies, the mean GDP per capita is $4218.48 with a standard deviation of $4945.91, from which we calculated a standard error of $1105.94. We can substitute these values into our t equation:

$$t = (\bar{X} - \mu)/s.e.$$

$$= (\$4218.48 - \$13,307.87)/\$1105.94$$

$$= -8.22$$

Remember that we interpret a t-score as the number of standard errors a sample mean is from a population mean. Because in this case we have a negative t-score, we know that the sample mean is less than the population mean. So the African nations are 8.22 standard errors below the mean for all former British colonies. From our experience with z-scores, we know that the probability of being more than two standard deviations from the mean is fairly slim. But z-distributions are not identical with t-distributions, so we need to complete our analysis. We want to know the probability of getting a t-score this large due to random error, so we look this up on the t table. (See Appendix 4, p. 445.)

The first thing you should notice on the table is the bottom line within the table. Under degrees of freedom, it has an infinity sign ∞. This means that the values on this line correspond with an infinitely large sample size. As you move across this row, you will find values for t-scores. You'll notice that they are all positive. As with z-scores, we know that we can ignore the negative sign—it just indicates that the sample mean is less than the population. In the bottom row, because of the large sample size, the t-scores actually correspond with z-scores. So the first element in the row is 0.674. If you look up a z-score of 0.67 on a z table, you will see a probability of 0.2486. This would round up to 0.25. Now look back at the t table to the 0.674. Follow that column up to its heading. It says ".25." So the probability of having a t-score of 0.674 or larger is 25 percent. If we translate this into math, we would write $P(t = 0.674) = 0.25$. The thing you have to remember with the t table is that it is giving you the probability of being in the tail, whereas the z table is giving you the probability of being between the mean and the value of X. So you actually need to subtract the probability in the z table from 0.50 to make it correspond with the bottom row of the t table. Look across to where the t-score = 1.960. The heading at the top of the column says 0.025. So the probability of being in the tail is 2.5 percent. Now look up a z-score of 1.96 in Appendix 3. The probability of being between the mean and that z-score is 0.4750. So the probability of being in the tail is $0.5000 - 0.4750 = 0.0250$, or 2.5 percent—just the same as corresponded to the t-score for a large sample. The conclusion you can draw from this is that if you have a large sample, the t-distribution is the same as the z-distribution, so you can just as easily look it up in a z table.

Because we normally want our results to be statistically significant at the 0.05 level, we will normally look at the column headed 0.05. Looking at the bottom row, we can see that for a very large sample, we would need a t-score of 1.645. So the sample mean would need to be 1.645 standard errors away from the population mean before we would conclude with confidence that what we are seeing is probably not due to chance. But that is for a very large sample. As our sample size decreases, we would move up the column and see that we need a larger and larger t-score. If we had a sample of 1001 (d.f. = 1000), we would need a t-score of 1.646. For a sample of 101 (d.f. = 100), we would need a t-score of 1.660. That is fairly close to the t-score you needed for an infinitely large sample. Comparing those bottom three rows, you can see that the t-distribution for a sample size of 100 is actually fairly close to the z-distribution found in the bottom line. But as the sample size gets smaller, the t-score you need in order to be statistically significant will increase. As you move all the way up to the top of the 0.05 column, you will see that if you have a sample size of only two (d.f. = 1), the entry says you need a t-score of 6.314. This means that if you have a sample size of two, you would need to see a sample mean that is 6.314 standard errors away from the population mean before you could be confident that something was actually going on.

So let's use the t table to decide whether we can be confident that African countries have different GDPs from other former British colonies. First, we find the appropriate row for the sample size. In this case, our sample size was twenty. Our degree of freedom is one less than the sample size: $20 - 1 = 19$. So we go down to the row for 19 degrees of freedom. Each of the columns is labeled with a probability—and this probability is connected to the picture at the top of the table. We want relationships that are statistically significant at the 0.05 level. So we want the probability of getting a sample mean in the tail that is less than 0.05. If we move to the right from where the row is labeled d.f. = 19 to the column with the heading 0.05, that cell will tell us what t-score will be statistically significant at the 0.05 level. The number in that cell is 1.729. This means that with twenty cases, we need a t-score of 1.729 or higher to be statistically significant at the 0.05 level. Because 1.729 is less than 8.22 (the standard error for former British colonies in Africa), we know that our relationship is statistically significant at the 0.05 level. It is very unlikely that the lower GDPs experienced in Africa are due to random factors. We can be fairly confident that the experience in Africa was different from that in other British colonies.

Proportions as a Special Case: Showing Employment Discrimination

In the United States, means testing is used as evidence in discrimination cases. The reason the U.S. Census Bureau keeps asking questions about the race and ethnicity of people who live in this country even though that information is no longer relevant to either taxation or representation is so that we have a measure of the racial composition of the population against which we can compare samples. If those samples do not look like the population, we can conclude that something is happening to cause the disparity.

From these data, the Equal Employment Opportunity Commission, the federal agency charged with enforcing laws against employment discrimination, has calculated the racial, ethnic, and gender makeup of various sectors of private industry. For example, of 803,349 total executive-level managers in the country, 91,845 are minorities.[10] Race is a nominal-level variable, and so technically, you can't calculate a mean. But suppose you have a variable X for which you were to code all minority executives as "1" and all other executives as "0." (We call a dichotomous variable that has been coded this way a dummy variable.) If you were to use these data as if they were interval, you could then add up all the values of X to get the summation. You would find that $\Sigma X = 91,845$, or the total number of minority executives in the United States. If you then divided by N (or the total number of executives), you would get 91,845/803,349 = 0.1143. You wouldn't interpret this in the normal way you would a mean; it doesn't make sense to say that executives are 0.1143 minority.

Table 6.4 Racial Composition of Executives

Race X	Frequency f_x	Xf_x	$X - \mu$	$(X - \mu)^2$	$(X - \mu)^2 f_x$
0 White	711,504	0	−0.1143	0.0131	9320.70
1 Minority	91,845	91,845	0.8857	0.7845	72,052.40
	$N = 803,349$	$\Sigma X = 91,845$		$SS = \Sigma(X - \mu)^2 f_x = 81,373.10$	
		$\mu = (\Sigma X)/N$		Variance $= SS/N = 0.1013$	
		$= 91,845/803,349$		$\sigma = \sqrt{0.1013} = 0.3183$	
		$= 0.1143$			

Source: "2009 Job Patterns for Minorities and Women in Private Industry," *2009 EEO-1 National Aggregate Report*, 2009, www1.eeoc.gov/eeoc/statistics/employment/jobpat-eeo1/2009/index.cfm.

But it does make sense to interpret that number as a proportion. From it, you know that 11.43 percent of executives are minorities.

Table 6.4 shows these data in tabular form. I've worked out the standard deviation for this population in the same way we've done for tabular data in prior chapters. As in Chapter 4, the population size ends up being irrelevant when calculating the standard deviation for a proportion. I've used N in the equations as appropriate, but I could just as easily have used the following equation, which uses only p (or the proportion) and doesn't use *N* at all:

$$\sigma = \sqrt{p*(1-p)}$$
$$= \sqrt{0.1143*(1-0.1143)}$$
$$= \sqrt{0.1143*(0.8857)}$$
$$= \sqrt{0.1013}$$
$$= 0.3183$$

As in Chapter 4, the finding is hidden in the algebra that I used in calculating the standard deviation in Table 6.4. Simply by the nature of using proportions, the *N*s end up cancelling out in the end.

These census data give us the population mean—which, in the case of this dummy variable, is the population proportion. We can then compare the employment records of any given company with the population proportion to determine if that company has hiring practices outside the range that is normal for the country. And that is precisely what the EEOC does. So suppose you get called in to examine the hiring practices of a company and find that of twenty total executives, this company has one minority executive. Your sample proportion (or mean) is 1/20 = 0.05. We found above that when calculating the standard deviation of proportion of a population, the *N*s cancel out and all we need is the proportion. This algebra holds for sample standard deviations as well. So for samples, I would normally need to divide by n − 1 instead of *N* for the standard deviation, but if I'm dealing with a sample proportion, I don't need to account for the size of the sample at all. The equation for population standard deviation "σ" is identical for the sample standard deviation "s":

$$\sigma = \sqrt{p*(1-p)}$$
$$s = \sqrt{p*(1-p)}$$
$$= \sqrt{0.05*(1-0.05)}$$
$$= \sqrt{0.05*0.95}$$
$$= \sqrt{0.0475}$$
$$= 0.2179$$

Our question is whether the proportion of minority executives in this company is exceptional or not. Given the population mean of 11.43 percent minority executives, what is the probability of observing a company with only 5 percent minority executives as shown in Table 6.5? If the probability is less than 0.05, we would conclude that something is going on at this company and it might be reasonable to prosecute it for racist hiring practices. In order to answer that question, we need to look at the standard error of the mean to see if a mean of 0.05 is an outlier. From the sample standard deviation of 0.2179, you can calculate the standard error:

Table 6.5 Racial Composition of Executives at a Company

Race X	Frequency fx	Proportion
0 White	19	0.95
1 Minority	1	0.05
	n = 20	

$$\text{s.e.} = s / \sqrt{n}$$
$$= 0.2179 / \sqrt{20}$$
$$= 0.0487$$

We now have all the numbers we need to calculate a t-score:

$$t = (\bar{X} - \mu)/\text{s.e.}$$
$$= (0.05 - 0.1143)/0.0487$$
$$= -1.32$$

Finally, we look up this t-score on the t table in Appendix 4. Keep in mind that the t table will have only positive values, so just ignore the negative sign. Because our sample size was twenty, we look at the line for nineteen (one less than the sample size) degrees of freedom. If we want our results to be statistically significant at the 0.05 level, we look at the t-score in that row under the column labeled 0.05. That cell tells us that t needs to be at least 1.729. Because our t-score of −1.32 has a magnitude that is less than what was needed, we cannot reject the possibility that what we are seeing in this company could just be due to the fact that they don't have very many executives. They may not have as many minority executives as the industry average, but that could just be due to chance.

CONFIDENCE INTERVALS: TWO-TAILED DISTRIBUTIONS

In the previous chapter, we used our knowledge about the distribution along the normal curve to calculate the probability of being in particular ranges. In much the same way, we can use the Student's t distribution to calculate confidence intervals around an estimated mean. Recently, a *Forbes* article reported that the most affluent community in America is Westlake,

Texas, where the musicians the Jonas Brothers live.[11] The article was based on estimates made by the Census Bureau in its American Community Survey. The original census data reported that the mean household income for the town was $446,317. This estimate was accompanied by a margin of error of ±$179,565. When we read "margin of error," we should translate it to say that if we sandwich the mean in a range that extends a distance of the margin of error on each side, we are 95 percent confident that the actual mean falls in that range. If we assume that income in Westlake is normally distributed, then we can concluded that the actual mean is within $179,565 of the estimated $446,317. This confidence interval is pictured in Figure 6.3. It shows a normal curve around the estimate that is shaded on each side of the estimate a distance of that margin of error. If we are 95 percent confident that the real mean is in that range, then there must be a 5 percent chance that the real mean is in the tails past those two points. Because the two tails total a probability of 0.05, and they are symmetrical, each tail must contain half that much. So there must be a probability of 0.025 that it is less than $266,752. Conversely, there must be a probability of 0.025 that it is greater than $625,882.

Figure 6.3 Estimated Income of Westlake, Texas

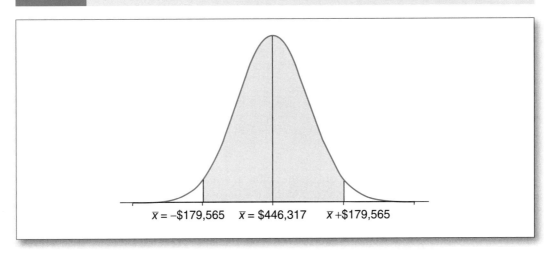

The Census Bureau was kind enough to give us the margin of error for its estimate of the income in Westlake, but that isn't always the case. As long as we are given the mean and either the standard error or the standard deviation and sample size, we can calculate the margin of error around our estimate of the mean. To calculate the margin of error, we take the equation for the t-score and make some algebraic adjustments to it. We know that we want to get both a positive and a negative t-score so that we can get the range on each side of the mean. So we begin with the equation for the t-score and adjust it to get the population mean (μ) on one side of the equation:

$$\pm t = (\overline{X} - \mu)/\text{s.e.}$$

$$\pm t(\text{s.e.}) = (\text{s.e.})(\overline{X} - \mu)/\text{s.e.}$$

$$\pm t(s.e.) = \bar{X} - \mu$$
$$\mu \pm t(s.e.) = \bar{X} - \mu + \mu$$
$$\mu = \bar{X} \pm t(s.e.)$$

From the sample data, we can find the sample mean and standard error.

Figure 6.4 Curve for 95% Confidence Interval

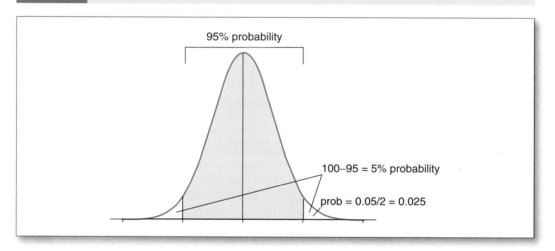

To find our t-score, we need to use a bit of logic. Remember how, with z-scores, the first step was always to draw a picture of a normal curve and insert the information that we are given? Let's do that now in Figure 6.4. We know that we want to be 95 percent confident, which means that we want to use the t-score for which 95 percent of the distribution is symmetrical around the mean and 5 percent of the distribution is in the two tails. Because the tails are symmetrical, we can split that 5 percent in half to find the probability of being in one tail. So we want the t-score for which one tail has a probability of 0.025. If the sample size is large, then t = 1.96. So for a large sample, we can rewrite the confidence interval as

$$\mu = \bar{X} \pm 1.96(s.e.)$$

All we need then are the mean and standard error for any large dataset and we can calculate the confidence interval.

The National Center for Education Statistics is charged with collecting data on the quality of education in the United States, and it collects a wide variety of data. For example, it collects average test scores in math as well as other subject areas. In 2009, the average math score for twelfth graders was 153 on a scale from 0 to 300.[12] Along with the mean, they reported a standard error of 0.7. With that information, we can calculate a 95 percent confidence interval because we already know that a 95 percent confidence interval has a t-score of 1.96.

$$\mu = \bar{X} \pm t(s.e.)$$
$$\mu = \bar{X} \pm 1.96(s.e.)$$
$$\mu = 153 \pm 1.96(0.7)$$
$$\mu \leq 153 \pm 1.372$$
$$153 - 1.372 \leq \mu \leq 153 + 1.372$$
$$151.628 \leq \mu \leq 154.372$$

We can be 95 percent confident that the actual mean score for twelfth graders is somewhere between 151.628 and 154.372. This window is very narrow because the sample size is so large—it makes the standard error miniscule in relationship to the magnitude of the scores.

Confidence Intervals with Proportions

Confidence intervals are particularly important during election years. When you hear about pre-election polls, they are always framed in terms of a margin of error, which we should be careful to translate into a confidence interval. In the 2000 presidential race, the final Gallup poll before the election predicted that George W. Bush would receive 49.0 percent of the vote. The standard deviation of the poll is entirely dependent on that proportion:

$$s = \sqrt{p*(1-p)}$$
$$= \sqrt{0.49*(1-0.49)}$$
$$= \sqrt{0.49*(0.51)}$$
$$= \sqrt{0.2499}$$
$$= 0.4999$$

But although the standard deviation depends only on the proportion, the standard error depends on the sample size. Suppose that they interviewed 1,400 people. Then the standard error would be

$$s.e. = s/\sqrt{n}$$
$$= 0.4999/\sqrt{1400}$$
$$= 0.0134$$

Remember that for a confidence interval of 95 percent, you would use a t-score of 1.96. In that case, the margin of error would be t(s.e.) = 1.96(0.0134) = 0.0263. This yields a 95 percent confidence interval of

$$\mu = \bar{X} \pm t(s.e.)$$

$$\mu \leq .49 \pm 0.0263$$

$$.49 - 0.0263 \leq \mu \leq .49 + 0.0263$$

$$.4637 \leq \mu \leq .5163$$

So Gallup pollsters were 95 percent confident that Bush would get between 46 and 52 percent of the vote. The problem with that election was that it was within the margin of error for Bush to either win or lose the election. So although Bush was polling better than Gore, the pollsters were usually careful to say that the election was too close to call. In other words, they were not 95 percent confident that Bush would receive more votes than Gore.

CHOOSE A SAMPLE SIZE

Pollsters actually use their understanding of the role of sample size in determining the margin of error to decide how many people to survey. If pollsters know an election is close, they can increase their sample size to decrease their margin of error in order to make a better prediction of who will win the election. If we substitute the equation of the standard error into the margin of error, we get

$$\text{Margin of error} = 1.96(s.e.)$$

$$= 1.96(s/\sqrt{n})$$

Suppose we're a pollster and want to decrease the margin of error to 1 percent. How many people do we need to survey?

$$0.01 = 1.96(s/\sqrt{n})$$

$$0.01\sqrt{n} = \sqrt{n}\; 1.96(s/\sqrt{n})$$

$$0.01\sqrt{n} = 1.96s$$

$$0.01\sqrt{n}/0.01 = 1.96s/0.01$$

$$\sqrt{n} = 196s$$

All you need is the standard deviation and you can find out what sample size you need.

The problem is that you can't know the standard deviation until after you've done the survey. So what do you do? It sounds like a conundrum until you remember that for a dichotomous variable, the standard deviation is entirely dependent on the proportion. You can find the maximum possible standard deviation and just use that. As a result, you will have the sample size you need for the worst-case scenario. Table 6.6 shows the possible standard deviations for various possible proportions. As you can see, the standard deviation keeps increasing until you get to a 50-50 split and then it folds back onto itself. So the

Table 6.6 Standard Deviations Associated with Various Proportions

p	$1-p$	$p(1-p)$	$s = \sqrt{p(1-p)}$
.1	.9	0.09	0.30
.2	.8	0.16	0.40
.3	.7	0.21	0.46
.4	.6	0.24	0.49
.5	.5	0.25	0.50
.6	.4	0.24	0.49
.7	.3	0.21	0.46
.8	.2	0.16	0.40
.9	.1	0.09	0.30

maximum standard deviation is at p = 0.50, when the election is a dead heat. At this point, s = 0.50, so we can substitute this maximum standard deviation into our equation and get the most conservative estimate of the necessary sample size. In the end, we might not need to survey that many people, but we know we are safe if we do:

$$\sqrt{n} = 196s$$
$$= 196(0.50)$$
$$= 98$$
$$n = 98^2$$
$$= 9604$$

So if you have a sample size of 9604, you are guaranteed of having 95 percent confidence with a margin of error of only 1 percent. Unfortunately, a survey of that size would be very expensive, so most pollsters survey fewer people with the consequence that the margin of error is much larger than 1 percent.

SUMMARIZING THE MATH: SAMPLING A POPULATION

When we are able to measure a full population, we can describe the parameters of a variable with its mean and standard deviation. But once we start looking at smaller samples, we insert uncertainty into our description. In this chapter, we compared a population mean with the mean of a smaller sample to see whether the subgroup was different from the population. Just as diagnosticians worry about making false-positive and false-negative diagnoses, as statisticians we worry about making Type I and Type II errors. Our biggest concern is that we will make a Type I error and incorrectly conclude that a difference is not random when it really is. So we normally set a threshold of 0.05 and say a difference is statistically significant when there is only a 5 percent chance that we are making a Type I error. We use a means test to compare the population and sample means to determine whether or not the difference is statistically significant.

Means Testing

We use different symbols when describing samples instead of populations.

1. The mean is calculated the same way as for the population but is called $\bar{X}$, (pronounced X-bar) instead of μ (mu):

$$\bar{X} = (\Sigma X)/n$$

2. The standard deviation is also given a different symbol (s instead of σ), but its equation changes to compensate for the fact that a sample will always have a narrower distribution than the population from which it is taken. So instead of dividing by N, you divide by $n - 1$.

$$s = \sqrt{(\Sigma(X - \bar{X})^2)/(n-1)}$$

3. From our estimate of the standard deviation, we calculate the standard error of the mean:

$$s.e. = s/\sqrt{n}$$

4. We then calculate a t-score, which compares the population mean with the sample mean:

$$t = (\bar{X} - \mu)/s.e.$$

5. Finally, we look up the t-score on a t table in Appendix 4 to find the probability that the difference we observed between our subgroup and the population is due to chance. We find the row on the table corresponding to our sample size:

$$d.f. = n - 1$$

Next, we move across that row to find the cell in which the t-score is just under the t-score for our sample. We look up at the heading for that column and see the probability of getting that mean if the population mean holds true for our sample. If our t-score is greater than the value under $p = 0.05$, we say that the difference between the population and sample means is statistically significant at the 0.05 level. If it is greater than the value under $p = 0.01$, our findings are statistically significant at the 0.01 level.

A Political Example: Soviet Corruption. The Soviet Union kept a very tight rein on behavior within its component states. With the breakup of the Soviet Union, the post-Soviet states faced increasing freedom, which led to the possibility of increasing corruption. A common measure of corruption is produced by Transparency International, where states with low scores are more corrupt than states with high scores. Worldwide, the average score is 4.00. Are the post-Soviet states actually more corrupt than other countries of the world? Do a means test on the Corruptions Perception Index to answer the question. (Note that a score of 10 means there is no corruption and a score of 0 means a country is totally corrupt.)

1. Calculate the sample mean according to the equation:

$$\bar{X} = (\Sigma X)/n$$

Table 6.7 is my work table. At the bottom of the second column, I total the scores and divide by the number of cases to get an average score of 2.99.

Table 6.7 Corruption Perception Index for Post-Soviet States

Country	CPI*	$X - \bar{X}$	$(X - \bar{X})^2$
Armenia	2.6	−0.39	0.15
Azerbaijan	2.4	−0.59	0.35
Belarus	2.4	−0.59	0.35
Estonia	6.4	3.41	11.63
Georgia	4.1	1.11	1.23
Kazakhstan	2.7	−0.29	0.08
Kyrgyzstan	2.1	−0.89	0.79
Latvia	4.2	1.21	1.46
Lithuania	4.8	1.81	3.28
Moldova	2.9	−0.09	0.01
Russia	2.4	−0.59	0.35
Tajikistan	2.3	−0.69	0.48
Turkmenistan	1.6	−1.39	1.93
Ukraine	2.3	−0.69	0.48
Uzbekistan	1.6	−1.39	1.93
		SS = 24.50	
	$\Sigma X = 44.8$	Variance = 24.50/(15 − 1) = 1.75	
n = 15	$\bar{X} = (\Sigma X)/n = 2.99$	$s = \sqrt{1.75} = 1.32$	

Source: Transparency International, *Corruption Perceptions Index 2011*, www.transparency.org/cpi2011/in_detail.

*Note that states with low scores are more corrupt.

2. Calculate the sample standard deviation using the equation:

$$s = \sqrt{(\sum(X - \bar{X})^2)/(n-1)}$$

In the third column of my worktable, I subtract the mean from the value in the second column. In the fourth column, I square the difference found in the third column. At the bottom of the fourth column, I get the sum of squares, which I then divide by n – 1, which is 14 in this case, to get the variance. Then I take the square root of the variance to get the standard deviation for my sample, which is 1.32.

3. From the standard deviation, calculate the standard error of the mean:

$$s.e. = s/\sqrt{n}$$

The standard error is $1.32/\sqrt{15} = 0.34$.

4. Calculate a t-score comparing the population mean with the sample mean:

$$t = (\bar{X} - \mu)/s.e.$$

$t = (2.99 - 4.00)/0.34 = -1.01$.

5. Finally, look up the t-score on a t table to find the statistical significance. Remember that you find the row by d.f. = n – 1.
I ignore the negative sign and look to see if the entry in the t table is less than 1.01. For this sample, d.f. = 15 – 1 = 14. In that row, to be statistically significant at the 0.05 level, you would need a t-score of 1.761, which is greater than 1.01. Although the post-Soviet states appear to be more corrupt than the world average, this difference might well be due to chance.

Means Testing with Proportions

Mathematically, the rules for means testing are the same for proportions as for interval-level data except that the means of the population and sample are the proportion of cases with the attribute.

1. Identify the population mean and sample mean.

2. For proportions, the equation for the sample standard deviation is based entirely on the sample proportion, where $p = \bar{X}$:

$$s = \sqrt{p*(1-p)}$$

3. Estimate the standard error in the same way:

$$s.e. = s/\sqrt{n}$$

4. Calculate a t-score that compares the population mean with the sample mean:

$$t = (\bar{X} - \mu)/s.e.$$

5. Finally, look up the t-score on a t table, choosing the row corresponding to our sample size:

$$d.f. = n - 1$$

Find the cell in that row that is just less than our t-score. Looking at the top of that column, we see the probability that the difference we observed between our subgroup and the population is due to chance.

A Political Example: "Driving While Black". In 2010, economist Timothy Bates published a study of the "crime" of driving while black. He analyzed all of the traffic tickets given in Eastpoint, Michigan, from 1996 to 1999. He did a means test to see whether the treatment of blacks and whites in that Detroit suburb could have been the result of chance.[13] In order to measure the racial makeup of drivers on the streets studied, the author actually counted cars over a four-month period. He estimated that on the mostly white streets of Nine Mile Road and Ten Mile Road, 5.3 percent of the drivers were black. Of the 2,270 tickets given on those two roads, 7.8 percent of them went to blacks. Do a means test. What is the probability that this difference is due to chance rather than racial profiling by police?

1. Identify the population mean and sample mean.

$$\mu = 0.053, \; \bar{X} = 0.078$$

2. For proportions, the equation for the sample standard deviation is based entirely on the sample proportion, where $p = \bar{X}$:

$$s = \sqrt{p*(1-p)}$$

I use the sample proportion to calculate the standard deviation. $p = \bar{X} = 0.078$, so $s = \sqrt{0.078(1 - 0.078)} = 0.268$.

3. Estimate the standard error in the same way:

$$s.e. = s/\sqrt{n}$$

s.e. $= 0.268/\sqrt{2270} = 0.0056$

4. Calculate a t-score that compares the population mean with the sample mean:

$$t = (\bar{X} - \mu)/s.e.$$

$t = (0.078 - 0.053)/0.0056 = 4.46$

5. Finally, look up the t-score on a t table, where d.f. = n − 1.
   d.f. = 2270 − 1 = 2269 = ∞. Our t-score of 4.46 is larger than 2.576, the largest value on that row. So the probability of getting this disparity in ticket rates due to chance is less than 0.005. We can conclude that the disparity is statistically significant at the 0.01 level.

Confidence Intervals

A confidence interval gives us the range of possible values around our sample mean between which we are confident the actual mean lies. The general equation is

$$\mu = \bar{X} \pm t(s.e.)$$

1. First we find the standard deviation.
   a. If it is an interval-level variable, the standard deviation is given by

   $$s = \sqrt{(\sum(X-\bar{X})^2)/(n-1)}$$

   b. But if it is a proportion, the standard deviation is

   $$s = \sqrt{p*(1-p)}$$

2. Second, find the standard error from the standard deviation:

   $$s.e. = s/\sqrt{n}$$

3. Third, we find the t-score associated with the confidence we want, remembering that this is a two-tailed test, so after we subtract our confidence from 1, we then divide it by 2 so that the probability of being wrong is evenly divided between the two tails. If we know that we want to be 95 percent confident in this interval, then we substitute a t-score of 1.96 because that is the t-score that corresponds to having a probability of 0.025 in each of the two tails.

4. Finally, we substitute our sample mean along with the appropriate t-score and standard error into the equation:

   $$\mu = \bar{X} \pm t(s.e.)$$

A Political Example: Final Poll 2008. In the final poll of the 2008 election, Gallup surveyed 2,458 people. Among likely voters, 53 percent indicated that they planned to vote for Barack Obama.[14] Find the confidence interval within which pollsters were 95 percent sure Obama's vote would fall.

1. Because we are dealing with proportions, the standard deviation is

   $$s = \sqrt{p*(1-p)}$$

   Because p = .53, s = √.53*(1−.53) = 0.499.

2. Find the standard error from the standard deviation:

$$s.e. = s/\sqrt{n}$$

The s.e. = $0.499/\sqrt{2458} = 0.010$.

3. For a 95 percent confidence interval, substitute the sample mean and standard error into the following equation:

$$\mu = \bar{X} \pm 1.96(s.e.)$$

$$\mu = .53 \pm 1.96(0.010) = .53 \pm 0.0196$$

$$0.53 - 0.02 < \mu < 0.53 + 0.02$$

$$0.51 < \mu < 0.55$$

Gallup was more than 95 percent confident that Obama would win the election.

Choosing a Sample Size

If you know the size you want your margin of error to be, you can work backward to calculate the size of sample you would need to get that confidence interval.

1. Begin with the equation for the margin of error:

$$\text{Margin of Error} = t(s.e.)$$

$$= t(s/\sqrt{n})$$

2. You substitute the appropriate t-score for the level of confidence you want. If you want to be 95 percent confident that the actual mean is in the range, you substitute a t-score of 1.96:

$$\text{Margin of Error} = 1.96(s/\sqrt{n})$$

3. Next, you know that in a survey, the highest a standard deviation can be for a proportion is where the results are evenly split. So you substitute the maximum standard deviation of 0.50 into the equation:

$$\text{Margin of Error} = 1.96(0.50/\sqrt{n})$$

$$= 0.98/\sqrt{n}$$

4. But you want to move the variables around so that the equation is given for n, the sample size:

$$\sqrt{n} = 0.98/\text{Margin of Error}$$

$$(\sqrt{n})^2 = (0.98/\text{Margin of Error})^2$$

$$n = (0.98/\text{Margin of Error})^2$$

A Political Example: Do Your Own Poll. Suppose you want to find how the students at your university will vote during an election. How big of a sample do you need to be 95 percent confident with a 5 percent margin of error?

I substitute the 5 percent margin of error into the equation: $n = (0.98/\text{Margin of Error})^2 = (0.98/0.05)^2 = 384$.

USE SPSS TO ANSWER A QUESTION WITH MEANS TESTING

As political scientists, we occasionally find that a subgroup has a different mean from the population on a given variable. Although we know that will happen occasionally simply due to random error, when the two means are very different, we wonder whether there is some underlying factor leading to the difference. Assuming that the underlying variable is normally distributed, we want to know the probability that the subsample's distribution could be due to chance. We call this the statistical significance. In general, we want the difference to be statistically significant at the 0.05 level, reducing the probability of it being due to chance to 5 percent. We use a means test to compare the population mean to the distribution of our sample.

In SPSS, after cleaning your data, the first step is to find the population mean. Although you can do this by getting a frequency of the variable and requesting a mean on the "statistics" page, this can be cumbersome. If you have a truly interval-level variable, there may be as many categories of your variable as you have cases. If that is true, the frequency would have that many rows. It is easier to find the mean using the "Analyze," "Descriptive Statistics," and "Descriptives" commands. As in Chapter 5, you will bring your variable over to the "Variable" box. But this time, instead of wanting to save a standardized version of the variable, you will click on "Options." There are many descriptive statistics you can request here. For our purposes, you definitely want the mean, but you might also want the minimum and maximum just to do a double check that the data are clean.

Once you have the population mean, you want to compare it to the sample distribution. You will want to select only those cases that belong to the subgroup you are comparing to the population. Use the "Data," "Select If" command to indicate the value of a particular variable that identifies your subgroup. Remember here that in the box, you will indicate a variable along with the value(s) it is equal to, less than, or greater than (using the appropriate operator). You can check to make sure the command worked properly by entering the data window; there will be slashes through the case number of those cases that are not part of the subgroup.

BOX 6.2 How to Use SPSS to Do a Means Test

After opening your data in SPSS and cleaning your variables, get the population mean for the variable.

>Analyze
>>Descriptive Statistics
>>>Descriptives
>>>>"*Variable*"→Variable
>>>>\>Options
>>>>>\>Mean
>>>>>\>Continue
>>>>\>OK

Select the cases that belong to the group you want to compare to the population:

>Data
>>Select Cases
>>>\>If condition is satisfied
>>>\>If
>>>>"*GroupIdentifier*"→"*GroupIdentifier=appropriate value*"
>>>>\>Continue
>>>\>OK

Compare the group to the population mean for the variable:

>Analyze
>>Compare Means
>>>\>One-Sample T-Test
>>>>"*Variable*"→Test Variable
>>>>Test Value="*population mean*"
>>>>\>OK

Finally, you do the means test. Under "Analyze," one of the options is "Compare Means." You want to choose the "One-Sample T-Test" option. Bring your variable into the "Test Variable" box. Then, for the "Test Value," enter the population mean you found above. Get SPSS to process your request by clicking on "OK." The SPSS commands are summarized in Table 6.8.

SPSS will provide you with two tables of output. The first one is labeled "One-Sample Statistics." This will give you the statistics of the variable for your sample, including the sample mean. You can compare the sample mean to the population by finding the difference between them. But you want to know if this difference is statistically significant. The second table is labeled "One-Sample Test," which you gave in the "test value" box. This will compare the distribution for the sample with the mean for the population. It gives you the t-score, but you don't really need that because in the next column, it gives you the level of statistical significance for the difference. If this is smaller than 0.05, you say that the difference is statistically significant at the 0.05 level. If this is smaller than 0.01, it is statistically significant at the 0.01 level. On occasion, SPSS will give a value of 0.000 for the significance, but the probability is never actually zero. Any result might be due to chance; SPSS has just rounded this to zero because of the limited number of decimal points. In the right-hand column, the table gives you a confidence interval for the difference between the sample and population means. You are 95 percent confident that the difference lies in this range. Table 6.9 summarizes the full process of doing a means test.

BOX 6.3 How to Do a Means Test

A. Find the population mean.

B. Identify your subgroup.

C. Compare the sample mean to the population mean.

D. Write your memo describing your conclusions.

A Political Example: Civil War and Infant Mortality

The year is 2009. Los Angeles is hosting a meeting of fourteen African first ladies focusing on health care. You are interning in the office of Maria Shriver, the first lady of California. On the first day of the conference, she gives the welcoming address. When she returns to the office later that day, she asks you to look into something. She had been having a conversation with Sia Koroma, the first lady of Sierra Leone, in which the issue was raised about the impact of civil war on infant mortality. Ms. Koroma argued that civil war doesn't just kill people during the battles; it also has a long-term impact on life expectancy because it destroys various kinds of infrastructure.[15] Ms. Shriver is thinking about addressing this issue in future speeches, but first she wants to check out the facts. She asks you to collect data and write a memo analyzing whether those nations involved in an internal conflict actually do have a higher infant mortality rate.

1. Open your dataset in SPSS and make sure that the variables "InfantMortality" and "InternalWar" are clean using the "Define Variable Properties" command.
   InfantMortality has a possible value of "−9," which I set to "Missing." The other values range between 1.80 and 121.63, which seem quite reasonable. For

InternalWar, I assign the value labels "no internal war" for a value of 0 and "internal war" for a value of 1.

2. **Find the population mean.**
   Using the "Descriptives" command, I get a mean infant mortality rate of 29.287979 per thousand live births for the population of all countries.

3. **Select your subgroup.**
   I select those countries that are involved in a civil war by using the "Select Cases" command.

4. **Compare the subgroup mean to the population mean.**
   Figure 6.5 shows the location for the command for a means test. Figure 6.6 shows how to enter the data for the test.

Figure 6.5 Finding the Command for a Means Test

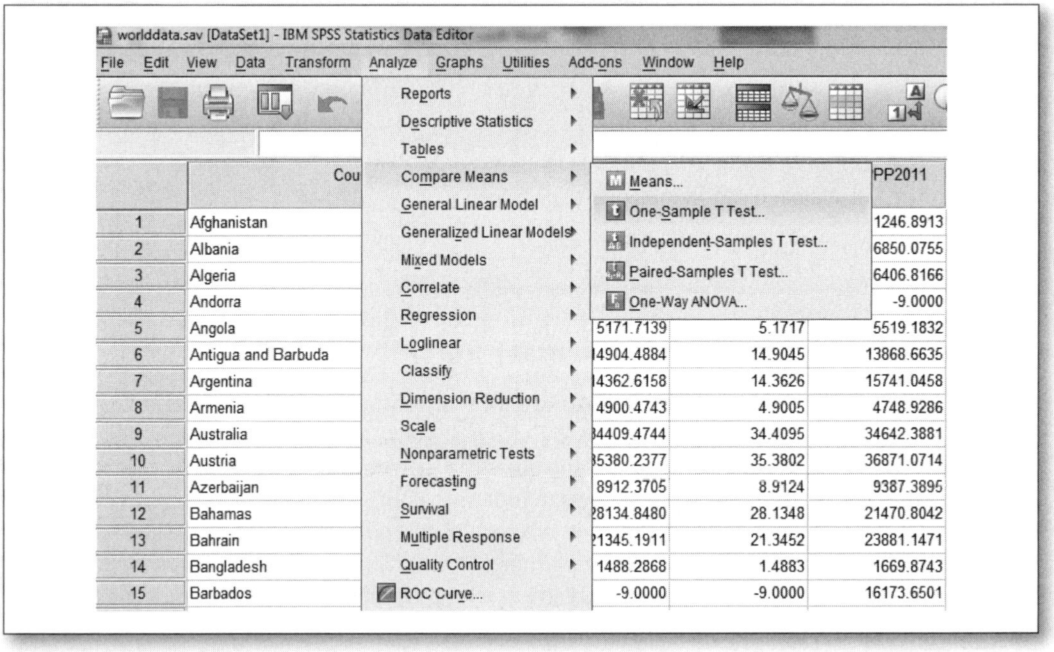

5. **Analyze the output and write your memo.**
   Figure 6.7 shows the resulting output. I can see that those nations involved in an internal war have an infant mortality rate of 48.055, which is twenty deaths more per thousand live births than the population mean. The table reports a significance of 0.003, so this difference is statistically significant at the 0.01 level. I am 95 percent confident that nations involved in internal wars have infant mortality rates that are at least 6.7 deaths per thousand live births higher than countries as a whole, and maybe as much as 30.6 deaths higher.

Figure 6.6 Requesting a Means Test in SPSS

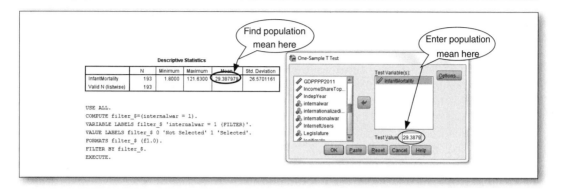

Figure 6.7 Output from SPSS for a Means Test

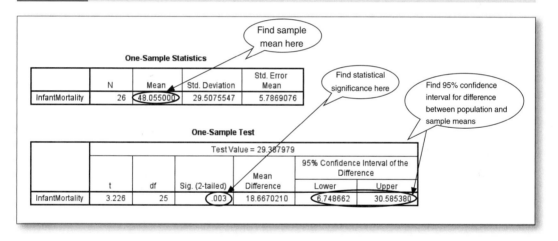

Memo

To: First Lady Maria Shriver
From: T. Marchant-Shapiro, intern
Date: April 22, 2009
Subject: Impact of Civil War on Infant Mortality

You asked me to research the long-term impact of purely internal wars on the infant mortality rate. In order to answer this question, I collected the infant mortality rates (in terms of number of infants under the age of one per thousand live births) from the CIA World Factbook. I found that Sia Koroma was correct when she told you that these countries have higher infant mortality rates.

On average, the nations of the world experience twenty-nine deaths of infants under the age of one for every thousand live births. Among the twenty-six nations that were involved in a solely internal military conflict between 2000 and 2009, the average was about twenty deaths higher than that—with an average infant mortality rate of forty-eight. There is less than a 1 percent chance that this difference is due to chance. We can be 95 percent confident that civil war increases the infant mortality rate by at least 6.7 deaths per thousand live births, and possibly as many as 30.6 deaths.

You can feel comfortable in the future repeating Sia Koroma's claim that countries that have been involved in internal war are left with higher infant mortality rates. This statistical analysis does not answer the question of whether she was correct in asserting that this difference occurs because of the destruction of infrastructure, although that seems like a reasonable hypothesis. Regardless of the causal mechanism, decreasing the infant mortality rate in these countries would be a worthy goal.

Your Turn: Means Testing

YT 6.1 The expectation is that African countries not only have worse economies than other former colonies, they also have less stable governments. Table 6.8 gives the instability index for each of the former British colonies in Africa.

Table 6.8 Instability Measure for Former British Colonies in Africa

Country	Instability Index X	$X - \bar{X}$	$(X - \bar{X})^2$
Botswana	3		
Cameroon	16		
Egypt	13		
Gambia	15		
Ghana	14		
Kenya	12		
Lesotho	11		
Malawi	16		
Mauritius	1		
Nigeria	17		
Sierra Leone	19		
Somalia	25		
South Africa	8		
Sudan	24		
Swaziland	8		

Source: Marshall, Monty G., and Benjamin R. Cole, *Polity IV Project* 2012. www.systemicpeace.org/polity/polity4.htm.

1. Calculate the mean, standard deviation, and standard error for these data.

2. Compare the sample mean to the population mean for twentieth-century British colonies as a whole (μ = 10.30). Calculate the t-score. Are you 95 percent sure that African countries are less stable politically than other British colonies?

YT 6.2 In the study of "Driving while Black," the author was interested in whether police were more likely to attribute bad intent to drivers who "don't belong." As a result, he also analyzed data on Eight Mile Road, which bordered on a black community, in order to analyze whether whites were disproportionately ticketed there because they "didn't belong." On this road, he found that 60.7 percent of the drivers were white. Of the 3,896 drivers ticketed on this road, 65 percent were white. What is the probability that this difference is due to chance?

YT 6.3 In February 2012, President Obama began to do battle with Catholic Church leaders over a requirement that health plans cover the cost of birth control. During the initial hubbub, Gallup did a national poll of Catholics to see whether their approval of the president had changed as a result of it.[16] Earlier, the president's approval rating among Catholics had been 0.49. The survey of 755 Catholics after the announcement gave him an approval rating of 0.46. Compute the 95 percent confidence interval for the survey. Based on the survey results, are we 95 percent certain that Obama's approval ratings changed? (Is the earlier 0.49 rating within this range?)

YT 6.4 In 2004, President George W. Bush defeated John Kerry by 2.4 percent of the vote. Prior to the election, how large a sample would a survey have needed in order to predict with 95 percent confidence that Bush would win within that 0.024 margin of error?

Apply It Yourself: Assess Maternal Mortality Rate Increases

It is still April 2009. First Lady Maria Shriver is intrigued by your finding that nations that have been involved in an internal war have higher infant mortality rates than other countries. She asks you to research a follow-up question. Sierra Leone's first lady Sia Koroma indicated that another aftereffect of these kinds of conflict is that maternal mortality rates increase. Is that true? In SPSS's "WorldData," the variable "MaternalMortality" includes data supplied by UNICEF of the number of maternal deaths per hundred thousand live births.

1. Open your dataset in SPSS and make sure that the variables "MaternalMortality" and "InternalWar" are clean using the "Define Variable Properties" command. Be sure to check for missing values as well as any unusual maternal mortality rates.

2. Find the population mean.

   >Analyze

   >Descriptive Statistics

>Descriptives

    "MaternalMortality"→Variable

    >Options

      >Mean

      >Minimum

      >Maximum

      >Continue

    >OK

3. Select your subgroup.

>Data

>Select Cases

    >If condition is satisfied

    >If

      "InternalWar"→"InternalWar=1"

      >Continue

    >OK

4. Compare the subgroup mean to the population mean.

> Analyze

  >Compare Means

    >One-Sample T-Test

      "*MaternalMortalilty*"→Test Variable

      Test Value=*population mean*

      >OK

5. Write your memo.

Key Terms

Dummy variable (p. 157)

Parameter (p. 149)

Standard error of the mean (p. 153)

Statistical significance (p. 171)

Type I error (p. 146)

Type II error (p. 146)

CHAPTER 7

Hypothesis Testing

Examining Relationships

Frequently, when we think of the Industrial Revolution, we think of assembly line production, where jobs were divided into their component parts. Each task was performed faster if each person on the assembly line did a single task repeatedly, passing the product on to the next worker in the line to perform the next task. In contrast to the artisan tradition of having a single individual make each item from beginning to end, assembly line production resulted in a faster production of goods at a lower price. Karl Marx critiqued industrialization as being dehumanizing, but no one argued that it introduced a high level of efficiency into the production process.

After implementation of assembly line production during the Industrial Revolution, management experts turned to other aspects of work life to increase productivity. One study, conducted outside of Chicago in the Hawthorne Western Electric plant, studied the effects of various physical characteristics of the work environment (lighting, cleanliness, and humidity) as well as interpersonal dynamics (privacy, supportive supervisors, and work breaks) in an attempt to maximize productivity. Because the results of the experiments are so important, a myth has grown up around what happened in the plant. The mythical version of the experiment tells the story that the experimenters dimmed the lights in the factory, did a means test comparing productivity before and after, and found that productivity increased in a way that couldn't be due to random error. They then dimmed the lights again; once again, productivity increased at a statistically significant level. Eventually, the experimenters had dimmed the lights so much that the equipment couldn't even be seen. Every change produced increases in productivity that couldn't be explained by chance. What the researchers finally realized was that although the change in productivity was not due to chance, it was also not due to the lighting. Rather, the workers knew they were being watched and so worked harder. Although this description of the illumination studies is mythical, the bulk of the research done at the Hawthorne Plant did support the finding that being studied does affect behavior.[1] This is called the **Hawthorne Effect**.

One of the consequences of the Hawthorne Effect is the heavy emphasis that researchers now place on experimental designs. We understand that it is not sufficient to simply do a means test comparing the pre-test mean with the sample mean after the experimental stimulus. Instead, we very carefully do pre- and post-tests on both an experimental group and a control group—without any member of either sample knowing in which group he or she is a member. This comparison of two (or more) groups requires a statistical test slightly different from the means test. In examining the means test in the previous chapter, we looked at the possibility that the attributes of subgroups are different from the population. By assuming that the attribute was normally distributed across the population, we were able to calculate the probability that the difference we observed is simply due to chance. On occasion, we saw that the probability of it being due to chance was so small that we felt confident coming to the conclusion that something was going on. In this chapter, we will compare means of multiple sample subgroups using a technique called analysis of variance, or ANOVA. This is a technique particularly well-suited to experimental research designs and so it is used more often by psychologists than other social scientists. But it also makes a nice transition from our understanding of probabilities into a discussion of causal relationships. We begin this chapter with a discussion of hypothesis testing and then move to ANOVA as an example of one technique used to do hypothesis testing.

HYPOTHESIS TESTING

In this chapter, we take our knowledge of probability distributions and use it to further answer the "why" questions that we find so interesting. As we do so, we can continue to think about the world in terms of causation. The fact that the attribute about which we are concerned varies across the population immediately raises the question of why there is that variation. Marx saw variations in national wealth and asked, Why are some countries richer than others? He also saw variations in personal income and asked, Why do some individuals earn more than others? Marx also saw variations in which groups had more political power and asked, Why do some groups have more power than others?

Like Marx, after asking a "why" question, we proceed to use our understanding of the world to come up with a story to explain the variation. Implicit in the story we tell is our understanding of human nature and interpersonal dynamics. We use that understanding to develop a theory about how the world works in that specific context. Our theory is the story we tell to explain why there is variation. A theory may well be based on our experience with how the world works. But rather than being empirical in nature, a theory posits a general explanation of the causal connections between important factors in the dynamic we observed. Although we create theories to be consistent with our observations, theories are tentative in nature. Normally, we use them to make predictions about how the world should work in a situation we have yet to observe. We state these predictions as hypotheses about what we expect to see if the theory is correct in an as yet unobserved situation.

BOX 7.1 Numbers in the News

The Bill and Melinda Gates Foundation tries to channel funds toward solving problems that are not glamorous. One of the "down and dirty" problems facing the children of the world is moderate to severe diarrhea (MSD), a cause of 800,000 deaths per year. In 2006, the Gates Foundation funded the Global Enteric Multicenter Study (GEMS) to study MSD in children in sub-Saharan Africa and South Asia. During the study, which lasted for three years, one of the questions the researchers asked was about the long-term impact of MSD. Because malnutrition is also a problem among children in these regions, GEMS set up an experimental design that included over 9,000 test subjects suffering from MSD and over 13,000 control subjects who were not. The study found that the test subjects were eight times more likely to die in the next two months than the control subjects.[2] In addition to highlighting the problem as an important public policy issue, the study was also able to identify the most common causes in order to focus research attention on finding cures.

Hypotheses

As scientists, we want more than a story of how the world works. We actually want a hypothesis that we can use to evaluate how well our theory describes reality. Any hypothesis needs to do three things in order to be useful:

1. A hypothesis needs to be based on a theory.

2. A hypothesis needs to state a relationship between an independent variable (what we think of as the cause) and a dependent variable (the effect about which we asked the "why" question).

3. A hypothesis needs to be testable.

Only when a hypothesis meets these three requirements can we use it to collect and analyze data that can potentially increase our understanding of how the world works.

Let us take the example of national wealth. We can tell a story about the wealth of a nation being dependent on the nature of its economy: Industrial countries are going to be richer than agrarian countries. The causal relationship I expect to see is that level of industrialization (my independent variable) ought to affect the wealth of a nation (my dependent variable). It is sometimes helpful to set up the relationship between my two variables as a model, or a word picture connecting two concepts with an arrow that indicates the direction of the relationship. The model looks something like this:

Independent Variable → Dependent Variable

If you think in terms of cause and effect, it wouldn't make sense to draw a picture that places the effect before the cause. Similarly, a variable that is dependent on another has to come after it. Sometimes, you are lucky enough to be describing two variables that occurred chronologically at different times. In that case, it's pretty easy to say that the event that occurred first must be independent; the variable occurring second, dependent.

To help my students remember which is which, I suggest they think of their parents as independently wealthy and so able to pay for their children's education. If that is the case, odds are that the independently wealthy parents will claim their children as dependents for tax purposes. These independent parents were the cause for their children's existence; the children's birth was an effect that was dependent on the parents' partnership. The independent parents lead to the dependent children.

By their nature, some variables do not have any (socially interesting) cause and so must be independent variables. For example, race and sex are almost always independent variables. Almost never will they be modeled as dependent variables because they are attributes with which an individual is born and so cannot be influenced by social factors that occur later in life. Characteristics with which we are born are almost always independent variables.

Most other social variables, though, interact in society in complex ways and so could be either an effect or a cause, depending on the setting. In the example of wealth and industrialization, what came first is not clear cut. At any given point in time, you can measure both a nation's wealth and its industrialization. So, to identify which variable is independent and which is dependent, I need to look at the story I told about the relationship between the two. I said that as a nation develops industries, it will become wealthier as a result. Because I framed it that way, industrialization is the independent variable upon which wealth is dependent. So I would draw my model as follows:

Industrialization → National Wealth

In order to test my hypothesis, I have to be able to measure these two variables. For my dependent variable, gross domestic product is usually a good measure of the overall wealth of a nation. But if I use that measure, a very large country will be wealthy even if its citizens are poor. To eliminate that bias in measurement, it makes sense to control for the size of the population. To do that, I operationalize national wealth as the GDP per capita. For my independent variable, when I talk about industrial countries, I am thinking in broad terms of developed versus developing countries.

Now I need to phrase my hypothesis in such a way as to be testable. Hypotheses take certain forms. Usually, the dependent variable is mentioned first and the independent variable is mentioned second. The way in which the two variables are related is also mentioned. This is relatively easy if the categories of both variables have order. If we believe that increases in the independent variable (the cause) lead to increases in the dependent variable (the effect), we say that there is a positive (or direct) relationship. For example, if we believe that wealthy people get more education, we would hypothesize: There is a positive relationship between education (the dependent variable) and wealth (the independent variable). If we believe that increases in the independent variable lead to decreases in the dependent variable, we say that there is a negative (or inverse) relationship. For example, if we believe that increases in the unemployment rate lead to decreases in presidential approval ratings, we would hypothesize: There is a negative relationship between presidential approval and the unemployment rate.

If I had a continuous measure of industrialization, I could hypothesize: There is a positive relationship between GDP and industrialization. But in the absence of a continuous variable for industrialization, I chose to measure it in terms of two categories, developed versus developing nations. So this statement of the hypothesis isn't quite right. The problem here is that if one or both of the variables are nominal, stating the hypothesis gets a little tricky. If there is no order to the categories, it wouldn't make sense to talk about a positive or negative relationship because you can't have increases or decreases in a nominal-level variable. In that case, you would separate the independent variable into two groups and state what value of the dependent variable is more connected to the first group than the second. Suppose you think that the race of an individual affects his or her income. In this case, the independent variable (race) is nominal and the dependent variable (income) is interval. To write your hypothesis, you would separate your independent variable (race) into two groups (whites and blacks) and contrast them on income: Whites earn more than blacks. This format is a better fit for my hypothesis about the impact of industrialization on wealth: Developing countries will have lower GDP per capita than developed countries.

If your dependent variable is nominal and your independent variable is ordered, you still divide your independent variable into two groups. Suppose you think that richer people are more likely to go to college. In this case, wealth (your independent variable) is ordered, whereas college attendance (your dependent variable) is not. Divide your independent variable into two groups (richer and poorer) and contrast which group is more likely to belong to one of the categories of your dependent variable (college): Rich people are more likely to attend college than poor people.

The Null Hypothesis

As we translate our theory into a testable hypothesis, we need to keep the following questions at the forefront: If our theory is correct, what do we expect to see? How will the world look different if it is correct than if it is incorrect? When we observe variation in salaries, we may theorize that the respect that society gives a particular group may affect how much members of that group are paid. What do I expect to see if that theory is correct? Holding qualifications for a job constant, I expect members of respected demographic groups to be paid more than members of non-respected groups. To state a testable hypothesis, we need to decide what our population is and then identify who is respected and who is not. I could be thinking about all Americans and distinguish on the basis of race: Whites earn more than blacks. I could be thinking about whites and distinguish between women and men: Among whites, men earn more than women. Conversely, I could be thinking about blacks and distinguish based on gender: Among blacks, women earn more than men. The point is to make the hypothesis clear enough that we know exactly what we are looking for once we collect the data.

One aspect of having a testable hypothesis is that we know what it looks like if we are wrong. For a hypothesis to be testable, we cannot define it in such a way as to guarantee that we are right. The term we use to describe this aspect of a testable hypothesis is that it must be falsifiable. This means that as soon as we write a hypothesis, we need to also write a null hypothesis. Where the hypothesis states that there is a relationship between the dependent and independent variables, the **null hypothesis** states that there is no relationship between them. Normally, a hypothesis is designated by H_1 and the null hypothesis is designated by H_0. Look at how the hypotheses stated above correspond with null hypotheses:

H_1 = There is a positive relationship between education and wealth.

H_0 = There is no relationship between education and wealth.

H_2 = There is a negative relationship between presidential approval and the unemployment rate.

H_0 = There is no relationship between presidential approval and the unemployment rate.

H_3 = Whites earn more than blacks.

H_0 = Whites do not earn more than blacks.

H_4 = Rich people are more likely to attend college than poor people.

H_0 = Rich people are not more likely to attend college than poor people.

Remember that with tests of statistical significance, we will actually be asking the question "What is the probability of observing this outcome simply due to chance?" Keeping this in mind, it becomes clear that we are actually testing the null hypothesis, rather than the hypothesis of what we think will happen if our theory is correct. Measures of statistical significance are sometimes called "**p values**" to remind us that they are probabilities: **Statistical significance** is the probability that the null hypothesis is correct. If you are rooting for your hypothesis to be correct, that means that you want your p value to be small.

In tables, statistical significance can be labeled in different ways. Sometimes it is labeled "Stat. Sig.," sometimes, just "Sig.;" sometimes "Prob.," sometimes, just "p." It is always given as a decimal between zero and one. Normally, we just look to see if it is greater or less than 0.05 because the standard is, stick with the null hypothesis unless we find that there is less than a 5 percent probability of our observation being due to chance. In that case, we say that the relationship is statistically significant at the 0.05 level. But sometimes, the actual p value will be reported, in which case you just remember that it is the probability of the null hypothesis being correct. It is very important to keep in mind that there still might be a relationship between the two variables even if we fail to meet the 5 percent threshold. Sometimes you might even want to report it. For example, if you don't have very many cases, you might still want to discuss a relationship with a p value of 0.10 because that means that there is only a 10 percent chance of the null hypothesis being correct. Because it is always possible that there is a relationship we just happened to miss, we never say "We proved the null hypothesis." Instead, if a relationship is not statistically significant, we say, "We cannot reject the null hypothesis." Similarly, if we do meet the 5 percent threshold, it is still possible that there is no relationship between our two variables. That is simply the point at which we feel confident concluding that something is probably going on. We never say, "We proved our hypothesis." Instead, once we've concluded that there is a sufficiently small chance that the relationship we observed is just random, we say, "We can reject the null hypothesis."

In Chapter 6, I described Type I and Type II errors in an intuitive way. I connected Type I errors to a false positive diagnosis in medical terms and Type II errors with a false negative diagnosis. Technically speaking, though, the definitions of Type I and II errors are connected to the null hypothesis. Technically, a Type I error is when you reject the null hypothesis when you shouldn't. A Type II error is when you fail to reject the null hypothesis when you should. For example, before entering the Iraq war, U.S. security analysts hypothesized that Iraq had weapons of mass destruction (WMDs). They gathered evidence (which later proved faulty) in order to reject the null hypothesis that Iraq did not have any WMDs. They made a Type I error. Conversely, after World War II, U.S. leaders worried about the possibility of the Soviets developing nuclear weapons. Although security analysts tested the hypothesis that the Soviets had nuclear capacity, they never collected enough evidence to reject the null hypothesis. It wasn't until 1949, when planes picked up enough atomic debris off the coast of the Soviet Union to convince scientists that the Soviets had tested a bomb (on which Americans bestowed the moniker "Joe One" after Joseph Stalin) that the analysts realized that previously they had made a Type II error.[3]

Analysis of Variance

Once we have a well-stated hypothesis, it is simply a matter of collecting the data for our dependent and independent variables. Our analysis is similar to the means testing we conducted in Chapter 6. The difference is that in the previous chapter, we compared the sample with the population. In this chapter, we are comparing two (or more) subsamples. Analysis of variance (or ANOVA—ANalysis Of VAriance) is the technique that we use when our dependent variable is on a continuous scale (it has an interval level of measurement) and our independent variable has a limited number of categories. Our hypothesis is that membership in each of those categories leads to different values of the dependent variable.

Where in means testing, we compared a sample mean with a population mean, with ANOVA, we compare the means of subsamples with each other and with the full sample. The full sample has its own mean and its own distribution around that mean, shown in Figure 7.1. When we talk about that distribution, we first calculate the distance for each case from the mean. We then square that distance. We then sum all those squared distances. At this point, if we were calculating the standard deviation, we would then divide by n − 1 to get the sample variance and then take the square root to get the standard deviation. But with ANOVA, we actually stop at the sum of squared distances. We call this the "Total Sum of Squares." Picture this Total Sum of Squares (TSS) as the result of a wide range of values of our dependent variable. The distribution is centered on our sample mean, but it is dispersed around it. Soon, I'll walk you through a problem in which we will calculate the Total Sum of Squares using a work table, but for now, just keep a picture in your mind of all the data points and how they are all different distances from the mean.

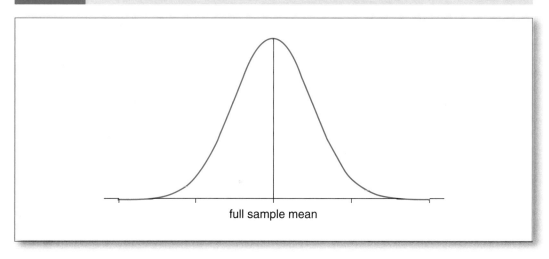

Figure 7.1 Distribution of Full Sample

full sample mean

Now picture this distribution as a composite of what you would get if you took two separate smaller distributions and added them together. Each of the two distributions has a different mean with a different distribution around the mean, as shown in Figure 7.2. The two distributions may overlap, but they really do peak at different places. If we didn't know that they came from two different groups, though, we would have combined all the cases together and it would look like the single distribution of Figure 7.1. But because we do know that they come from different groups, it is easy to see that the cases are actually distributed around their group means. For each of the groups, we can calculate the ubiquitous sum of squares around that group's mean. You can imagine that these two parallel distributions are much tighter than the original one was. There is still variation in each of the curves where each data point is a certain distance from the group mean. We call that the "Within-group Sum of Squares." We haven't explained that variation. But the difference between the Total Sum of Squares and the Within-group Sum of Squares is the amount of variation we were actually able to explain by knowing to which group each case belonged. We call that the "Between-group Sum of Squares." In equation form, these three different sums of squares relate as follows:

Total Sum of Squares = Within-group Sum of Squares + Between-group Sum of Squares

TSS = WSS + BSS

Because the TSS is a constant for a particular variable from any given sample, we know that WSS and BSS must be inversely related. The higher the variance around the group means, the less variance that has been explained by group membership. Conversely, the tighter the distribution around the group means, the more variance explained by membership in the groups.

Figure 7.2 Distribution of Two Sub-Samples

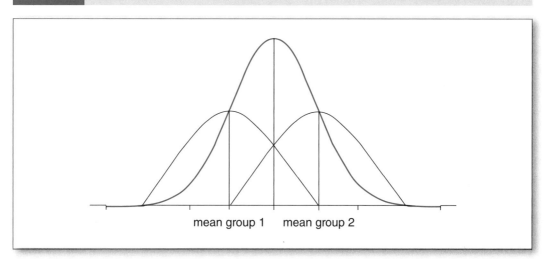

Let's go back to the example we used in the previous chapter of the level of corruption found in each of the post-Soviet states. Half of those countries are found in Europe and half are in Asia. Is it possible that when we look at a country's region, we would find a different pattern of corruption? I hypothesize that among post-Soviet states, European countries have a lower level of corruption than Asian countries. The null hypothesis is that the European countries do not have a lower level of corruption. The data are found in Table 7.1.

Practically speaking, we begin by using all of the data and calculating the Total Sum of Squares by going through the same process we did to find the standard deviation but stopping after we find the sum of squares and before we find the variance. We then calculate the Within-group Sum of Squares by doing the same thing for each of the subgroups and adding them together. Finally, we subtract the Within-group Sum of Squares from the Total Sum of Squares to get the Between-group Sum of Squares:

$$BSS = TSS - WSS$$

To compare the countries from Asia and Europe, we need to separate the countries into these two groups. Begin the work table by putting all the cases from the first group (Asia) above all the cases from the second group (Europe) with two extra rows between the two categories, as in Table 7.2. In the second column, place the case identifier—in this instance, the name of the country. Put the dependent variable (remember that we are using an inverse measure of corruption: the Transparency Index) into the third column. Below each group, calculate its mean by totaling the values on the index and dividing by the number of cases in that group. At the very bottom, find the total mean by adding the group totals and dividing by total number of cases. In the fourth column, subtract the total mean from each value of X given in column 3. In the fifth column, square the differences. Total the fifth column to get

Table 7.1   Corruption among Post-Soviet States by Region

Country	Region	Transparency Index*
Armenia	Asia	2.6
Azerbaijan	Asia	2.4
Belarus	Europe	2.4
Estonia	Europe	6.4
Georgia	Asia	4.1
Kazakhstan	Asia	2.7
Kyrgyzstan	Asia	2.1
Latvia	Europe	4.2
Lithuania	Europe	4.8
Moldova	Europe	2.9
Russia	Europe	2.4
Tajikistan	Asia	2.3
Turkmenistan	Asia	1.6
Ukraine	Europe	2.3
Uzbekistan	Asia	1.6

Source: Transparency International, *Corruption Perceptions Index 2011*. www.transparency.org/cpi2011/in_detail.

*Zero equals most corrupt.

the Total Sum of Squares. In this case, TSS = 24.50. Once you've found the Total Sum of Squares, you need to repeat the process for the individual groups. In the sixth column, subtract the appropriate group mean from the value of X found in the third column. In the last (seventh) column, square the difference. At the bottom of the seventh column, sum all the squares from both groups to get the Within-group Sum of Squares; here, WSS = 19.09. Finally, you get the Between-group Sum of Squares by subtracting the Within-group Sum of Squares from the Total Sum of Squares according to the equation:

$$BSS = TSS - WSS$$

$$= 24.50 - 19.09$$

$$= 5.41$$

Describe the Pattern. Now that we've done the math, we can begin to analyze the relationship between our dependent and independent variables. The first step is to describe the

Table 7.2 Work Table for ANOVA: Corruption Level for Post-Soviet States by Region

Region	Country	Transparency Index	$X - \bar{X}_{Total}$	$(X - \bar{X}_{Total})^2$	$X - \bar{X}_{Group}$	$(X - \bar{X}_{Group})^2$
Asia	Armenia	2.6	2.6 − 2.99 = −0.39	$(-0.39)^2 = 0.15$	2.6 − 2.42 = 0.18	$(0.18)^2 = 0.03$
Asia	Azerbaijan	2.4	−0.59	0.35	−0.02	0.00
Asia	Georgia	4.1	1.11	1.23	1.68	2.81
Asia	Kazakhstan	2.7	−0.29	0.08	0.28	0.08
Asia	Kyrgyzstan	2.1	−0.89	0.79	−0.33	0.11
Asia	Tajikistan	2.3	−0.69	0.48	−0.13	0.02
Asia	Turkmenistan	1.6	−1.39	1.93	−0.83	0.68
Asia	Uzbekistan	1.6	−1.39	1.93	−0.83	0.68
$n_{Asia} = 8$		$\Sigma X_{Asia} = 19.4$				
	$\bar{X}_{Asia} = \Sigma X_{Asia}/n_{Asia} = 2.42$					
Europe	Belarus	2.4	−0.59	0.35	2.4 − 3.63 = −1.23	$(-1.23)^2 = 1.51$
Europe	Estonia	6.4	3.41	11.63	2.77	7.67
Europe	Latvia	4.2	1.21	1.46	0.57	0.32
Europe	Lithuania	4.8	1.81	3.28	1.17	1.37
Europe	Moldova	2.9	−0.09	0.01	−0.73	0.53
Europe	Russia	2.4	−0.59	0.35	−1.23	1.51
Europe	Ukraine	2.3	−0.69	0.48	−1.33	1.77
$n_{Europe} = 7$		$\Sigma X_{Europe} = 25.4$			WSS = $\Sigma(X - \bar{X}_{Group})^2 = 19.09$	
	$\bar{X}_{Europe} = \Sigma X_{Europe}/n_{Europe} = 3.63$					
$n_{Total} = 15$	$\Sigma X_{Total} = \Sigma X_{Asia} + \Sigma X_{Europe} = 44.8$		TSS = $\Sigma(X - \bar{X}_{Total})^2 = 24.50$			
	$\bar{X}_{Total} = \Sigma X_{Total}/n_{Total} = 2.99$					

Source: Transparency International, *Corruption Perceptions Index 2011.* www.transparency.org/cpi2011/in_detail.

pattern that we see. For analysis of variance, the pattern is the comparison between the different means that we've calculated. We know that among all post-Soviet states, the average score on the Transparency Index is 2.99 on a 10-point scale, where zero is the most corrupt a nation can be. Is it possible that this mean is masking an underlying pattern in which Asian countries have higher levels of corruption than European countries? Separating the two sets of countries, we find that it looks like our hypothesis is correct: Asian countries have an average of 2.42, whereas European countries have an average of 3.63. Although the pattern appears consistent with our hypothesis, we still need to find out whether the pattern we see is statistically significant or whether it might plausibly be due to chance.

Statistical Significance. The second step of any analysis is to determine whether it is likely that a relationship exists. In order to determine whether or not the difference between the groups is due to chance, we calculate the F-statistic. The **F-statistic** is always proportional to the ratio of the between-group variation to the within-group variation. If you are comparing only two groups, F is actually the number of degrees of freedom multiplied by the Between-group Sum of Squares divided by the Within-group Sum of Squares:

$$F = (d.f.)BSS/WSS$$

As the spread around the group mean gets wider, the denominator gets larger. Simultaneously, the numerator gets smaller because BSS is inversely related to WSS. Both of these two factors decrease the F-statistic. But if the spread within the sub-groups gets narrower, the opposite would happen, increasing the F-statistic. As it works out, if we are comparing only two groups, the F-distribution is identical to the t-distribution we used in the previous chapter except that you need to take the square root of F to get the corresponding value of t. So as in that chapter, the higher value of F is associated with the means being distinct. A large F-statistic means that the difference in the means of the two groups is probably not due to chance. With a small F-statistic, though, it is quite possible that the difference we observed is due to chance. In contrast to the means test of the previous chapter, though, ANOVA is not a one-tailed test. Because we are comparing two different sample distributions, ANOVA is a two-tailed test. This means that when we look it up on our table, we need to adjust the columns we look at to find the level of statistical significance.

As with t tables, F has corresponding degrees of freedom that are dependent on the sample size as well as the number of categories of our independent variable. But unlike a t-test, the degrees of freedom do not equal n − 1. You'll recall that for a means test, we lost a degree of freedom because we had to calculate the mean. But for ANOVA, we have to calculate a separate mean for each of the groups we are analyzing. This means that we lose a degree of freedom for each of those means. As a result, for ANOVA, the d.f. = n − k, where k is the number of categories in our independent variable. To make the math more manageable, we are doing ANOVA for only two groups. This means that in the math portion of this chapter, our degrees of freedom will always be n − 2. Limiting our comparison to only two groups also allows us to use the t table to find the statistical significance. With more than two groups, use SPSS to calculate both the F-statistic and its statistical significance. For now, we can solve for the F-statistic in our comparison of Asian and European post-Soviet states:

$$F = (d.f.)BSS/WSS$$

$$= (13)5.41/19.09$$

$$= 3.684$$

Because we're going to look this up on a t table, we need to translate the F-statistic into a t-score by taking its square root.

$$t = \sqrt{F}$$

$$= \sqrt{3.684}$$

$$= 1.919$$

We can look this up on our t table knowing that we have $15 - 2 = 13$ degrees of freedom. Because ANOVA is a two-tailed test, in this row, the t-score necessary for the difference to be statistically significant at the 0.05 level is $t = 2.160$, which unfortunately is larger than the t-score of 1.919 we found from our data. Although the relationship is not statistically significant at the 0.05 level, with this few cases it is not unreasonable to look at the column for a 0.10 probability. Here we see a t-score of 1.771, which is less than what we found with our data. We can conclude that the difference between the two groups is statistically significant at the 0.10 level. We can reject the null hypothesis with 90 percent confidence.

Thus far, we have been focusing our discussion entirely on questions of statistical significance—what is the likelihood of observing a particular difference just because of chance? We say that a relationship is statistically significant at the 0.05 level if there is only a 5 percent probability of observing it due to chance. Because we are actually testing the null hypothesis, we want a small number for statistical significance because a small number indicates that it is very unlikely that the null hypothesis is correct. Statistical significance only makes us confident that there is a relationship between our dependent and independent variables. But being confident that there is a relationship leaves us wondering how strong it is.

Substantive Significance. Third, we make our best guess of how strong the relationship appears to be. I call this substantive significance. To determine whether a relationship is substantively significant, we use various measures of association. Measuring the strength of an association can be tricky because how we calculate it depends on how much information is contained in our data. In the remainder of this book, we will be examining various measures of association that are appropriate at different times, depending on the level of measurement of our variables.

With ANOVA, we have an independent variable with a limited number of categories (this limited information has a nominal or ordinal level of measurement) connected to a dependent variable that is measured on a scale (that increased level of information has an interval level of measurement). The appropriate measure of association is eta-squared. (We are back to Greek again: The Greek letter "η" may look like a calligraphy version of the letter "n" but it is actually the lower-case version of the Greek letter eta and is pronounced "ayta."). This measure of association uses the same information we collected above about the Total, Between-group, and Within-group Sum of Squares of the total distribution. But this time, instead of getting the ratio of the Between-group Sum of Squares compared to the Within-group Sum of Squares, as we did in calculating the F-statistic, we compare the Between-group Sum of Squares to the Total Sum of Squares.

$$\eta^2 = \text{Between-group Sum of Squares/Total Sum of Squares}$$

$$\eta^2 = BSS/TSS$$

$$= 5.41/24.50$$

$$= 0.22$$

The question here is, How much of all the variation we originally saw in the distribution of our dependent variable can be explained by knowing to which group each individual belongs? As group membership explains more variance, η^2 will approach one. If the distribution for each group is identical, then the BSS approaches zero and so η^2 approaches zero as well.

Usually, as non-statisticians, when we talk about the significance of a relationship, we do not mean statistical significance. Actually, one of the hardest things you'll have to do this semester is get yourself in the habit of thinking about statistical significance as answering the question: Is there a relationship between these variables? This is difficult to learn because intuitively when we ask "Is this a significant relationship?" what we really mean is "How strong is the relationship?" In contrast to statistical significance, I call this substantive significance. To answer it, we need a measure of association rather than a measure of statistical significance. In ANOVA, the appropriate measure of association is eta (which is actually the square root of eta-squared).

Measures of association usually follow certain patterns. It is fairly standard for measures of association to range between 0 (a nonexistent relationship) and 1 (a perfect relationship). We can judge how strong a relationship is by where on the continuum from zero to one the measure of association falls. Table 7.3 gives you a rule of thumb to use in evaluating the relative strength of different values of association. In the hard sciences, you might expect much higher measures of association than are given in this table because you are measuring variables that have direct physical causal relationships. In the social sciences, though, we are measuring human interactions, which are much more nebulous. As a result, we are very happy with a measure of association over 0.30 and call it a strong relationship. At the other extreme, when testing a hypothesized relationship between two variables, we may have been thrilled to have been able to reject the null hypothesis because it is statistically significant at the 0.05 level. But that relationship may still have a measure of association that is less than 0.10. If that is the case, although we've concluded

Table 7.3 How to Interpret Measures of Association

Measure of Association (X)	Qualitative Interpretation
$0 \leq X < 0.10$	Very Weak
$0.10 \leq X < 0.20$	Weak
$0.20 \leq X < 0.30$	Moderate
$X \geq 0.30$	Strong

that there probably is a relationship, we would describe it as very weak. For the relationship between region and corruption, we calculated an eta-squared of 0.22. If we take the square root of that, we get eta = 0.47. We can conclude that there is a strong relationship between corruption and region.

On occasion, when you look at the relationship between two variables, you will find that it has a strong measure of association but is not statistically significant. The standard is that if the statistical significance does not allow you to reject the null hypothesis, you do not ask how strong the relationship is. It doesn't make sense to indicate how strong a relationship is if you don't think a relationship exists. But this is unusual, so it is worth asking why it occurred in the instance you are analyzing. It could be that (as with the example of corruption in the post-Soviet states) you have too few cases, and if you were able to increase your sample size, the relationship would become statistically significant. It could be that you have one or more outliers that are boosting up the relationship artificially. Or it could be that something else weird is going on. If you run across a relationship that has a nice measure of association but is not statistically significant, it is worth looking at the data to see what happened. Maybe you can fix the problem. But if you can't, just report that because it is not statistically significant, you cannot reject the null hypothesis.

Eta-squared is member of a special class of measures of association called Proportional Reduction in Error, or PRE, measures. **PRE measures** give us a measure of how well we can predict the value of the dependent variable if we know the value of the independent variable. For eta-squared, we know that group membership accounts for a particular proportion of the variance of the dependent variable. So if $\eta^2 = 0.70$, we can say that the independent variable explains 70 percent of the variance of the dependent variable. For the corruption in post-Soviet states, we calculated $\eta^2 = 0.22$. We can conclude that region explains 22 percent of the variance of the corruption found in post-Soviet states. Usually, you will not get an eta-squared that is very high. Think about it: You have a full distribution of values of your dependent variable similar to Figure 7.3. If you draw lines at two or three points along that continuum, even if those points are widely spaced, the clusters around those points would have to spread fairly wide. A limited number of categories simply cannot explain all the variance of a continuous variable. If two categories can explain even 20 percent of the variance, that's pretty impressive.

SUMMARIZING THE MATH: HYPOTHESIS TESTING AND ANOVA

As scientists, we develop theories about how the world works. In order to make those theories as accurate as possible, we posit hypotheses that make predictions about what we should observe if the theories are correct. With an experimental research design, we frequently use ANOVA because it compares an interval-level variable across different groups. With ANOVA, we describe the pattern by comparing the means for the groups. We identify the statistical significance by using the F-statistic to determine whether there appears to be a relationship. We evaluate the substantive significance by using eta, interpreted as a standard measure of association, and eta-squared, interpreted as a PRE measure.

Hypothesis Testing

When using the scientific method, we delineate a hypothesis to help us evaluate the empirical evidence.

1. A hypothesis is a theoretically based statement of the relationship between a dependent variable (the effect) and an independent variable (the cause) that can be empirically tested.

2. How to properly state a hypothesis depends on the level of measurement of your variables.

    a. If both variables have order (ordinal or interval), you say that there is a positive/direct or negative/inverse relationship between your dependent variable and your independent variable.

    b. If one of the variables doesn't have order (nominal level of measurement), you divide the independent variable into two groups and contrast the values of the dependent variable for those two groups.

3. As soon as you state your hypothesis, you state the null hypothesis, which is that there is no relationship between the dependent and independent variables.

4. With an empirical test of the hypothesis, you actually are testing the null hypothesis.

    a. If there is more than a 5 percent chance that your results are random, you say, "We cannot reject the null hypothesis."

    b. If there is less than a 5 percent chance that your results are random, you say, "We can reject the null hypothesis."

A Political Example: Software Piracy. U.S. software companies are concerned that many nations do not enforce copyright protection of computer software. This can lead us to ask the question, What kinds of countries have higher piracy rates?

1. Suppose you think that poor people are more likely to install illegal software. In that case, the lower a country's GDP per capita, the higher the rate of piracy you

would expect to see. Properly state a hypothesis and its corresponding null hypothesis.

In this relationship, income would be the independent variable that leads to the piracy rates, the dependent variable. Both piracy rates and income (measured as GDP per capita) are interval-level variables. Because both variables have order, I can use the language of a positive or negative relationship.

H_1 = *There is a negative relationship between a country's rate of software piracy and its GDP per capita.*

H_0 = *There is no relationship between a country's rate of software piracy and its GDP per capita.*

2. Suppose you think that membership in the European Union would influence piracy because of the EU's strict oversight of copyright and patent protections. Properly state a hypothesis and its corresponding null hypothesis.

In this relationship, the independent variable is membership in the EU, which is a nominal-level variable. That means that I cannot use the language of a directional relationship. Instead, I divide the independent variable into two groups and compare them on the dependent variable.

H_1 = *EU nations have a lower rate of software piracy than non-EU nations.*

H_0 = *EU nations do not have a lower rate of software piracy than non-EU nations.*

ANOVA

When we want to see if group membership affects an interval-level dependent variable, we use analysis of variance. Within ANOVA, we calculate the Total Sum of Squares (TSS). This is the number we got right before finding the variance, where we totaled all of the squared differences from the mean. The TSS is an indication of how spread out the dependent variable is—how much variance needs to be explained. We do the same calculation for each subgroup to get the Within-group Sum of Squares (WSS). The WSS is how much variability in the dependent variable is left unexplained by membership in a group. The Between-group Sum of Squares (BSS) is the difference between the two: BSS = TSS − WSS. The BSS is how much of the variability of the dependent variable is explained by group membership.

To analyze a relationship, we always follow the three-step process of answering the following three questions: Do the data follow a pattern consistent with the hypothesis? Does it look like a relationship exists? How strong does it look like the relationship is? For ANOVA, we follow the process in the following way.

1. First, we describe the pattern we see by comparing the average scores of the dependent variable for each of our groups.

2. Second, we identify the statistical significance with an F-test. We need to calculate three sums of squares that we will use later. We begin with the TSS,

which is the total squared difference between the case value and the population mean. Then, within each group, we find the squared difference between the case value and the group mean and then sum those group squared differences for all the cases to get the WSS. Finally, we take the difference between the TSS and the WSS to get the BSS.

To find the statistical significance for the difference between group means, we find the F-statistic. If we are comparing only two groups, we get the F-statistic by finding the ratio between the BSS and the WSS and multiplying it by the degrees of freedom (where d.f. = n − 2): F = (n − 2)BSS/WSS. In this case, we can look up the square root of the F-statistic on a t table using d.f. = n − 2. If the two-tailed probability is greater than 0.05, we know that the difference in means that we observed might be due to chance and we do not reject the null hypothesis. If the probability is less than 0.05, we conclude that the difference is probably not due to chance and we reject the null hypothesis.

3. We then make our best guess regarding how strong the relationship is. For ANOVA, the measure of association is eta (η) where η^2 = BSS/TSS. If η (or any other measure of association) is less than 0.10, we say that it is very weak; if it is more than 0.30, we say that the relationship is strong.

4. Because eta-squared (η^2) is a PRE measure, we can interpret it as the proportion of variance of the dependent variable explained by membership in the groups defined by your independent variable.

A Political Example: Software Piracy and EU Membership among Eastern European States. If EU nations are more likely to have lower rates of software piracy controlling for wealth, we would expect to see that pattern among the generally poorer nations of Eastern Europe. Table 7.4 shows the piracy rates among the Eastern European Group, an unofficial UN regional association. Describe the pattern you see between these two groups. Is it statistically significant? How strong is the relationship?

Table 7.4 Piracy Rates among Eastern European States

Region	Country	Piracy Rate (%)
Non-EU	Albania	75
Non-EU	Armenia	88
Non-EU	Azerbaijan	87
Non-EU	Bosnia and Herzegovina	66
Non-EU	Croatia	53
Non-EU	Macedonia	66

Region	Country	Piracy Rate (%)
Non-EU	Moldova	90
Non-EU	Russia	63
Non-EU	Serbia	72
Non-EU	Ukraine	84
EU	Bulgaria	64
EU	Czech Rep.	35
EU	Estonia	48
EU	Hungary	41
EU	Latvia	54
EU	Lithuania	54
EU	Poland	53
EU	Romania	63
EU	Slovakia	40
EU	Slovenia	46

Source: Business Software Alliance, "Shadow Alliance: 2011 Global Software Piracy Study," 9th ed., May 2012. http://portal.bsa.org/globalpiracy2011/downloads/study_pdf/2011_BSA_Piracy_Study-Standard.pdf.

1. Describe the pattern by comparing the average scores of the dependent variable for each group.

   To find the group means, I need to complete a work table for computing an analysis of variance. This is shown in Table 7.5. Among non-EU nations, the average piracy rate is 74.4 percent. In contrast, 49.8 percent of software installations among EU members are pirated. We can conclude that among Eastern European countries, EU membership leads to a drop in piracy of 24.6 percentage points.

2. Second, we determine the statistical significance with an F-test where $F = (n - 2)$ BSS/WSS. Look up the square root of the F-statistic on a t table using d.f. = n − 2. From the worktable, we find that TSS = 5271.8 and WSS = 2246.0. From this, we find BSS:

$$BSS = TSS - WSS$$
$$= 5271.8 - 2246.0$$
$$= 3025.8$$

Table 7.5 Work Table for ANOVA: Software Piracy for Eastern European States by EU Membership

Region	Country	Piracy Rate (%)	$X - \bar{X}_{Total}$	$(X - \bar{X}_{Total})^2$	$X - \bar{X}_{Group}$	$(X - \bar{X}_{Group})^2$
Non-EU	Albania	75	12.9	166.41	0.6	0.36
Non-EU	Armenia	88	25.9	670.81	13.6	184.96
Non-EU	Azerbaijan	87	24.9	620.01	12.6	158.76
Non-EU	Bosnia and Herzegovina	66	3.9	15.21	−8.4	70.56
Non-EU	Croatia	53	−9.1	82.81	−21.4	457.96
Non-EU	Macedonia	66	3.9	15.21	−8.4	70.56
Non-EU	Moldova	90	27.9	778.41	15.6	243.36
Non-EU	Russia	63	0.9	0.81	−11.4	129.96
Non-EU	Serbia	72	9.9	98.01	−2.4	5.76
Non-EU	Ukraine	84	21.9	479.61	9.6	92.16
$n_{Non-EU} = 10$	$\Sigma X_{Non-EU} = 744$					
$\bar{X}_{Non-EU} = \Sigma X_{Non-EU}/n_{Non-EU} = 74.4$						
EU	Bulgaria	64	1.9	3.61	14.2	201.64
EU	Czech Rep.	35	−27.1	734.41	−14.8	219.04
EU	Estonia	48	−14.1	198.81	−1.8	3.24
EU	Hungary	41	−21.1	445.21	−8.8	77.44
EU	Latvia	54	−8.1	65.61	4.2	17.64
EU	Lithuania	54	−8.1	65.61	4.2	17.64
EU	Poland	53	−9.1	82.81	3.2	10.24
EU	Romania	63	0.9	0.81	13.2	174.24
EU	Slovakia	40	−22.1	488.41	−9.8	96.04
EU	Slovenia	46	−17.35	301.02	−3.8	14.44
$n_{EU} = 10$	$\Sigma X_{EU} = 498$					$WSS = \Sigma(X - \bar{X}_{Group})^2 = 2246.0$
$\bar{X}_{EU} = \Sigma X_{EU}/n_{EU} = 49.8$						
$n_{Total} = 20$	$\Sigma X_{Total} = \Sigma X_{Non-EU} + \Sigma X_{EU} = 1242$		$TSS = \Sigma(X - \bar{X}_{Total})^2 = 5271.8$			
$\bar{X}_{Total} = \Sigma X_{Total}/n_{Total} = 62.1$						

Source: Business Software Alliance, "Shadow Alliance: 2011 Global Software Piracy Study," 9th ed., May 2012. http://portal.bsa.org/globalpiracy2011/downloads/study_pdf/2011_BSA_Piracy-Study-Standard.pdf.

$$F = (n - 2)BSS/WSS$$
$$= (20 - 2)(3025.8)/2246.0$$
$$= 24.25$$
$$t = \sqrt{F} = \sqrt{24.25} = 4.92$$

For this set of data, d.f. = n − 2 = 18. Looking at a t table, we can see that for a two-tailed test, we need t = 2.878 for prob. = 0.01. Because 4.92 > 2.878, the pattern we saw was statistically significant at the 0.01 level. We can reject the null hypothesis.

3. Measure the strength of the relationship with eta (η) where η^2 = BSS/TSS.

$$\eta^2 = BSS/TSS$$
$$= 3025.8/5271.8$$
$$= 0.574$$
$$\eta = .758$$

The eta of 0.758 indicates that there is a very strong relationship between software piracy and EU membership. The eta-squared indicates that EU membership explains 57.4 percent of the variance in the rate of software piracy among Eastern European nations.

USE SPSS TO ANSWER A QUESTION WITH ANOVA

Frequently, ANOVA is used with experimental research designs, but it is equally useful whenever you have an interval-level dependent variable and an independent variable with a limited number of categories. It is easiest to think of ANOVA as answering the question of whether membership in a group affects the distribution of that group on a scale. I limited the math in the chapter to calculating ANOVA with only two categories because it is easiest to have SPSS do the calculations with more than two categories and with lots of cases.

You begin your data analysis in the same way you always do: by opening your data in SPSS and checking the variables you are going to use to make sure that they are clean. By using the "Define Variable Properties" command, you first see whether the values for the variables all make sense and set the appropriate values to missing. You also want to make sure that each of the values of your independent variable is labeled by the name of the group it identifies. Third, if appropriate, add any relevant information about the variable into the variable label.

Sometimes you will want to include all the cases in your dataset for your analysis, but sometimes you want to limit it to a subset. For example, I have the Transparency Index for all the countries of the world, but in Table 7.1, I was interested only in post-Soviet states. In order to get that subset, I used the "Select Cases" command to make that limitation. Similarly, if you want to analyze only a subset of cases, select them by indicating the values of the identifying variable you want. As usual, you can check to make sure you communicated what you wanted to SPSS by checking the "Data Window" to make sure the cases you want excluded from analysis have a slash through their case number. This process is summarized in Box 7.2.

BOX 7.2   How to Use SPSS to Select a Subset of Cases

After opening your data in SPSS and cleaning them:

>Data
>>Select Cases
>>>If condition is satisfied
>>>If
>>>>"Variable"→"mathematical operators for condition"
>>>>>For example: "PartyID < 4"
>>>Continue
>>OK

At this point, if you enter the data window, you will see slashes through the case numbers for all cases not meeting your criteria.

BOX 7.3   How to Use SPSS to Conduct an Analysis of Variance

After opening your data in SPSS, cleaning them, and selecting the appropriate cases:

>Analyze
>>Compare Means
>>>Means
>>>>"Dependent Variable" → Dependent Variable
>>>>"Group Variable" → Independent Variable
>>>>Options
>>>>>ANOVA Table and eta
>>>>>Continue
>>>>OK

Once you have everything set up properly, you request the ANOVA. Under "Analyze," choose "Compare Means" and "Means." To get the information you need for the ANOVA, you will need to go into the "Options" window. Once it is open, you'll want to request the ANOVA table and eta. This process is summarized in Box 7.3.

Once you have the results from SPSS, you are in a position to analyze them. We always follow a three-step process in any analysis. First, you describe the pattern you see. Does it look like the data support your hypothesis? The report table gives the means for your groups. Comparing those means, are they different? Second, identify the statistical significance. What is the probability that the difference you just described is due to chance? In the ANOVA table, the last column is labeled "Sig." This is the statistical significance. If this value is less than 0.05, the relationship is statistically significant at the 0.05 level and you can reject the null hypothesis. Third, if you've concluded that there probably is a relationship between your two variables, evaluate the strength of the relationship. In the Measures of Association table, you'll see "eta" and "eta-squared," and you should interpret both of these. First, interpret the strength of the relationship by eta: If eta is less than 0.10, the relationship is very weak; less than 0.20, weak; less than 0.30, moderate; and greater than 0.30, strong. Second, give the PRE interpretation of eta-squared. Multiply the decimal by 100 and say that the independent variable explains that percent of the variance of the dependent variable.

As part of writing up the results of your ANOVA, you will want to include a table summarizing your results. This will be similar to the Means table, which was part of the SPSS output, except that you will want to append to it the information you obtained regarding the statistical and substantive significance of the relationship between the two variables. As usual, you begin with a descriptive title. Then you insert a table that will look similar to the frequency tables we've already created, except that there will be two columns and the number of rows will be the number of categories of your independent variable plus four. The first column will contain the categories of your independent variable; the second, each group's mean value on the dependent variable. After the group mean, you give the total mean. Then you need to display the relevant statistics. Under the first column, you'll show the number of cases; under the second, the values of eta and eta-squared. To indicate the level of significance, we normally use stars next to the measure of association. Normally, one star indicates that the relationship is statistically significant at the 0.05 level; two stars, at the 0.01 level. The overall format of the table looks similar to a frequency table: Use the "Borders and Shadings" command to place lines before and after the table as well as after the column headings. Although you can still follow the table with information about the data in a smaller font, the first thing you will give after the table is a key to the stars indicating statistical significance. The process of making a table for your ANOVA is summarized in Box 7.4.

BOX 7.4 How to Create a Professional-Looking Table for ANOVA

1. Give it a title describing its contents.
2. Insert a table of the appropriate size: columns = 2, rows = 4 + number of categories of independent variable.

    \>Insert
    > \>Table

(Continued)

(Continued)

>Insert Table
Number of columns = 2
Number of rows = 4 + number of categories of independent variable

3. In the first row, label the two columns: *Independent Variable Label*, "Average," *Dependent Variable Label*.
4. In the first column, list the categories of the independent variable, followed by the word "Total."
5. In the second column, give the means of those categories.
6. In the row under "Total," indicate your sample size "n = ".
7. Under the second column, identify eta and eta-squared. Place stars next to these values to designate the level of significance of the relationship.
8. Draw the appropriate lines.

Highlight the entire table:
>Borders and Shading *(the arrow, not the icon)*
>Borders and Shadings
Settings
>None
Style
The single line should be highlighted.
Preview
Click above and below the table to add lines there.
>OK

Highlight the first line (with column labels):
>Borders and Shading *(the arrow, not the icon)*
>Bottom Border

9. Below the last line, use a smaller font to first give a key for the stars indicating the level of significance "*prob. < 0.05." Second, add the data source and any other clarifying information.

A Political Example: Partisanship and Support for the President

It is 2008, and you are volunteering for the McCain presidential campaign. As the Republican nominee, normally John McCain would rely on the sitting Republican president as a source of support. But this year, he is wondering whether campaigning by President Bush might hurt him in the eyes of the independent voters who will be necessary to win the election. Although support for President Bush was uniformly high in the aftermath of 9/11, over the next few years, opposition to the war in Iraq and, later, unhappiness over the state of the economy led to greater variation in support for him.

Nicolle Wallace, the senior adviser on message, asks you to evaluate how partisanship has affected individuals' evaluations of President Bush. Your dependent variable is placement of President Bush on a 100-degree feeling thermometer, and your independent variable is self-identification as a Republican, Independent, or Democrat. She is pretty confident that Republicans are more likely to feel warmly toward President Bush than Democrats, but she is particularly curious about how self-labeled Independents (who are likely to have the most impact on this election) evaluate the president.

1. Open your data in SPSS and clean them.

    I am going to use the 2008 American National Election Study data.[4] My independent variable is Party Identification (PartyID), and my dependent variable is Feeling Thermometer for President Bush (FTBush). When I open the data, I look at these two variables in "Define Variable Properties." The feeling thermometer looks fine; its missing categories have already been set to missing and the variable ranges from 0 to 100 as it should. Under PartyID, I see that there are values set to missing. But I also see that the categories are not set up the way I expect (see Figure 7.3). The value 1 = Democrat, 2 = Republican, 3 = Independent, 4 = Other party, and 5 = No preference. I really don't want those last two categories included in my analysis.

Figure 7.3 Defining Variable Properties for Party Identification

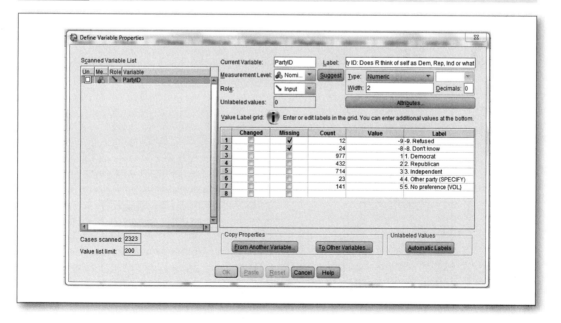

2. If necessary, use the "Select Cases" command to limit your analysis to the desired cases.

Because I want to include only Democrats, Republicans, and Independents, I use the "Select Cases" command to limit the cases to the ones I want. In Figure 7.3, I see that these labels are connected to the values 1, 2, and 3, and I want to exclude the values 4 and 5. Figure 7.4 shows the mathematical operation within the Select Cases command to do this.

Figure 7.4 Selecting only Democrats, Independents, and Republicans

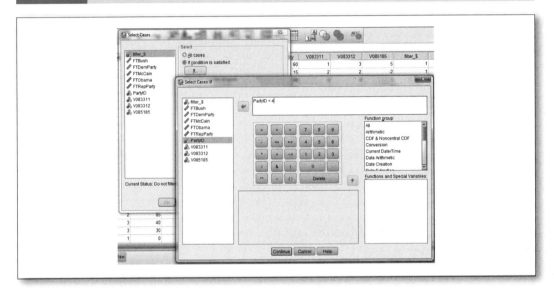

After processing my request, I get a frequency of PartyID (shown in Figure 7.5) to make sure that this variable now looks right. It has only the three desired categories, so I must have done it correctly.

Figure 7.5 Partisan Identification

J1. Party ID: Does R think of self as Dem, Rep, Ind or what

		Frequency	Percent	Valid Percent	Cumulative Percent
Valid	1. Democrat	977	46.0	46.0	46.0
	2. Republican	432	20.3	20.3	66.4
	3. Independent	714	33.6	33.6	100.0
	Total	2123	100.0	100.0	

CHAPTER 7 Hypothesis Testing: Examining Relationships

3. Request the analysis of variance.

I use the "Means" command, making sure to request both an ANOVA table and eta. The location of the commands is shown in Figure 7.6. Figure 7.7 shows the actual commands.

Figure 7.6 Finding the Commands for an Analysis of Variance

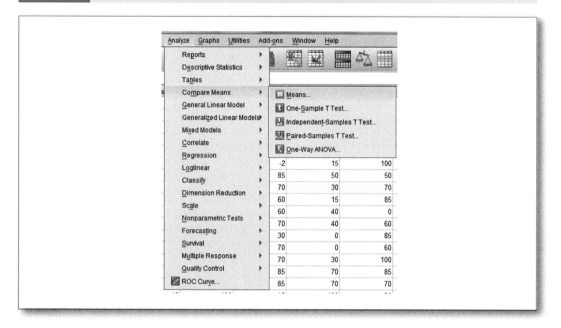

Figure 7.7 Getting an ANOVA for Bush Feeling Thermometer by Party ID

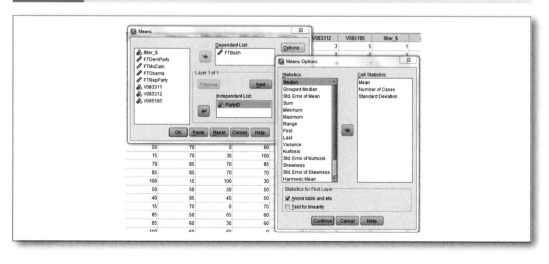

Figure 7.8 Output for ANOVA of Bush Feeling Thermometer

Report

B1a. Feeling Thermometer: President

J1. Party ID: Does R think of self as Dem, Rep, Ind or what	Mean	N	Std. Deviation
1. Democrat	24.69	973	24.661
2. Republican	65.93	432	24.619
3. Independent	37.20	712	27.257
Total	37.31	2117	29.882

Group and Total Means

ANOVA Table

		Sum of Squares	df	Mean Square	F	Sig.
B1a. Feeling Thermometer: President * J1. Party ID: Does R think of self as Dem, Rep, Ind or what	Between Groups (Combined)	508784.712	2	254392.356	389.524	.000
	Within Groups	1380621.526	2114	653.085		
	Total	1889406.237	2116			

Statistical significance

Measures of Association

	Eta	Eta Squared
B1a. Feeling Thermometer: President * J1. Party ID: Does R think of self as Dem, Rep, Ind or what	.519	.269

Eta and Eta-squared

4. Analyze the results.

   a. Describe the pattern.

   Figure 7.8 shows the ANOVA output. In the Report table, I can see that, nationwide, Americans do not feel very warmly toward President Bush because the total mean is 37.31°, which is 13° under the neutral temperature of 50°. My expectation was that Republicans would feel warmly toward the president and Democrats would feel cold. This is correct: Republicans place him at a warm 66°, and Democrats place him at a cool 25°. But it does not appear that Independents are neutral. They place President Bush at about the national average, with a Feeling Thermometer rating of 37°.

   b. Can you conclude with confidence that a relationship exists? In other words, is this relationship statistically significant?

   In the ANOVA table, the significance is reported at .000. I know that this does not mean that there is a 0 percent probability that the pattern I saw is due to chance, but it does indicate that the relationship is statistically significant at the 0.01 level. I can reject the null hypothesis.

   c. How strong does it look like the relationship is?

   In the Measures of Association table, I see that eta is 0.519, which is well above the 0.3 threshold to call this a strong relationship. I also see that eta-squared is 0.269, which indicates that partisanship explains 26.9 percent of the variance in how Americans place President Bush on a feeling thermometer.

5. Write your memo.

Memo

To: Nicolle Wallace
From: T. Marchant-Shapiro, volunteer
Date: September 1, 2008
Subject: Partisan evaluation of President Bush

As we enter the general election portion of the campaign, you asked me to evaluate how continued campaigning by President Bush is likely to affect support for Senator McCain among self-identified Independents. In order to answer the question, I used data from the American National Election Study, which asked respondents to place the president on a feeling thermometer from 0 (very cold) to 100 (very warm). Independents feel somewhat cold toward the president.

The table below shows how warmly different partisans feel toward President Bush. On average, Americans feel somewhat cold toward the president, placing him at 37.31°. As might be expected, Republicans feel warmly toward the president (66°), and Democrats feel cold (25°). But Independents are not neutral. They, like the nation as a whole, feel somewhat cold toward the president, placing him at 37.2°. There is less than a 1 percent chance that this pattern is due to chance. There is a strong relationship between feelings for the president and partisanship. In fact, partisanship explains 26.9 percent of the variance in support for the president.

Based on these findings, I would not recommend using President Bush to campaign for Independent voters—this would be more likely to lose us support than to gain it. Perhaps it might be helpful to have him campaign among party regulars because they do feel warmly toward him. But in doing that, we would need to be careful to limit his appeals, because the cool feelings of Democrats and Independents more than offset the warmth among Republicans.

Feeling Thermometer for President Bush by Partisanship

Party Identification	Average Feeling Thermometer
Republican	65.93
Independent	37.20
Democrat	24.69
Total	37.31
N = 2117	Eta = 0.519**
	Eta-Squared = 0.269**

Source: American National Election Study, ANES 2008 Time Series Study. www.electionstudies.org/studypages/2008prepost/2008prepost.

*prob. < 0.05.

**prob. < 0.01.

Question Wording: *I'd like to get your feelings toward some of our political leaders and other people who are in the news these days. I'll read the name of a person and I'd like you to rate that person using something we call the feeling thermometer. Ratings between 50 degrees and 100 degrees mean that you feel favorable and warm toward the person. Ratings between 0 degrees and 50 degrees mean that you don't feel favorable toward the person and that you don't care too much for that person. You would rate the person at the 50-degree mark if you don't feel particularly warm or cold toward the person: President George W. Bush.*

Your Turn: Hypothesis Testing

YT 7.1 For each of the following relationships: Properly state (1) a hypothesis and (2) a null hypothesis.

1. You think that the number of years of education a person has is likely to increase the income that they earn.
2. You think that the higher taxes are, the fewer people businesses hire.
3. You think that incumbents (as opposed to challengers) raise more campaign contributions.
4. You think that the more popular presidents are, the more likely they are to be reelected.

YT 7.2 When I hypothesized that those post-Soviet states located in Asia would have more corruption, I was thinking that the European countries would be more likely to be part of the democratic tradition that looks askance on corruption. If European countries are more democratic, I would also expect that the European countries would have more political rights. Table 7.6 shows the political rights score (on a 40-point scale where a 40 indicates the most rights) produced by Freedom House.

1. Complete the work table for the analysis of variance.
2. Describe the pattern by comparing the means.
3. Calculate the F-score and look up its square root on the t table with d.f. = n − 2. Is the relationship statistically significant?
4. Calculate eta-squared and eta. How strong is the relationship?

Apply It Yourself: Examine Partisanship's Effect on Feelings toward the Democratic Party

It is 2011, and you are interning at the Democratic National Committee. Your boss, the political director, is concerned about preparing for the 2012 presidential and congressional elections. In the 2010 midterm elections, the proportion of Americans who identified themselves as Democrats declined dramatically. The director is afraid that the decline in partisanship will lead voters to evaluate the Democratic Party more harshly and thus make them less likely to vote Democratic in the election. She has asked you to evaluate the impact that partisanship has on the views of the parties. The "ANES2008" data is an SPSS save file from the 2008 American National Election Study. Using these data, find out whether partisanship affects how warmly Americans feel toward the Democratic Party.

Table 7.6 Work Table for ANOVA: Political Rights for Post-Soviet States by Region

Region	Country	Political Rights	$X - \bar{X}_{Total}$	$(X - \bar{X}_{Total})^2$	$X - \bar{X}_{Group}$	$(X - \bar{X}_{Group})^2$
Asia	Armenia	11				
Asia	Azerbaijan	6				
Asia	Georgia	22				
Asia	Kazakhstan	6				
Asia	Kyrgyzstan	16				
Asia	Tajikistan	8				
Asia	Turkmenistan	1				
Asia	Uzbekistan	0				
$n_{Asia} =$		$\Sigma X_{Asia} =$				
$\bar{X}_{Asia} = \Sigma X_{Asia}/n_{Asia} =$						
Europe	Belarus	4				
Europe	Estonia	39				
Europe	Latvia	33				
Europe	Lithuania	37				
Europe	Moldova	28				
Europe	Russia	7				
Europe	Ukraine	23				
$n_{Europe} =$		$\Sigma X_{Europe} =$			$WSS = \Sigma(X - \bar{X}_{Group})^2 =$	
$\bar{X}_{Europe} = \Sigma X_{Europe}/n_{Europe} =$						
$n_{Total} =$	$\Sigma X_{Total} = \Sigma X_{Asia} + \Sigma X_{Europe} =$		$TSS = \Sigma(X - \bar{X}_{Total})^2 =$			
	$\bar{X}_{Total} = \Sigma X_{Total}/n_{Total} =$					

Source: Freedom House, "Freedom in the World: Aggregate Scores, 2003-2011," 2012. www.freedomhouse.org/report/freedom-world-aggregate-and-subcategory-scores.

1. Open your dataset in SPSS and clean it. Your dependent variable is the feeling thermometer for the Democratic Party (FTDemParty) and your independent variable is party identification (PartyID). Make sure that all the appropriate values are set to missing.

2. If necessary, use the "Select Cases" command to limit your analysis to the desired cases. Notice that it includes two categories in which you are not interested—"Other Party" and "No Preference." You do not want to include these cases in your analysis, so proceed to select only those cases for which "Democrat," "Republican," and "Independent" are coded for partisanship. That occurs when this variable is less than or equal to 3.

   >Data
   >>Select Cases
   >>>If Condition is Satisfied
   >>>If
   >>>"PartyID<4"
   >>>Continue
   >>>OK

3. Request the analysis of variance. Be sure to save these tables to a Word document so that you can use them as you write your memo. (Warning: The ANOVA table is very wide, so switch page layout so that the orientation is "landscape"—you can then move the table's column lines over to make the table narrower so that you can switch back to "portrait.")

   >Analyze
   >>Compare Means
   >>>Means
   >>>>"FTDemParty" → Dependent Variable
   >>>>"PartyID" → Independent Variable
   >>>>Options
   >>>>>ANOVA Table and eta
   >>>>>Continue
   >>>>OK

4. Now write your memo: Will those people who became Independents have colder feelings about the Democratic Party? How about those who became Republicans? Be sure to include the four parts of a memo: heading, introduction, body, and conclusion. Within the body, be sure to go through the three steps of analysis: describe the pattern, identify its statistical significance, and evaluate the strength of the relationship. At the end of the memo, include a table with the results of the ANOVA. For grading purposes, attach your SPSS output: "Report," "ANOVA Table," and "Measures of Association."

Key Terms

Analysis of variance (ANOVA) (p. 185)
Dependent variable (p. 181)
Empirical (p. 180)
Falsifiable (p. 183)
F-statistic (p. 190)
Hawthorne effect (p. 179)
Hypothesis (p. 181)
Independent variable (p. 181)
Model (p. 181)

Negative relationship (p. 182)
Null hypothesis (p. 184)
Positive relationship (p. 182)
PRE measures (p. 193)
p values (p. 184)
Statistical significance (p. 184)
Substantive significance (p. 191)
Theory (p. 180)

CHAPTER 8

Describing the Pattern

What Do You See?

In the 2012 State of the Union, President Barack Obama instituted an innovation: Whereas television viewers saw the traditional view of the president facing members of Congress in the chamber of the House of Representatives, viewers on the White House's webpage had the option of an "enhanced" version. It contained a blue sidebar to the traditional image, presenting PowerPoint-style images to supplement the president's words.[1] One of the images was of particular interest to those of us who are *How to Lie with Statistics* aficionados. As the president spoke the words "My administration has put more boots on the border than ever before,"[2] the visual next to him showed the footprints of two boots. The slide indicated that in 2004, there were 10,000 border patrol agents, whereas in 2011, there were 21,444. To reinforce the numerical doubling in agents, below the numbers, the second bootprint was twice as long as the first.[3] This visual falls into the classic fallacy of "The One Dimensional Picture." In this fallacy, a one-dimensional number is depicted with a two-dimensional picture. The problem with it is that to make the bootprint look normal, if you double the length, you also have to double the width, which is exactly what was done here. The result is that the area covered by the larger bootprint is not double the area of the first bootprint; it is actually four times as large, leaving a visual impression that instead of doubling, the size of the border patrol had quadrupled.

Perhaps President Obama decided to attach visual images to his speech because he understands the truth behind the adage "A picture is worth a thousand words." Certainly, his use of the FBI's most wanted poster for Osama bin Laden with a big red "X" through it communicated very clearly that bin Laden was no longer a threat to the United States. But the bootprint slide (in addition to a few of the graphs) is a clarion call to us that in order to keep the numbers honest, we need to keep the visuals honest as well.

In academic work, we do not normally use pictures like the bootprint to communicate information. We normally have two kinds of visuals in our writing: figures and tables. Please resist the temptation to call them charts; we simply do not use that term. **Figures** present information in some kind of picture. In a statistical context, this will frequently refer to graphs or diagrams. Tables refer to information in the form of numbers or words that are presented in rows and columns. In this chapter, we'll discuss how to present information in figures (specifically graphs) and tables (specifically contingency tables). Our goal here is present the information in a clear way in order to aid us in the first step of the analysis process: describing the pattern.

CHOOSING THE APPROPRIATE FORM OF PRESENTATION

In the previous chapter, we saw that the correct formulation of the hypothesis depended on the level of measurement of the variables. Similarly, the level of measurement determines the proper form of the visual presentation of data. When you have two variables with an interval level of measurement, you can graph their values to show the relationship. In most cases, though, a contingency table is a better way to communicate the information. For example, if one of the variables has a limited number of categories, a graph becomes meaningless. Similarly, even when you have two interval-level measures, if you have many data points, a graph gets too messy. In both instances, the data would be communicated more clearly in a contingency table in which you have collapsed the interval-level data into ordered categories. Finally, whenever you have categorical data (whether nominal or ordinal), your best bet is to present the data in a contingency table.

GRAPHS: RELATIONSHIPS AND SCALES

When we state a hypothesis for ordered data, we use the following language: There is a positive/negative relationship between (the dependent variable) and (the independent variable). This language reflects our intuitive understanding of what relationships look like. If I were to state, "There is a positive relationship between weight and height," you would immediately picture a graph in your mind that looks something like Figure 8.1. You almost intuitively remember from algebra that a line with a positive slope increases as you move to the right. Conversely, if I were to hypothesize, "There is a negative relationship between weight and exercise," you might picture a graph like Figure 8.2. Another memory from

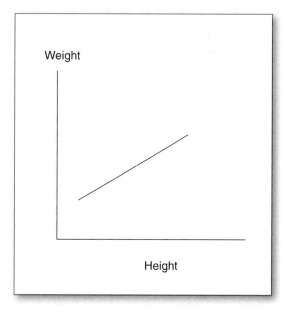

Figure 8.1 The Relationship between Weight and Height

algebra is that a negative slope decreases as it moves to the right. We borrowed that intuition about positive and negative slopes when we used the language of positive and negative relationships in stating hypotheses.

Not all relationships can be graphed. Although we use the language of positive and negative relationships for any hypotheses where both the dependent and the independent variables have order, we can only graph interval-level data because the axes have scales. Perhaps if we are using ordinal-level data, it would be more appropriate to use the language of direct and inverse relationships. Certainly, you should not try to graph ordinal-level data. In addition to needing interval-level data, if you want to create an honest graph, the following rules apply.

1. The independent variable corresponds to the x-axis; the dependent variable, to the y-axis.

2. You should label the axes both by the names of the variables and by the units with which they are being measured.

3. The axes should be marked with intervals that are evenly spaced and make sense.

4. Mark each of your data points with a dot, making sure that they are all in the correct spot.

5. Normally, you draw a smooth line (or curve) going through the space containing the dots in such a way as to be as close as possible to the data points.

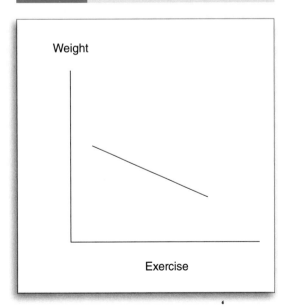

Figure 8.2 The Relationship between Weight and Exercise

Suppose we hypothesize that there is a negative relationship between voter turnout and time. (Or more simply put: Voter turnout has been declining for the past fifty years.) In Table 8.1, I have the percent of turnout during all the presidential elections since 1960. My thought is that something has been happening over time to American voters, making them less likely to vote. My dependent variable is turnout; my independent, time. The first step, then, is to set up a grid where time forms the x-axis and turnout forms the y-axis. I am measuring time in years and so I give the x-axis the label "Year." I am measuring turnout according to the percent of the voting age population that voted for president, so I give the y-axis the label "Turnout for Presidential Election in Percent of Voting Age Population." These axes are found in Figure 8.3.

The next step is to mark the axes with intervals that are evenly spaced and make sense.

I'm going to be graphing fifty years' worth of data. Normally, you want to have some space before the data begins and some space after it ends so I actually begin at 1950 and end at 2020. I chose to make my intervals at every ten years. I could have made it every four years to connect with the presidential elections, but I decided that it would clutter the grid to have that many intervals.

The interval for the y-axis is a little trickier. It makes for a prettier graph if the data points fill up the space. Some computer programs deliberately choose to place the origin of the graph at a point just below the minimum values for the two variables. They then extend each of the axes just past the maximum values for each. For the independent variable, that is not usually

Table 8.1 Voter Turnout for Presidential Elections since 1960

Year	Turnout
1960	62.77
1964	61.92
1968	60.84
1972	55.21
1976	53.55
1980	52.56
1984	53.11
1988	50.15
1992	55.23
1996	49.08
2000	51.30
2004	55.27
2008	57.48

Source: Wooley, John, and Gerhard Peters, "Voter Turnout in Presidential Elections: 1828-2008." *The American Presidents Project*, 2012. www.presidency.ucsb.edu/data/turnout.php.

Figure 8.3 Axes for Graphing Turnout by Year

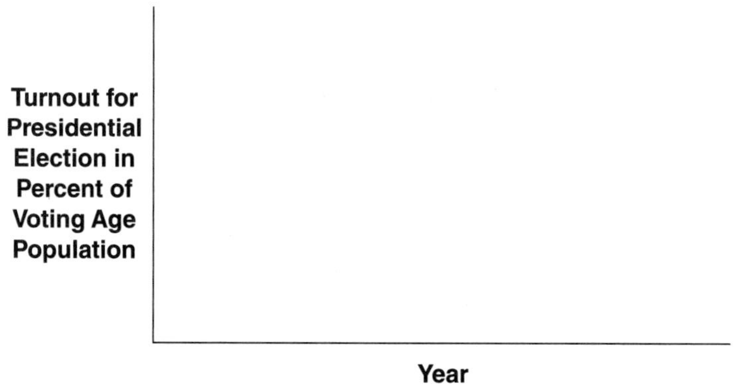

Turnout for Presidential Election in Percent of Voting Age Population

Year

a problem unless you deliberately choose to leave out data that would contradict your hypothesis. But if you do it for the dependent variable, you may end up with what *How to Lie with Statistics* calls a "Gee-Whiz Graph." If you have a trend, no matter how slight, it will always look like a large change if you adjust the grids so that the data fill the window. Although many of the graphs shown in the enhanced version of the State of the Union included the origin, there were a couple of "Gee-Whiz Graphs" as well. These tended to be the graphs depicting an economic recovery since President Obama's election. For example, one graph was labeled "Recent Gains in Manufacturing Employment." The x-axis begins very conveniently in November (rather than January) 2009 (presumably when the gains began rather than when Obama came into office). More problematic is the y-axis, which ranges from a low of 11.40 million to a high of 11.80 million jobs.[4] By making the data fill the graph, the creator of this slide made a 3.5 percentage point increase in manufacturing jobs appear like a 600 percent increase. It could well be that a 3.5 percent increase is reason to cheer, but that's an argument to be made logically with words, not emotionally with pictures. Occasionally, there are good reasons to limit the graph to a range around the data points. If that is the case, try to highlight the fact that your y-axis does not begin at y = 0. For example, sometimes the axis is broken by two hash marks indicating that a section of the axis has been removed. The State of the Union graphs dealt with this problem by not including a line for the x-axis. As I graphed the trend in turnout, though, I chose to begin the y-axis at the origin and label every 10 percent of turnout above that. My grid is shown in Figure 8.4.

My next step is to mark each data point on the grid. The key here is to make sure that you have entered the data correctly. All of us make mistakes, so double check. As a consumer of data, be sure to also double check the data points on figures you see in the news. Journalists are human, too, and sometimes the patterns they talk about are actually incorrectly graphed data points. My data points are graphed in Figure 8.5.

Figure 8.4 Turnout by Year Grid

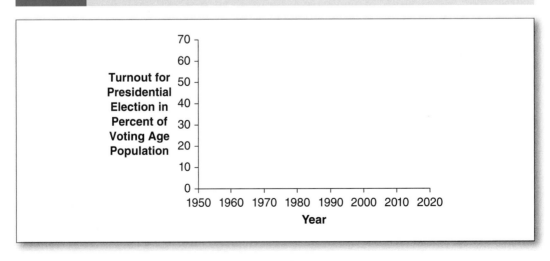

Figure 8.5 Turnout in Presidential Elections since 1960

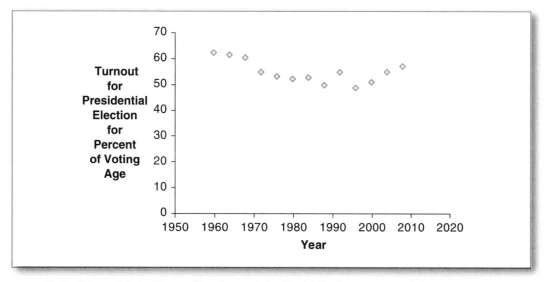

Source: Wooley, John, and Gerhard Peters, "Voter Turnout in Presidential Elections: 1828-2008." *The American Presidents Project*, 2012. www.presidency.ucsb.edu/data/turnout.php.

Finally, show the trend by drawing a well-fitting line. This does not mean playing dot-to-dot. Rather, superimpose a single line so that the data points are evenly spaced above and below the line. My line is shown in Figure 8.6. You can see that the line does not connect the dots. Rather, it is a straight line that is placed in such a way as to minimize how far the dots are above and below it. In Chapter 12, we'll learn how to use a procedure called regression to mathematically find the line that best fits the data, but for now, just keep the goal of minimizing the distance from the data points in mind as you draw your own line.

Once we've drawn a well-fitting line, we can begin to describe the pattern that we see in the data. The line I've drawn definitely has a negative slope, which suggests that there has been a downward trend in voter turnout since 1960. At this point, we cannot give the slope of the line (although the regression analysis in Chapter 12 will estimate the slope), but we can give a general feel for what has happened. If I look at my line, in 1960, turnout was about 60 percent. By 2008, my line had decreased to about 52 percent. In describing this pattern, I would say that turnout decreased by about 8 percentage points between 1960 and 2008.

Occasionally, we will hypothesize a curvilinear, rather than linear, relationship. For example, I know that the races of the twenty-first century have been very exciting and so have had fairly high turnouts. I might want to hypothesize that after four decades of decline in voter turnout, it is now on the rise again. If I were to make that hypothesis, I would want to fit a curve, not a line, onto the data in my graph. I would end up with Figure 8.7. This curve begins with a turnout of about 64 percent in 1962, declines to its nadir at 48 percent in 1996, and then rises again to 58 percent in 2008. I would describe the pattern by saying

Figure 8.6   Voter Turnout with a Line Showing Trend

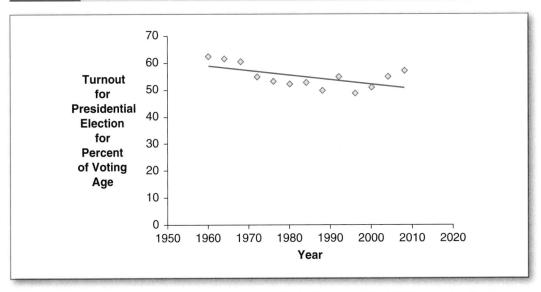

Source: Wooley, John, and Gerhard Peters, "Voter Turnout in Presidential Elections: 1828-2008." *The American Presidents Project*, 2012. www.presidency.ucsb.edu/data/turnout.php.

Figure 8.7   Voter Turnout with a Curve Showing Trend

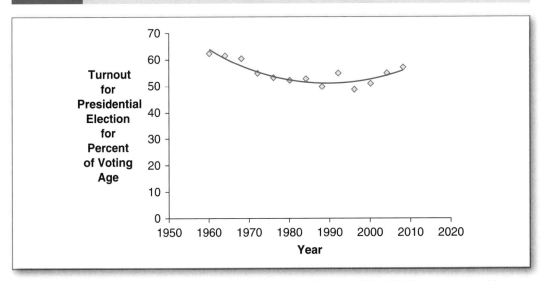

Source: Wooley, John, and Gerhard Peters, "Voter Turnout in Presidential Elections: 1828-2008." *The American Presidents Project*, 2012. www.presidency.ucsb.edu/data/turnout.php.

that after 1960 voter turnout declined by about 16 percentage points over the course of three decades, but since the early 1990s, it has rebounded by about 10 percentage points.

Sometimes, we may want to compare the trends for two variables. When I talked about voter turnout decreasing over time, I discussed my assumption that something was happening in our society to make Americans less likely to vote. If I were to posit another hypothesis that identified that cause, I might want to include a second variable on the graph. For example, I know that education is highly correlated with voter turnout. Suppose I hypothesize that declining turnout is the result of declining education. If I measure education in terms of percent of adults who have graduated from high school, I might want to include the graduation rate on my graph. Because both variables are measured on the same scale (percent), I can just superimpose the second variable on the first with a key to distinguish between the two. This is shown in Figure 8.8. It looks like graduation rates have increased over 45 percentage points, whereas turnout rates have declined 8 percentage points. The pattern does not support my hypothesis that declining turnout rates are the result of declining graduation rates.

Sometimes, I want to superimpose a second variable that does not use the same scale as the first. In that case, you would put a second grid to the right of the graph with the

Figure 8.8 Two-Variable Graph with Same Scale

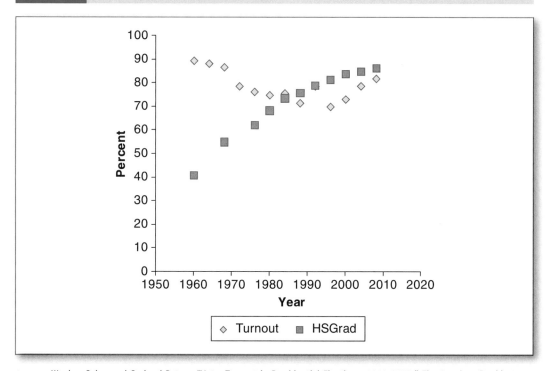

Sources: Wooley, John, and Gerhard Peters, "Voter Turnout in Presidential Elections: 1828-2008." *The American Presidents Project*, 2012. www.presidency.ucsb.edu/data/turnout.php; "Digest of Education Statistics." *National Center for Education Statistics*, 2011. http://nces.ed.gov/programs/digest/d10/tables/dt11_008.asp.

appropriate scale for the second variable. When I saw the upsurge in turnout after 2000, I discussed it in terms of how hotly fought those races were. A reasonable proxy for tight races is the amount of campaign spending. Suppose I compare the trend for turnout with the trend for campaign spending in presidential elections. The scale for spending is in millions of dollars rather than percent, so I need a different scale, which I place to the right of the graph. This is shown in Figure 8.9. The pattern shown here is that campaign spending is increasing dramatically over time—both in recent years, when turnout increased, and in early years, when turnout declined. This pattern does not support my hypothesis that turnout rates are a result of the excitement of a race.

Figure 8.9 Two-Variable Graph with Different Scales

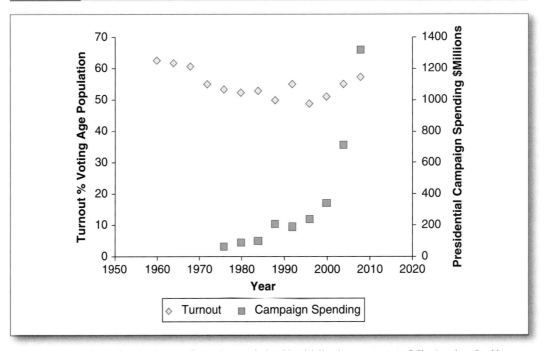

Sources: Wooley, John, and Gerhard Peters, "Voter Turnout in Presidential Elections: 1828-2008." *The American Presidents Project*, 2012. www.presidency.ucsb.edu/data/turnout.php; "Presidential Fundraising and Spending, 1976-2008," *Open Secrets: Center for Responsive Politics*, 2009. www.opensecrets.org/pres08/totals.php.

VISUALIZING A RELATIONSHIP: CONTINGENCY TABLES

Although graphs are helpful in showing trends, they are less useful when you have lots of data points that may or may not overlap. If I were analyzing survey data, I might have thousands of data points. A graph of all those data points would be too messy to be useful. Somehow, I would need to summarize the data. In addition, I can only graph interval-level data. As a result of these two factors, more often than not, I will present my data visually in a contingency table.

The original census in Sweden collected data about multiple variables and used them to create cross tabulations. With those data, they were not really testing hypotheses about the causal relationship between variables, but they were able to find anomalies. For example, they found that young women tended to die of infections after childbirth. That observation then led them to ask, Why are there differences in death rates? As they answered this question, the Swedes were subsequently able to improve public health. In this chapter, I am going to use the term "contingency table" rather than "cross tabulation" because I am interested in testing my hypothesis of whether my independent variable affects my dependent variable. Both "cross tabulation" and "contingency table" refer to a bivariate frequency distribution, but "contingency table" does a better job of describing what we are using the table for statistically.

Set up a Contingency Table

Table 8.2 shows the relationship between domestic war and political competition. Our hypothesis is that politically competitive nations are less likely to go to war than less competitive nations. We will test that hypothesis in the next chapter, but for now, we want to present the data in a way that allows us to describe whether the data have a pattern that appears to support the hypothesis.

Notice that Table 8.2 is currently set up so that the independent variable is in the columns and the dependent variable is in the rows. This is standard. At the top of each column is the label for each category of the independent variable. At the left of each row is the label

Table 8.2 Domestic War by Political Competition, Raw Data

Involved in Domestic War?	Political Competition					
	Repressed	Suppressed	Factional	Transitional	Competitive	Total
No	13	14	26	40	35	128 (84.2%)
Yes	1	2	9	11	1	24 (15.8%)
n =	14	16	35	51	36	152
Percent of Total	(9.2%)	(10.5%)	(23.0%)	(33.6%)	(23.7%)	(100.0%)

Sources: Polity Project IV, "PARCOMP," *Political Regime Characteristics and Transitions*, 1800-2010, Integrated Network for Societal Conflict Research, 2010. www.systemicpeace.org/inscr/inscr.htm; Gleditsch, Nils Petter, Peter Wallensteen, Mikael Eriksson, Margareta Sollenberg and Håvard Strand "Armed Conflict 1946–2001: A New Dataset," *Journal of Peace Research* 39 (2002): 615–637. www.pcr.uu.se/research/ucdp/datasets/ucdp_prio_armed_conflict_dataset/.

for each category of the dependent variable. This is standard as well. The cells contain the number of cases that jointly exhibit those particular values of the dependent and independent variables. Finally, the table includes the marginals—or total number of cases—for both the columns and the rows. Again, this is standard.

Table 8.3 Domestic War by Political Competition, Partial Calculation

Involved in Domestic War?	Political Competition					Total
	Repressed	Suppressed	Factional	Transitional	Competitive	
No	13/14 = .929 = 92.9%	14	26	40	35	128 (84.2%)
Yes	1	2	9/35 = .257 = 25.7%	11	1	24 (15.8%)
n =	14	16	35	51	36	152
Percent of Total	(9.2%)	(10.5%)	(23.0%)	(33.6%)	(23.7%)	100.0%

Sources: Polity Project IV, "PARCOMP," *Political Regime Characteristics and Transitions, 1800-2010*, Integrated Network for Societal Conflict Research, 2010. www.systemicpeace.org/inscr/inscr.htm; Gleditsch, Nils Petter, Peter Wallensteen, Mikael Eriksson, Margareta Sollenberg and Håvard Strand "Armed Conflict 1946–2001: A New Dataset," *Journal of Peace Research* 39 (2002): 615–637. www.pcr.uu.se/research/ucdp/datasets/ucdp_prio_armed_conflict_dataset/.

Unfortunately, this presentation of the data makes it hard to see whether the data support the hypothesis or not. In each of the categories of political competition, there are lots of nations that are not involved in an internal war. But because each column has a different number of cases, we cannot simply compare the raw numbers across the levels of competition. In order to do that, the columns need to be standardized. That means that we are less concerned with the number of cases that belong in each cell than in the proportion in the cell in comparison to the total for the column. For each cell, I divide the number of cases in that cell by the number of cases in that column to find the proportion of countries with that level of competition that are involved in a domestic war. I then multiply the proportion by 100 to get the percent. In Table 8.3, I have done that for two of the cells. Thirteen of the fourteen repressed countries were not involved in an internal war between 2000 and 2010. This means that 92.9 percent of the repressed countries avoided internal war. Of the thirty-five factional countries, nine (or 25.7 percent) were involved in an internal war.

Table 8.4 shows the completed contingency table for the relationship between war and level of competition. Although the structure of the table has stayed the same (the dependent variable is still in the rows; the independent, in the columns) the interior looks different. I have replaced the cell counts with the column percent. Although many scholars

Table 8.4 Domestic War by Political Competition, 2000–2010

Involved in Domestic War?	Political Competition					Total
	Repressed	Suppressed	Factional	Transitional	Competitive	
No	92.9%	87.5%	74.3%	78.4%	97.2%	84.2%
						(128)
Yes	7.1%	12.5%	25.7%	21.6%	2.8%	15.8%
						(24)
Total	100.0%	100.0%	100.0%	100.0%	100.0%	100.0%
n =	14	16	35	51	36	152

Sources: Polity Project IV, "PARCOMP," *Political Regime Characteristics and Transitions, 1800-2010*, Integrated Network for Societal Conflict Research, 2010. www.systemicpeace.org/inscr/inscr.htm; Gleditsch, Nils Petter, Peter Wallensteen, Mikael Eriksson, Margareta Sollenberg and Håvard Strand "Armed Conflict 1946–2001: A New Dataset," *Journal of Peace Research* 39 (2002): 615–637. www.pcr.uu.se/research/ucdp/datasets/ucdp_prio_armed_conflict_dataset/.

include the column percent with the cell count below it in parentheses (in the way I have done for the row marginals), I prefer the visual simplicity of including only the column percent in the cells. It is much easier to scan and compare the percentages when they are not divided by cell counts. Normally, we don't like to lose data, but in this case, we actually don't lose any information because we could re-compute any cell count by simply multiplying the column percent for the cell by the number of cases in the column. For the marginals, though, I include both the column percent and the raw count below it in parentheses. I want the raw count both so I can re-compute the cell counts if needed and also because sample size affects the statistics we'll be calculating in later chapters. Because an imbalance in the relative size of the categories can affect our results, we need to be able to see the marginal frequencies. I want the column percent in the marginals at the right of the table so I can compare any given category with the average for all the cases. It may seem silly to include the 100 percent at the bottom of all the columns, but it serves as a useful cue to your reader that the table is correctly percentaged by column. Add up the column percents for each cell in a column to get the total percent in the marginals at the bottom of the table. This should always total 100.0 percent. Going to the trouble of adding the percentages in each column to get the total percent at the bottom serves as a good check on your math. Be careful, though. If you have more than two categories in your dependent variable, the total at the bottom could be off by 0.1 percent just due to rounding error. Don't worry about that beyond making a note at the bottom of the table that the difference is due to rounding error. But if the total is off by more than 0.1 percent, definitely check your math. The steps I followed in creating this contingency table are summarized in Box 8.1.

BOX 8.1 How to Create a Contingency Table

1. Give the table a title in the form of the name of the *"dependent variable"* "by" the *"independent variable."*

2. Insert a table with as many rows as it has categories in the dependent variable plus three. It has as many columns as it has categories in the independent variable plus two.

3. In the first row, type the name of the dependent variable in the first column, each of the categories of the independent variable in the next columns, and the word "Total" in the last column.

4. Below the name of the dependent variable in the first column, place each of the categories of the dependent variable. In the row third from the bottom, type "Total"; second from the bottom, "n"; and leave the first column of the bottom row empty.

5. Calculate the column percents for each cell by dividing the cell count by the column count and multiplying by 100.

6. Include marginals for both the rows and the columns—in both column percent and number of cases.

7. Place lines after the title, after the column headings and at the end using the "Borders and Shadings" command in Word.

8. Give any relevant information in a smaller font below the bottom line.

Describe the Pattern in a Contingency Table

With the contingency table properly constructed, we are now in a position to describe the pattern of the relationship between our dependent and independent variables. To get a feel for the data, in any given row, you can take the total percent in the right-hand column and compare it with the column percents for each of the categories. You should be looking for cells that are really different from the marginal percent—either higher or lower. Scanning the first row of Table 8.4, you find that 82.4 percent of all countries are not involved in a domestic war. Politically suppressed countries are fairly close to that, with 87.5 percent of them not being involved in war. But factional governments are much lower—only 74.3 percent of factional governments are not at war—as are transitional countries (78.4 percent). In contrast, repressed and competitive countries are much higher—92.9 percent and 97.2 percent, respectively. The second row shows the flip side of the same relationship: Factional and transitional countries are much more likely (25.7 percent and 21.6 percent) to be at war than the 15.8 percent of countries as a whole. In contrast, repressed and competitive countries are much less likely (7.1 percent and 2.8 percent) to be involved in a domestic war.

After getting a feel for how the data look, next, do a more specific analysis for your write-up. Because both rows tell essentially the same story from a different perspective, you will normally choose only one row to summarize in describing the pattern contained in the table. Think about how you've framed your hypothesis and choose the row accordingly. If you've hypothesized that competitive countries are much more likely to stay out of war than factional countries, describe the pattern in the "no war" row. If you've hypothesized that factional countries are much more likely to go to war than competitive countries, describe the "yes war" row. When I presented the issue, I talked about war being more likely in the middle range than on either end. It makes sense, then, to look at the "yes war" row. In order to make the comparison for this row, I would subtract the percentage of one cell from the percentage in the cell to which I want to compare it. Here, I can conclude that factional countries are 18.6 percentage points more likely to be involved in a domestic war than repressed countries (because $25.7 - 7.1 = 18.6$) and 22.9 percentage points more likely than a politically competitive country ($25.7 - 2.8 = 22.9$). In the next few chapters, we'll learn how to determine whether this relationship is statistically and substantively significant. But for now, we simply compare the percents to see whether cases in different categories of the independent variable appear to show a different pattern for their values of the dependent variable.

Collapse Data for a Contingency Table

The human brain is able to process only so much information at a time. If you have a surfeit of categories in a contingency table, it is difficult to see the pattern. This becomes very obvious if you try to create a contingency table with an interval-level variable, but is equally important if you are dealing with an ordinal- or nominal-level variable with too many categories. The table we've been looking at to analyze the relationship between domestic war and level of political competition pushes the size of a table to its limit. You really do not want more than five categories. In Chapter 4, when we were creating a frequency table for law schools, we went through the rules of collapsing. The same rules, summarized in Box 8.2, apply here.

BOX 8.2   How to Collapse Variables

1. It is best to have only three to five categories.
2. If you are collapsing an interval-level variable, keep the size (in terms of the variable's scale, not the number of cases) of the ranges equal (with the possible exception of the lowest and highest categories) and make sure both the size and the endpoints make sense.
3. If you are collapsing a categorical variable, you need to have a reason for which categories you group together. Most often, you will leave the largest categories (in terms of number of cases) and group the smallest categories together in an all-inclusive "Other" category. But it is preferable, if possible, to group similar categories together.
4. Define the categories in such a way as to avoid categories with zero cases.
5. Make sure that each case falls into exactly one category.

If you've ever seen a map of Red States and Blue States, the first thing you noticed was probably that the Blue States are mostly on the coasts and the Red States are in the heart of the country. The next thing you might have noticed is that the Blue States tend to be richer than the Red States. This seems odd, because the New Deal realignment reconfigured the two parties in terms of economic interest: Since then, Republicans have tended to be richer than Democrats. Is it possible that the New Deal realignment has been turned on its head? Reagan articulated a value-based vision of the Republican Party. Is it possible that politics is no longer a conflict between rich and poor, where the rich are represented by the Republicans and the poor are represented by the Democrats? I hypothesize that poor states are more likely to vote Republican than rich states.

The unit of analysis for the Red State/Blue State distinction is states. I will categorize the dependent variable, partisanship, into the state support for presidential candidates during the 2012 election, where Red States were safe states for the Republican presidential candidate, Blue States were safe for the Democratic candidate, and Toss-up States were in play during the election.

For my independent variable, income, I will use the Census Bureau's estimate of median family income for 2010. Because this is an interval-level variable, I need to collapse it into three to five categories. The median incomes for the states range from a low of $37,985 in Mississippi to a high of $66,707 in New Hampshire. Because I want evenly spaced ranges for my categories, I will group this variable into ranges of $10,000. This gives me four categories: $30-40,000, $40-50,000, $50-60,000 and $60-$70,000. You may have noticed that these categories have a common endpoint. For example, $40,000 is both the highest value for the first category and the lowest value for the second category. We need to keep in mind that Rule 5 on how to collapse variables requires us to make sure that each case falls into precisely one category. It is standard to write the ranges to have the overlapping endpoints in order to indicate that all amounts—regardless of how far out the decimal goes—are included the ranges. By convention, the range begins with its low point and goes up to, but does not include, its upper end point. For example, when we write "$30-40,000" what we mean is "$30,000.00-$39,999.99." You can see that the convention yields a tidier-looking range than carrying the decimal out as far as it needs to go to include all cases.

Once I've decided how to collapse my variables so that each has a reasonable number of categories, I follow the usual instructions for creating a contingency table. First, I set it up so that the independent variable is in the columns and the dependent variable is in the rows. I then make headings for each of the categories of both variables. Table 8.5 shows the raw data for the relationship between a state's partisanship and its income. When I inserted the table into my document, I based the number of rows and columns on the number of categories in the dependent and independent variables. I can see at a glance that my dependent variable has three categories. As a result, I included 3 + 3 = 6 rows. Because my independent variable has four categories, I included 4 + 2 = 6 columns. I placed the value labels for my independent variable in the first row and the value labels for the dependent variable in the first column. The last column and last two rows are for the marginals. At this point, I can place the total number of cases for each row in the last column and the total number of cases for each column in the last row.

Table 8.5 State Partisanship by Median Family Income, Raw Data

Partisanship	$30-40,000	$40-50,000	$50-60,000	$60-$70,000	Total
Republican	4	11	5	0	20
Toss-Up	0	9	2	3	14
Democratic	0	3	9	4	16
Total					
n =	4	23	16	7	50

Sources: Silver, Nate, "Five-Thirty-Eight." *The New York Times*, August 20, 2012. http://fivethirtyeight.blogs.nytimes.com/2012/08/20/aug-20-when-the-polling-gets-weird/; U.S. Census Bureau, "Median Household Income by State: Single-Year Estimates." *Current Population Survey, Annual Social and Economic Supplements* 2010. www.census.gov/hhes/www/income/data/statemedian/.

Table 8.6 shows a correctly percentaged contingency table for the data. In each cell, I divided the cell count by the total number of cases in each column and multiplied by 100 to get the column percent. In the last column, I divided the number of cases in that row by my total "n." To get the total percent in the marginals at the bottom, I added the percentages

Table 8.6 State Partisanship by Median Family Income

Partisanship	$30-40,000	$40-50,000	$50-60,000	$60-$70,000	Total
Republican	100.0%	47.8%	31.3%	0.0%	40.0% (20)
Toss-Up	0.0%	39.1%	12.5%	42.9%	28.0% (14)
Democratic	0.0%	13.0%	56.3%	57.1%	32.0% (16)
Total	100.0%	99.9%	100.1%	100.0%	100.0%
n =	4	23	16	7	50

Sources: Silver, Nate, "Five-Thirty-Eight." *The New York Times*, August 20, 2012. http://fivethirtyeight.blogs.nytimes.com/2012/08/20/aug-20-when-the-polling-gets-weird/; U.S. Census Bureau, "Median Household Income by State: Single-Year Estimates." *Current Population Survey, Annual Social and Economic Supplements* 2010. www.census.gov/hhes/www/income/data/statemedian/.

Note: Some columns fail to total 100.0% due to rounding error.

in the cells of each column. You'll notice that due to rounding error the second column totals 99.9 percent and the third totals 100.1 percent. Because these are within 0.1 percent of 100 percent, the difference is probably just rounding error rather than a mathematical mistake in need of fixing.

I am now in a position to describe the pattern contained in the table. When I discussed my hypothesis, I framed it in terms of poor states voting Republican, so I choose the "Republican" row to analyze. In this country, 40 percent of the states are categorized as solid red Republican. As I compare that baseline to the cells in the first row, I see that all of the poorest states are red, whereas none of the richest states are. In the two middle categories, I see that those states with median incomes in the $40,000 range are more likely to vote Republican than states in the $50,000 range by a difference of 16.5 percentage points. The data do have a pattern that is consistent with my hypothesis.

One of the things I need to be careful of, though, is to state my findings in such a way that they are consistent with my unit of analysis. My hypothesis and my data were talking about states, not individuals. I need to be careful not to conclude on the basis of state-level data that something is happening at the individual level. This table does not justify me concluding that individuals with incomes of $45,000 are more likely to vote Republican than individuals with incomes of $55,000. To draw that conclusion, I would need to look at individual-level data.

SUMMARIZING THE MATH: GRAPHS AND CONTINGENCY TABLES

The first step in analyzing data is describing their pattern. Descriptions are always more evocative when they are accompanied by some kind of visual presentation of data. For political scientists, this will normally entail presenting the data in either a figure or a table. In this chapter, we learned how to make professional-looking graphs that we can include as figures. We also learned how to make professional-looking contingency tables as one form that a statistical table can take.

Graphs

You can use graphs to describe either a linear or a curvilinear relationship between two interval-level variables. Be sure to state that relationship as a hypothesis so you can easily identify both the dependent and independent variables. Graphs can be particularly helpful in visualizing trends over time. They become more cluttered and so less useful when you have too many data points. If you have a limited number of cases that you want to compare on two interval-level variables, use the following procedures to create a graph:

1. The independent variable corresponds to the x-axis; the dependent variable, to the y-axis.

2. You should label the axes both by the names of the variables and by the units with which they are being measured.

3. The axes should be marked with intervals that are evenly spaced and make sense.

4. Mark each of your data points with a dot, making sure that they are all in the correct spot.

5. Normally, you draw a smooth line (or curve) going through the space containing the dots in such a way as to be as close as possible to the data points.

To describe the pattern that you see, find the high point on the line and subtract the low point. The difference is the change you observe between the corresponding change on the x-axis between those two points.

BOX 8.3 Numbers in the News

Three years in a row, the Senate failed to pass a budget. Three years in a row, Republican senators rose to their feet, graphics presentations in hand, to draw attention to how many days it had been since a budget had been approved by the upper house of the United States Congress. Finally, on March 23, 2013, the Senate approved a budget. Because the House had already approved its version of the budget, the next step would normally be the appointment of a conference committee consisting of a mixture of Democrats and Republicans from both the House and the Senate to work out a compromise version for final approval by both houses. But those same Tea Party Republicans who had complained about the lack of a budget feared that the compromise budget might contain an increase in the debt ceiling. So they began objecting every time Senate Democrats asked for unanimous consent to appoint the members of the conference committee. Democrats, having learned very well the power of the visual presentation of data over the previous three years, responded to the obstruction by presenting their own data, a visual presentation of how many days it had been since a budget had been passed . . . [5]

A Political Example: Solidarity Movement. In 1980, a noncommunist trade union organized in the Eastern Bloc country of Poland. Under the leadership of Lech Walesa, this organization, dubbed the Solidarity Movement, became politically powerful both in Poland and elsewhere as it inspired anti-communist activity elsewhere in the Eastern Bloc. It was central to the breakup of Soviet control over the region in 1989, after which Walesa was elected president of Poland. How did political freedoms change over this period of time? I hypothesize that from 1980 to 1999, political freedoms increased in Poland. Table 8.7 shows the final political rights score from Freedom House for Poland in this period of time.

To set up the graph shown in Figure 8.10, I follow the directions:

1. The independent variable corresponds to the x-axis; the dependent variable, to the y-axis.
   The independent variable is year, which I place on the x-axis; the dependent variable is political rights, which I place on the y-axis.

2. You should label the axes both by the names of the variables and by the units with which they are being measured.
   At the bottom of the graph, I place the label "Year"; at the left, I place the label "Freedom House Political Rights Score."

Table 8.7 Political Rights Score for Poland, 1980-1999

Year	Political Rights	Year	Political Rights
1980	6	1990	2
1981	7	1991	2
1982	6	1992	2
1983	6	1993	2
1984	6	1994	2
1985	6	1995	1
1986	6	1996	1
1987	5	1997	1
1988	5	1998	1
1989	4	1999	1

Source: Freedom House, "Freedom in the World Country Ratings: 1972-2011," 2012. www.freedomhouse.org/report-types/freedom-world.

3. The axes should be marked with intervals that are evenly spaced and make sense.
   For the x-axis, I divide the scale into five-year increments. For the y-axis, I know that the political rights score ranges from a free score of 1 to a not-free score of 7. We normally put some space above and below the possible range, so I begin at 0 and go to 8, marking each point on the scale.

4. Mark each of your data points with a dot, making sure that they are all in the correct spot.
   These are all marked.

5. Normally, you draw a smooth line (or curve) going through the space containing the dots in such a way as to be as close as possible to the data points.
   Because I know that a country's score cannot be more than 7 or less than 1, it makes sense to use a curve rather than a line to fit the data points. This will allow for the two ends to have very little change while the middle changes very rapidly.

Figure 8.10 Change in Political Rights in Poland

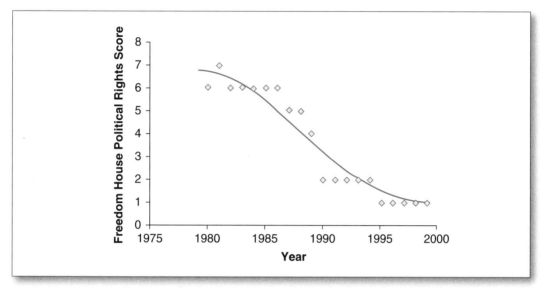

Source: Freedom House, "Freedom in the World Country Ratings: 1972-2011," 2012. www.freedomhouse.org/report-types/freedom-world.

Once I've set up the graph, I can describe the pattern. At the beginning of the Solidarity Movement, political freedoms changed very little, although there was a slight increase. Once the government began negotiating with Solidarity, political freedom increased dramatically. Within five years, Poland had achieved Freedom House's most politically free score. Political freedom did increase in Poland from 1980 to 1999.

Contingency Tables

Because graphs are useful only when looking at the relationship between two interval-level variables for which you have a limited number of cases, you are more likely to present your data in a contingency table. As with a graph, though, you need to have a well-stated hypothesis before you begin to build the table. You need to have clearly identified your dependent and independent variables. Also, you need to have clearly defined how both variables are measured. It is not a good idea to have more than five categories in either of your variables, so you may have to collapse them to achieve that goal. Make sure that the way you choose to collapse them makes sense. In particular, if you are collapsing an interval-level variable, create ranges that are of equal size. Finally, make sure that each case fits into one and only one category for each of the variables. Once you have two variables with a limited number of categories, take the following steps (summarized in Box 8.2):

1. Place the dependent variable in the rows, the independent variable in the columns. This means that the table will have as many rows as it has categories in

the dependent variable plus three. It will have as many columns as it has categories in the independent variable plus two.

2. Clearly label the categories of both variables: the categories of the dependent variable head each row; the categories of the independent variable, each column.

3. Calculate the column percents for each cell by dividing the cell count by the column count and multiplying by 100.

4. Include marginals for both the rows and the columns—in both percent and number of cases.

5. Give the table a title in the form of the name of the "*dependent variable*" "by" the "*independent variable*."

6. Place lines after the title, after the column headings and at the end using the "Borders and Shadings" command in Word.

7. Give any relevant information in a smaller font below the bottom line.

Once you have a properly formatted contingency table, you can describe its contents. Normally, you will choose a single row (whichever one is most connected to the way you phrased your hypothesis) to analyze. The marginal percent for that row is the baseline against which you compare the other percents in that row. Notice which columns have higher percents than the marginal and which have lower. Contrast those categories by finding the difference between the column percent for the row you chose to analyze.

A Political Example: Partisanship and Income at the Individual Level. We found that poorer states tend to vote Republican. To find out whether poorer people tend to vote Republican, we would need to look at individual-level data. We can answer that question using the 2010 General Social Survey from the National Opinion Research Center.

1. Place the dependent variable in the rows, the independent variable in the columns. This means that the table will have as many rows as it has categories in the dependent variable plus three. It will have as many columns as it has categories in the independent variable plus two.
   Because the income variable has twenty-five categories and the partisanship variable has seven, I collapsed both of them so that I could create a comprehensible contingency table with them. I grouped the income in ranges of $30,000, for a total of four categories. Because this is my independent variable, I will have 4 + 2 = 6 columns in my table. For partisanship, I decided to group the weak and strong partisans as Democrats and Republicans, and I grouped the Independents together regardless of whether they were leaning toward one party or the other. Because this is my dependent variable, these three categories mean that I will have 3 + 3 = 6 rows.

2. Clearly label the categories of both variables with headings for the labels of the variables.

   The dependent variable is Partisanship, so I place that in the rows and label each of the categories for it in the first column. The independent variable is Income, so I place that in the columns and label the top of each column with the categories. Cross tabulating these groupings gave the frequencies shown in Table 8.8.

Table 8.8 Partisanship by Income, Raw Data

Partisanship	Income				
	$0-30,000	$30-60,000	$60-90,000	$90,000+	Total
Republican	106	104	94	140	444
Independent	294	188	102	193	777
Democrat	257	171	101	133	662
Total					
n =	657	463	297	466	1883

Source: Smith, Tom W., Peter Marsden, Michael Hout, and Jibum Kim. *General Social Surveys, 2010* [machine-readable data file]. Principal Investigator, Tom W. Smith; Co-Principal Investigator, Peter V. Marsden; Co-Principal Investigator, Michael Hout; sponsored by National Science Foundation. NORC ed. Chicago: National Opinion Research Center [producer]; Storrs, CT: The Roper Center for Public Opinion Research, University of Connecticut [distributor], 2012.

3. Calculate the column percents for each cell by dividing the cell count by the column count and multiplying by 100.

   Next, I calculate the column percents by dividing the cell counts by the column "n." The correctly percentaged table is shown in Table 8.9. Some of the column total percents are not precisely 100.0 percent, but none is off by more than 0.1 percent so I'm fairly sure it is just rounding error.

4. Include marginals for both the rows and the columns—in both percent and number of cases.

   I already have the number of cases at the bottom of each column, and I also have totaled the column percentages up to 100 percent. I add in the total number of cases in each row, put that in parentheses, and then divide the number of cases in each row by the total number of cases to get the row percent, which I place above the row n in the last column for the row marginals.

5. Give the table a title in the form of the name of the "*dependent variable*" "by" the "*independent variable.*"

   I begin the title with "Partisanship" because that is the dependent variable and end it "by Income" because that is the independent variable.

6. Place lines after the title, after the column headings and at the end using the "Borders and Shadings" command in Word.

   The final product is shown in Table 8.9.

Table 8.9 Partisanship by Income

Partisanship	$0-30,000	$30-60,000	$60-90,000	$90,000+	Total
Republican	16.1%	22.5%	31.6%	30.0%	23.6% (444)
Independent	44.7%	40.6%	34.3%	41.4%	41.3% (777)
Democrat	39.1%	36.9%	34.0%	28.5%	35.2% (662)
Total	99.9%	100.0%	99.9%	99.9%	100.1%
n =	657	463	297	466	1883

Source: Smith, Tom W., Peter Marsden, Michael Hout, and Jibum Kim. *General Social Surveys, 2010* [machine-readable data file]. Principal Investigator, Tom W. Smith; Co-Principal Investigator, Peter V. Marsden; Co-Principal Investigator, Michael Hout; sponsored by National Science Foundation. NORC ed. Chicago: National Opinion Research Center [producer]; Storrs, CT: The Roper Center for Public Opinion Research, University of Connecticut [distributor], 2012.

Note: Some columns fail to total 100.0% due to rounding error.

7. Give any relevant information in a smaller font below the bottom line.
8. Describe the pattern by choosing a row and comparing the percentages across it.

    I originally hypothesized that poor people are more likely to vote Republican than rich people, so I am going to analyze the Republican row. I can see that 23.6 percent of the respondents identify as Republicans. If I compare that baseline to the percentage of respondents in each category of income, I can see that a higher percentage of those respondents who earn over $90,000 think of themselves as Republicans—30.0 percent. In contrast, only 16.1 percent of those who earn under $30,000 think of themselves as Republicans. I can conclude that the richest group of respondents is more likely to think of themselves as Republicans than those in the poorest category by a difference of 13.9 percentage points.

Why is there this contradiction between the state-level data and the individual-level data? Both results are accurate; the difference is in the unit of analysis. All we know at the state level is that poorer states tend to vote Republican. This does not mean that poorer people vote Republican. It just means that something is happening in those states to increase Republican voting. It is a fallacy to take aggregate-level data and draw individual-level conclusions. This fallacy can be called by two different names: the ecological fallacy or an aggregation bias. What we can see here is that although poor states tend to vote Republican, poor people still vote Democratic in the same way they did during the New

Deal realignment. The important message to take from this example is that as you describe the pattern you see in your data, be sure to phrase it in terms of the unit of analysis of your data. Do not draw conclusions about individuals from aggregate-level data.

USE SPSS TO ANSWER A QUESTION USING A CONTINGENCY TABLE

The first step in data analysis is always to describe the pattern that you see when you look at the relationship between two variables. Although you should always describe that pattern in words, it is also useful to present a visual representation of the data. Contingency tables are very useful tools to present data visually as you answer a question. Keep in mind that the point of the table is to present the numbers in such a way as to communicate information at a glance. For contingency tables, this means that you do not want to have too many categories because the human brain can process only so much information. So if either of your variables has more than five categories (especially if it is an interval-level variable), you should consider collapsing it into three to five categories for presentation in the table.

You already have all the skills you need to collapse variables and request a contingency table from SPSS. In Chapter 4, you learned how to collapse a variable using the "Recode into Different Variable" command. You'll want to first get a frequency of any variables you want to collapse so that you can choose equal-sized intervals that reflect the distribution of its variation in a logical way. Once you've chosen your intervals, recode the variable, remembering that for each category, the lower value of the range is included, but the upper end only goes up to, but does not include, the end point. As you are entering the commands into SPSS, be sure to include enough decimal points to make sure that each case falls into exactly one category.

If you use the same command a second time, you'll notice that SPSS leaves your previous entries in each of the boxes. This can be helpful if you make a mistake—you can just correct the errors without starting over completely. But when you want to collapse a second variable, you may want to just start fresh. In that case, you'll see the option "Reset" at the bottom of the initial screen to the right of the "OK" button that you'll click when you are completely finished. If you click "Reset" at the beginning, SPSS will clear everything so you can start fresh.

Once you've recoded your variables, use the "Define Variable Properties" command to assign value labels to your recoded variables. If you are recoding more than one variable, you'll need to watch out for a quirk in SPSS. After adding the value labels for the first variable, you'll click on the name of the second variable in the left-hand column. It will replace all the values of the first with the second. But for some reason, if the first variable has more categories than the second, SPSS will leave the labels for the extra values up on the screen. Unfortunately, if you try to delete them, it will delete them for the first variable even though the second is on the screen. The way to bypass this quirk is to make sure that you do not have those later labels open. So as you are inserting labels, begin at the bottom of the list and work your way up. If the last label you inserted was for the first category, then it will be replaced by the first category of the second variable, and the bottom labels will not show up. After you finish collapsing and labeling your variables, get a frequency to make sure they look right.

Once you have two variables with a limited number of categories, you can get SPSS to give you the information you need to create a contingency table. You want SPSS to give you both the raw count in each cell and its column percent. The cross tabulation command in SPSS allows you to request both the cell count and the column percent. As usual, you will begin the process of analyzing data by opening your dataset in SPSS and making sure that your data are clean. You then get a cross tabulation by clicking on "Analyze," "Descriptive Statistics," and "Crosstabs." Once in the Crosstabs window, the first step is to tell SPSS which variable to put in the columns and which to put in the rows. We will normally set these up so that the independent variable (the cause) is in the columns and the dependent variable (the effect) is in the rows. After specifying the variables, we then want to specify the contents of the cells. Click on the button labeled "Cells." In the "Counts" box, you'll see that "Observed" has already been checked. This will give you the actual number of cases in each cell. Now, under percents, check "column" to get the column percent. Click on "Continue" and "OK" to get SPSS to process your request. Box 8.4 summarizes this process.

BOX 8.4   How to Get a Contingency Table in SPSS

After opening and cleaning your data in SPSS:

>Analyze
>>Descriptive Statistics
>>>Crosstabs
>>>>"*Dependent Variable*"→Row Variable
>>>>"*Independent Variable*"→Column Variable
>>>>>Cells
>>>>>>Observed Count
>>>>>>Column Percent
>>>>>>Continue
>>>>>OK

With your SPSS output in hand, you are prepared to describe the pattern for the relationship between your dependent and independent variables. First, translate the output into a professional-looking table following the steps in Box 8.2. Within the body of the table, include only the column percent. In the marginals, place the cell count below the percentage. To describe the pattern, choose one row and compare the column percentages in it. Normally, you will identify the categories of your independent variable that have the highest and lowest column percentages in that row and calculate the difference between those two. This is how much more likely the first is than the second to fall in this row. Box 8.5 summarizes this process.

BOX 8.5 How to Use Crosstabs in SPSS to Answer a Question

1. Open your data in SPSS and clean them.

2. If you want a subset of cases, use the "Select Cases" command to do so.

3. Get a frequency of any variables you want to collapse.

4. Choose ranges to collapse the variables into three to five categories. The ranges should be equally sized and have logical end points. Remember that each case should be in exactly one category.

5. Use the "Transform" command to recode each variable into a different variable. Make sure that the lower end of each range includes the even number and the upper end is just below it to the correct number of significant digits. If you are going to recode more than one variable, after you finish with the first recode, be sure to "Reset" so that you can start fresh when you recode the second.

6. Use the "Define Variable Properties" command to assign variable and value labels.

7. Request your contingency table with the dependent variable in the rows and the independent variable in the columns. Using the "Cells" button, request the cell frequency and the column percent.

8. Translate the SPSS crosstab into a professionally constructed contingency table.

9. Analyze your results.

An SPSS Application: Life Expectancy and Female Literacy

You are interning at the Central Asia Institute, a nonprofit organization that builds schools in Pakistan and Afghanistan. Founder Greg Mortenson has long argued that the best way to decrease infant mortality is to educate girls. As he prepares to go on another speaking tour, he wonders whether female literacy would also increase the life expectancy of a nation. He asks you to get him the numbers. From Wikipedia, you get the list of developing countries for which to collect the appropriate data. For the female literacy variable, you are able to find in the CIA World Factbook a variable that gives the percent of women over age 15 who are able to read. For the life expectancy variable, GapMinder gives the life expectancy at birth in years.

1. Open your data in SPSS and clean them.
   My dependent variable is LifeExpectancy and my independent variable is FemaleLiteracy. When I look at these two variables with the "Define Variable Properties" command, I notice that both variables need to have "–9" set to missing, so I do that. The values for both variables look fine, although they both have too many categories to be included as is in a contingency table.

2. If you want a subset of cases, use the "Select Cases" command to do so.
   I actually collected the data for all the countries of the world. Because I want to include only the developing countries, I use the "Select Cases" command to select the cases when Development=1.

3. Get a frequency of any variable(s) you want to collapse.

4. Choose ranges to collapse the variables into three to five categories. The ranges should be equally sized and have logical end points. Remember that each case should be in exactly one category.
   FemaleLiteracy ranges from 12.6 to 100 percent. Because most of the cases are in the upper end, I am going to need more categories so that they don't all get bunched up in one category. I choose to collapse this variable into five categories of 20 percent each.
   LifeExpectancy ranges from 47.794 to 79.311 years. I decide to collapse this variable into four groups of ten years each.

5. Use the Transform command to recode each variable into a new variable. Make sure that the lower ends of your ranges include the even number while the upper end is just below it to the correct number of significant digits. If you are going to recode more than one variable, after you finish with the first recode, be sure to "Reset" so that you can start fresh when you recode the second.
   FemaleLiteracy has one decimal point, so I make sure the upper limits of the ranges end with "9.9." In my recoded new variable, I assign the range "Lowest thru 19.9" a value of 0; "20 thru 39.9" becomes 1, "40 thru 59.9" becomes 2, "60 thru 79.9" becomes 3, and "80 through highest" becomes 4. Figure 8.11a shows how to get the last interval. The option for the lowest interval is just above it.
   LifeExpectancy has three decimal points, so I make sure that the upper limits end with "9.999." Before entering the recoded values, I click on "Reset" because I want to clear all of the previous recodes. I recode the following intervals: lowest through 49.999 becomes 0; 50 through 59.999 becomes 1; 60 through 69.999 becomes 2; and 70 through highest becomes 3.

6. Use the "Define Variable Properties" to give your collapsed variable(s) value labels. Start with the biggest value of each and work your way up to the smallest.
   I work my way up cFemaleLiteracy, identifying the appropriate ranges for each value. With the label for the smallest value open (0 = 0-20%), I click on cLifeExpectancy and give its values labels.

7. Request your contingency table. Remember that the dependent variable should be in the rows and the independent variable should be in the columns. You want the cells to include both the cell count and the column percent.
   Figure 8.11b shows the location of the commands for a contingency table, or what SPSS calls a crosstab.
   Because I am hypothesizing that life expectancy is dependent on female literacy, my dependent variable is life expectancy and my independent is female literacy. I want life expectancy in the rows and female literacy in the columns. I then indicate that in the cells, I want both the observed count and the column percent. This is shown in Figure 8.12.

Figure 8.11a Identifying the Interval that Includes the Highest Values

Figure 8.11b Location of Crosstab Command

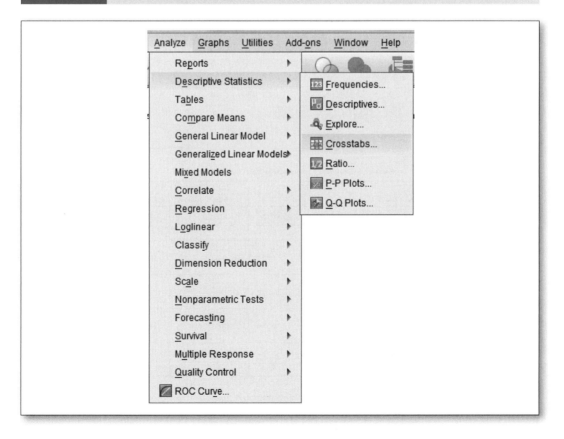

Figure 8.12 Requesting a Properly Percentaged Contingency Table

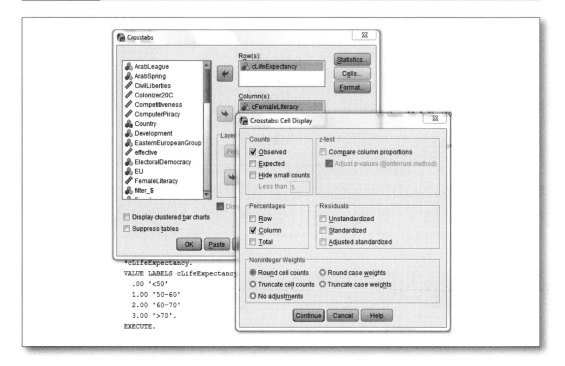

Figure 8.13 SPSS Output for Crosstab

collapsed Life Expectancy * collapsed Female Literacy Crosstabulation								
			\multicolumn{5}{c}{collapsed Female Literacy}		Total
			0-20%	20-40%	40-60%	60-70%	80%+	
collapsed Life Expectancy	<50	Count	7	2	3	1	6	19
		% within collapsed Female Literacy	50.0%	20.0%	13.6%	3.8%	7.3%	12.3%
	50-60	Count	2	6	10	6	5	29
		% within collapsed Female Literacy	14.3%	60.0%	45.5%	23.1%	6.1%	18.8%
	60-70	Count	2	2	8	7	17	36
		% within collapsed Female Literacy	14.3%	20.0%	36.4%	26.9%	20.7%	23.4%
	>70	Count	3	0	1	12	54	70
		% within collapsed Female Literacy	21.4%	0.0%	4.5%	46.2%	65.9%	45.5%
Total		Count	14	10	22	26	82	154
		% within collapsed Female Literacy	100.0%	100.0%	100.0%	100.0%	100.0%	100.0%

8. Translate the SPSS crosstab into a professionally constructed contingency table. The SPSS output is found in Figure 8.13. After typing the title for the table, I insert a table into my document. I look at the number of categories from my two variables to determine how many rows and columns to create. The independent variable has five categories, so my table will have 5 + 2 = 7 columns. The dependent variable has four categories, so I will make 4 + 3 = 7 rows.

In the SPSS table, I can see in the labels at the left that the first number in each cell is the cell count and the second number is the column percent. I translate this into a professional-looking table by including only the column percent in the main cells. I reverse the order for the marginal so that the percent is above the count. Finally, I place vertical lines at the top and bottom as well as after the column headings. This is shown in Table 8.10.

Table 8.10 Life Expectancy by Female Literacy

Life Expectancy	0-20%	20-40%	40-60%	60-80%	80-100%	Total
40-50	50.0%	20.0%	13.6%	3.8%	7.3%	12.3% (19)
50-60	14.3%	60.0%	45.5%	23.1%	6.1%	18.8% (29)
60-70	14.3%	20.0%	36.4%	26.9%	20.7%	23.4% (36)
70-80	21.4%	0%	4.5%	36.2%	65.9%	45.5% (70)
Total	100.0%	100.0%	100.0%	100.0%	100.0%	100.0%
n =	14	10	22	26	82	154

Sources: "Life Expectancy at Birth (Years)," *GapMinder*, 2011. www.gapminder.org/data/; "Literacy," CIA World Factbook, 2012. www.cia.gov/library/publications/the-world-factbook/fields/2103.html.

9. Write a memo.

Memo

To: Greg Mortenson
From: T. Marchant-Shapiro
Subject: Life Expectancy and Female Literacy
Date: August 27, 2012

You asked me to evaluate the impact of female literacy on the life expectancy in developing countries. I was able to find a measure of life expectancy on the GapMinder webpage and a measure of female literacy in the CIA's World Factbook. The data support the conclusion that those countries with higher female literacy rates have higher life expectancies.

The attached table shows the relationship between life expectancy and female literacy. For convenience in visualizing the relationship between the two variables, I've collapsed the two variables into ranges. It is interesting to look at the row with the highest life expectancies (70 to 80 years). You can see in that row that among the countries with 80 to 100 percent female literacy, 65.9 percent have residents that live on average past age 70. In contrast, only 21.4 percent of the countries with the lowest female literacy (under 20 percent) have a life expectancy greater than 70. This suggests that those countries that educate at least 80 percent of their women to read are more likely to have life expectancies of over 70 years than those countries in which less than 20 percent of the women can read, by a difference of 44.5 percentage points.

As you advocate for female education, you can add life expectancy as another of the advantages that you can expect in countries that teach women to read. You might, however, want to be careful about attributing all of the cause of this relationship to education. Because this is aggregate-level data, it is possible that there is some third variable, such as domestic war, that is affecting both education and life expectancy in these countries.

Life Expectancy by Female Literacy

Life Expectancy	0-20%	20-40%	40-60%	60-80%	80-100%	Total
40-50	50.0%	20.0%	13.6%	3.8%	7.3%	12.3% (19)
50-60	14.3%	60.0%	45.5%	23.1%	6.1%	18.8% (29)
60-70	14.3%	20.0%	36.4%	26.9%	20.7%	23.4% (36)
70-80	21.4%	0%	4.5%	36.2%	65.9%	45.5% (70)
Total	100.0%	100.0%	100.0%	100.0%	100.0%	100.0%
n =	14	10	22	26	82	154

Sources: "Life Expectancy at Birth (Years)," *GapMinder*, 2011. www.gapminder.org/data/; "Literacy," CIA World Factbook, 2012. www.cia.gov/library/publications/the-world-factbook/fields/2103.html.

Your Turn: Describing the Pattern

YT 8.1 — Do an Internet search for "Bad Graphs" to find examples of graphs that do not communicate the information honestly. Choose one to print and then analyze it: What makes it a bad graph?

YT 8.2 — One of the goals of the Iraq War was to displace Saddam Hussein in hopes of providing more political freedoms for the citizens of Iraq. Table 8.11 shows the raw scores that Freedom House uses to calculate its political rights index for Iraq from 2004 to 2011. This scale ranges from a not free score of "1" to a most free score of "40."

1. Graph the data.
2. Describe the pattern. Has political freedom increased?

Table 8.11 Political Rights in Iraq, 2003–2011

Year	Political Rights
2004	5
2005	5
2006	9
2007	11
2008	11
2009	11
2010	12
2011	12
2012	12

Source: Freedom House, "Freedom in the World: Aggregate Scores, 2003-2011," 2012. www.freedomhouse.org/report/freedom-world-aggregate-and-subcategory-scores.

YT 8.3

During the ratification of the Nineteenth Amendment giving women the right to vote, one issue debated was whether women would vote differently from men. Once women got the vote, for many years their voting patterns were quite similar to men. But beginning in the Reagan years, women began to be more supportive of the Democratic Party than men. This phenomenon has continued since then and has been dubbed the "gender gap." In the 2008 American National Election Survey, respondents fell into the following categories: 362 Democrat Men, 609 Democratic Women, 198 Republican Men, 232 Republican Women, 353 Independent Men, and 355 Independent Women.

1. Set up a properly formatted contingency table for these data.
2. Analyze the table: Does it appear that the gender gap continued during the 2008 election?

Apply It Yourself: Determine Stability across Legislative Systems

It is August 2012 and you are interning at the United Nations Mission in South Sudan (UNMISS). After years of ethnic conflict between northerners and southerners in Sudan, the residents of the South voted to secede and form their own country. Professor Akolda Man Tier, the chair of the National Constitutional Review Commission, has requested that UNMISS support the new nation in the creation of an original constitution. It will be easier to implement the new government if it is predominantly based on the government of Sudan to which everyone is accustomed. But because new countries always face the risk of instability, if there are certain changes that would increase the likelihood that the country will be stable, they are worth considering.

One of the age-old questions asked in creating governments is whether to adopt a unicameral or bicameral legislature. Sudan has a bicameral legislature, so that is the default plan for South Sudan. Hilde F. Johnson, head of the UNMISS delegation, has charged you with answering the question of whether countries with a unicameral system are more stable than those with bicameral systems. In "World Dataset," the variable "Legislature" is a dichotomous variable where a value of "1" indicates a unicameral system and "2," a bicameral system.[6] "StateFragilityIndex2010" is a measure of instability where stable countries have smaller values than unstable countries.[7]

1. Open your data in SPSS and clean them using "Define Variable Properties."
2. Get a frequency of "StateFragilityIndex2010" because you want to collapse it.
3. Choose ranges to collapse "StateFragilityIndex2010" into three to five categories. The ranges should be equally sized and have logical end points. Remember that each case should be in exactly one category.
4. Use the Transform command to recode "StateFragilityIndex2010" into a new variable named "Fragility." Make sure that the lower end of each range includes the even number and the upper end is just below it to the correct number of significant digits.
5. Use the "Define Variable Properties" command to give your collapsed variable labels for each category.
6. Use the "Analyze," "Descriptive Statistics," "Crosstabs" command to get the data for your contingency table. Remember that the dependent variable (fragility) should be in the rows and the independent variable (legislature) should be in the columns. Be sure to go into the "Cells" window to request both the cell count and the column percentage.

   >Analyze
   >Descriptive Statistics
   >Crosstabs
   "Fragility"→Row Variable
   "Legislature"→Column Variable
   >Cells
   >Observed Count
   >Column Percent
   >Continue
   >OK

7. Translate the SPSS crosstab into a professionally constructed contingency table.
8. Write a memo. Please attach the SPSS output of the crosstab for grading purposes.

Key Terms

Aggregation bias (p. 234)

Ecological fallacy (p. 234)

Figure (p. 212)

Table (p. 213)

CHAPTER 9

Chi-Square and Cramer's V

What Do You Expect?

The opening of the Olympic Games usually coincides with the quadrennial reopening of predictions of how many medals each country will win. Prognosticators can use many variables to predict how well each country will do. Frequently, they will include population size in their model because, presumably, the more people a country has, the more likely it is to have a large number of gifted athletes. Forecasters will also take the wealth of a nation into account because training even gifted athletes can be expensive. But although the human and financial resources may form a baseline of how well any given country will do, these augurs have noticed that there is always one country that does better than expected: the host country. By comparing their expectations before the games with what they observe during the games, analysts are able to identify anomalies. As they wonder why certain countries do better than expected, they are able to develop a more complete understanding of what determines success because their model includes an allowance for a home team advantage.[1]

As political scientists, we are also always on the lookout for events that defy our expectations. For example, up until the 1980s, exit polls had indicated that women normally voted the same way as men. So, going into the 1980 election, we expected that roughly the same proportion of women would vote for Ronald Reagan as men. However, the exit polls in 1980 indicated that 7 percentage points more men than women voted for Reagan that year.[2] This difference between what we expected (on the assumption that gender doesn't affect voting patterns) and our observations in 1980 led us to posit the theory that a gender gap had emerged that year in voting. Subsequent elections supported that conclusion, leading to a growing literature that asks the question, Why are women more likely to vote for Democrats than men?

This comparison, of what we expect to see if there is no relationship between our two variables with what we actually observe, underlies the statistical measure of chi-square—given by the lower-case Greek symbol χ^2 and pronounced kī square. **Chi-square** is a measure of statistical significance. Closely related to chi-square is **Cramer's V**, which measures substantive significance. These measures are appropriate when both variables are categorical—either nominal or ordinal—because neither assumes anything about the direction of a relationship.

They simply compare the observed number of cases in each cell of a contingency table with what you would expect to observe if the two variables were independent. We'll begin this chapter with a section on the probability of discrete events in order to identify the pattern we expect to see if two variables are independent. With that pattern, we can then learn how to calculate and interpret both chi-square and Cramer's V in order to reinforce the distinction between statistical and substantive significance.

THE PROBABILITY OF DISCRETE EVENTS

There are two basic principles of probability. First, probabilities always range between zero and one. A probability of zero means that there is no chance that an outcome could occur. A probability of one means that the outcome is certain to occur. Usually, a probability is somewhere between the minimum of zero and the maximum of one. Second, if you add up the probabilities of all the possible outcomes of the event, the total has to equal one. These two rules apply to two general types of probabilities: continuous and discrete. In Chapter 5, we addressed continuous probabilities with z-scores. In the variables addressed there, we looked at interval- and ratio-level measures for which the range of possible outcomes corresponded with a continuum of probabilities distributed in a normal curve. In this chapter, though, we are dealing with discrete probabilities where there are specific probabilities connected to each of a limited number of possible outcomes.

A simple example of a discrete event is flipping a coin. If you flip a coin, each of the two sides has a specific probability of landing on top. Usually a coin toss is a random event—heads and tails have equal likelihood of being tossed. Because the total likelihood of all possible alternatives is always one, the two possible outcomes of flipping a coin mean that each alternative has a likelihood of half of one (½ or 0.5). The probability of an outcome is written as P(X). For example, the probability of tossing a head (with a fair coin) is given as P(heads) = 0.5. Although probabilities are normally written as a decimal between 0.0 and 1.0, we sometimes multiply that decimal by one hundred to give the probability in terms of the percent chance of an outcome. In this case, we could say that there is a 50 percent chance of getting heads. Each time you flip the coin, the probability remains the same.

When the possible outcomes of an event are equally likely, as in the case of flipping a coin, determining the probability of each possibility is a simple matter of dividing one by the number of possible outcomes. We've already shown that the probability of tossing a particular side of a coin is ½. Similarly, to get the probability of rolling a particular side of a die, you would divide one by the number of sides of the die. If it is a standard six-sided die, you would divide one by six to get a probability of 1/6. If you had a fancy Dungeons and Dragons die—say, an eighteen-sided die—the probability would be 1/18.

But sometimes the possible outcomes are not equally probable. In the movie *The Terminal*, the Tom Hanks character is stuck in a New York airport terminal because his visa has been revoked due to a coup in his home country while he was in the air. Every day, he fills out the appropriate form to see if he can enter the United States, and every day, the immigration agent stamps the form "Denied." In one of my favorite scenes from the movie, the immigration agent finally asks him why he keeps filling out the form. He replies that she has two stamps, "Denied" and "Approved," so he has a fifty-fifty chance of being approved.[3] Unfortunately for the character, because the outcome is rule-based rather than random, the probability of being denied was closer to 1.0 than to 0.5. Similarly, another Hanks character from a different movie says, "My momma always said, 'Life was like a box of chocolates. You never know what you're gonna get.'"[4] That may be true if you have a box of creams, but if it were my box of chocolates, I would know with certainty that I was going to get a pecan turtle.

In the same way, political outcomes frequently do not always have equal probabilities. The U.S. Army Corps of Engineers, charged with maintaining the levees in New Orleans, knew that there was a very small but real probability of a Category 5 hurricane hitting New Orleans. They described that possibility as a one-in-five-hundred-year event.[5] That translates into a yearly probability of 1/500 = 0.002. Because probabilities of all the possible outcomes have to total one, that means that the probability that a Category 5 hurricane will not occur is 1.000 − 0.002 = 0.998.

Probabilities of Independent Events

Calculating probability gets more difficult when you are calculating the probability of a combination of two events. Suppose, for example, you are going flip a coin twice. We call probabilities **contingent** when the first outcome affects the probability of the second outcome. But with a fair coin, the outcome of the first toss doesn't change the probability of the second toss. We call probabilities **independent** when the two events are unrelated. Table 9.1 shows a cross tabulation for two tosses of a fair coin. For each individual toss, the probability of heads and tails remains 0.5 for each. But the total probability of the four possible outcomes can still total only 1.0, which means that the probability of each possible combination of two coin tosses will be lower than 0.5.

Table 9.1 Cross Tabulation of Tossing a Coin Twice

Second Toss	First Toss		Probability
	Heads1	Tails1	
Heads$_2$	P(Heads, Heads) = (0.5)(0.5) = 0.25	(Tails, Heads)	0.5
Tails$_2$	(Heads, Tails)	(Tails, Tails)	0.5
Probability	0.5	0.5	

When you have two independent events like two coin tosses, the probability of both outcomes occurring simultaneously is the product of the probability of the first and the probability of the second. So the probability of getting heads twice in a row is

$$P(Heads, Heads) = P(Heads_1) \times P(Heads_2)$$
$$= 0.5 \times 0.5$$
$$= 0.25$$

If you were to put that equation in general terms, you would say:

$$P(A \cap B) = P(A) \times P(B)$$

When you see the symbol "∩," you read the word "intersect." So in English, this math sentence would read, The probability of A intersect B equals the probability of A times the probability of B. The notion of intersect is the same as we have been discussing of both events occurring together. If we are thinking about the first *and* the second occurring simultaneously, we use the intersect rule by multiplying the probabilities of the two outcomes. If we are looking at a cross tabulation where the two variables are independent of each other, the probability of a case being found in one of the cells is given by multiplying the probability of being in the row by the probability of being in the column.

Statisticians frequently talk about games of chance because they are good examples of independent events where the outcome of one event does not change the probability of the outcomes of the other. Sometimes, in the political sphere, that assumption is also correct. For example, Table 9.2 shows a cross tabulation of region of residence and gender. If these two variables are independent of each other, we would expect to see half men and half women in every region of the U.S. Eyeballing the raw data, it appears that within each region and in the U.S. overall, there are roughly equal numbers of men and women.

But to find the probability that any given person belongs in any one cell of the table, you would focus on the marginals, multiplying the proportion in the row by the proportion in the column. These are shown in Table 9.3. If there is no relationship between gender and region, we would expect that the probability of being a woman in the Northeast would be the probability of being from the Northeast (0.181) multiplied by the probability of being a woman in the United States (0.507). So, assuming that gender and region are independent of each other, the probability of being a Northeastern woman is (0.181)(0.507) = 0.092.

Table 9.2 Population of Regions by Gender[6]

Region	Men	Women	Total
Northeast	26,740,485	28,184,294	54,924,779
Midwest	32,797,335	33,764,113	66,561,448
South	54,885,816	56,832,733	111,718,549
West	35,500,968	35,353,980	70,854,948
N =	149,924,604	154,135,120	304,059,724

Source: Howden, Lindsay M., and Julie A. Meyer, "Age and Sex Composition: 2010," U.S. Census Bureau, 2011. www.census.gov/prod/cen2010/briefs/c2010br-03.pdf.

Table 9.3 Proportional Population of Regions by Gender

Region	Men	Women	Total
Northeast		(0.181)(0.507) = 0.092	0.181 (54,924,779)
Midwest			0.219 (66,561,448)
South			0.367 (111,718,549)
West			0.233 (70,854,948)
P(X) (N)	0.493 (149,924,604)	0.507 (154,135,120)	1.000 (304,059,724)

Source: Howden, Lindsay M., and Julie A. Meyer, "Age and Sex Composition: 2010," U.S. Census Bureau, 2011. www.census.gov/prod/cen2010/briefs/c2010br-03.pdf.

Probabilities of Contingent Events

In the prior section, each of the events had outcomes that were independent of each other. Although one coin toss will not normally affect the outcome of a second toss, in political events, the first outcome will frequently affect the second. For example, the occurrence of a Category 5 hurricane affects the probability that the levees will be topped in New Orleans. Although in general, the probability of the levees being topped is only 1/300,[7] in the case of a Category 5 hurricane, the probability increases to nearly 1.0.

As social scientists, this understanding of independent probabilities forms the basis of our discussion of contingent probabilities. At the heart of political science is our desire to answer "Why?" questions. We assume that effects have causes, and we want to be able to identify them. The way that we do that is by showing that there is a difference between what we see and what we would expect to see if our effect were independent of our cause. If the probabilities do not appear to be independent of each other, we can begin to get answers to our "Why?" questions. Our effects appear to be contingent on our causes.

For example, the revolts that occurred in Tunisia, Egypt, and Libya in 2011 raised a question about why some countries are more likely to engage in internal domestic war than others. We might posit that the level of political competition affects the likelihood of internal conflict. On one extreme, among democracies, we would expect that the ability to participate in political dialogue should minimize the probability of civil war. At the other extreme, among repressed regimes, the control of the state should keep the people from being able to organize a revolt. The government need only confine the opposition because it is difficult to organize opposition from prison. In the middle, we would expect citizens to have more grievances than in democratic countries, but be more capable of revolting than in repressed countries.

Table 9.4 Domestic War by Political Competition, Marginals

Involved in Domestic War?	Repressed	Political Competition				Total
		Suppressed	Factional	Transitional	Competitive	
No						84.2% (128)
Yes						15.8% (24)
Percent of Total	9.2%	10.5%	23.0%	33.6%	23.7%	100.0%
N =	14	16	35	51	36	152

Sources: Polity Project IV, "PARCOMP," *Political Regime Characteristics and Transitions, 1800-2010*, Integrated Network for Societal Conflict Research, 2010. www.systemicpeace.org/inscr/inscr.htm; Gleditsch, Nils Petter; Peter Wallensteen, Mikael Eriksson, Margareta Sollenberg and Håvard Strand "Armed Conflict 1946–2001: A New Dataset," *Journal of Peace Research* 39 (2002): 615–637. www.pcr.uu.se/research/ucdp/datasets/ucdp_prio_armed_conflict_dataset/.

Table 9.4 addresses this question by returning to the cross tabulation found in Chapter 8 between the level of political competition and participation in an internal war. This table shows only the marginals so that we can determine the probabilities for each cell assuming that the two variables are independent of each other. In this case, we are basing the probability of being in any one category on how many countries have that attribute. Normally, we standardize the frequency by giving it as a percentage. To change that into the probability that any given country will have that attribute, we divide the percent of countries with the attribute by 100 to get the proportion. Looking at the marginals, you can see that 15.8 percent of the countries of the world were involved in a solely domestic war in the years between 2000 and 2010. Dividing that by 100, we know that there is a probability of 0.158 that any given country was involved in a domestic war in that period of time. Similarly, the probability of having a factional political order is 23.0/100 = .230.

Assuming the two variables are independent, we would expect that the probability of any country being in a particular cell is given by the equation of the intersection of the two outcomes:

$$P(A \cap B) = P(A) \times P(B)$$

If the two variables are independent we would expect that the probability of any given country of the world having a factional political system *and* being involved in an internal war is:

$$P(\text{domestic war} \cap \text{factional}) = P(\text{domestic war}) \times P(\text{factional})$$
$$= 0.158 \times 0.230$$
$$= 0.036$$

If the two variables are independent, for each of the cells we can calculate the probability of any given country being in it by multiplying the marginal proportion for the row by the marginal proportion for the column. These probabilities are given in Table 9.5.

Table 9.5 Domestic War by Political Competition, Cell Probabilities Assuming Independence

Involved in Domestic War?		Political Competition				
	Repressed	Suppressed	Factional	Transitional	Competitive	Total
No	(0.842)(0.092) = 0.077	0.088	0.194	0.283	0.200	0.842
Yes	0.015	0.017	0.036	0.053	0.037	0.158
Total	0.092	0.105	0.230	0.336	0.237	1.000
n =	14	16	35	51	36	152

Sources: Polity Project IV, "PARCOMP," *Political Regime Characteristics and Transitions, 1800–2010*, Integrated Network for Societal Conflict Research, 2010. www.systemicpeace.org/inscr/inscr.htm; Gleditsch, Nils Petter; Peter Wallensteen, Mikael Eriksson, Margareta Sollenberg and Håvard Strand "Armed Conflict 1946–2001: A New Dataset," *Journal of Peace Research* 39 (2002): 615–637. www.pcr.uu.se/research/ucdp/datasets/ucdp_prio_armed_conflict_dataset/.

We can also calculate the number of countries we expect to be in each cell by multiplying the probabilities of Table 9.5 by the total number of cases. For example, by multiplying the probability of a factional country being involved in a domestic war (0.036) by the total number of countries (152), we can calculate how many countries we expect to be factional and involved in domestic war, assuming that the two variables are independent of each other. The expected value is given as $E(X) = P(X)n = (0.036)(152) = 5.5$. In a similar way, we can calculate the expected number of cases in each cell, assuming that war is independent of political competition. These expected values are given in Table 9.6.

We began this example with the thought that countries with a moderate amount of political competition might be more likely to engage in domestic war than either repressed or competitive countries. If the level of political competition affects the probability of domestic war, then the actual frequency of each cell will be different from what we would expect if the two variables are independent. Table 9.7 shows the actual frequency for each cell. There actually are nine factional countries involved in some kind of domestic war—more than the 5.5 we expected if the two variables were independent. This suggests that there is actually a contingent, not independent, relationship between war and competition.

Table 9.6 Expected Frequencies of Domestic War by Political Competition

Involved in Domestic War?		Political Competition				Total
	Repressed	Suppressed	Factional	Transitional	Competitive	
No	(0.077)152 = 11.8	13.5	29.5	42.9	30.3	0.842
Yes	2.2	2.5	5.5	8.1	5.7	0.158
Total	0.092	0.105	0.230	0.336	0.237	1.000
n =	14	16	35	51	36	152

Sources: Polity Project IV, "PARCOMP," *Political Regime Characteristics and Transitions, 1800-2010*, Integrated Network for Societal Conflict Research, 2010. www.systemicpeace.org/inscr/inscr.htm. Gleditsch, Nils Petter; Peter Wallensteen, Mikael Eriksson, Margareta Sollenberg and Håvard Strand "Armed Conflict 1946–2001: A New Dataset," *Journal of Peace Research* 39 (2002): 615–637. www.pcr.uu.se/research/ucdp/datasets/ucdp_prio_armed_conflict_dataset/.

Table 9.7 Domestic War by Political Competition, Actual Frequencies

Involved in Domestic War?		Political Competition				Total
	Repressed	Suppressed	Factional	Transitional	Competitive	
No	13	14	26	40	35	128 (84.2%)
Yes	1	2	9	11	1	24 (15.8%)
n =	14	16	35	51	36	
Percent of Total	(9.2%)	(10.5%)	(23.0%)	(33.6%)	(23.7%)	

Sources: Polity Project IV, "PARCOMP," *Political Regime Characteristics and Transitions, 1800-2010*, Integrated Network for Societal Conflict Research, 2010. www.systemicpeace.org/inscr/inscr.htm. Gleditsch, Nils Petter; Peter Wallensteen, Mikael Eriksson, Margareta Sollenberg and Håvard Strand "Armed Conflict 1946–2001: A New Dataset," *Journal of Peace Research* 39 (2002): 615–637. www.pcr.uu.se/research/ucdp/datasets/ucdp_prio_armed_conflict_dataset/.

CHI-SQUARE

Although it appears that the actual distribution of domestic war in relationship to a country's level of political competition is different from what we would expect if the two variables were independent, we need to compare the differences between Tables 9.6 and 9.7 in a more systematic way before rejecting the null hypothesis. Even if two events are independent, there will be random fluctuations in the joint distribution. The extent of these fluctuations will vary depending on the size of the table and the number of cases contained in it. The fluctuations follow a pattern called a chi-square distribution. By comparing the chi-square value for a particular table to the general chi-square distribution, we can determine the statistical significance for the pattern contained in the table. If the null hypothesis is correct, we should observe a distribution within the table that reflects the distribution of two independent events as determined by the marginal proportions. If our observed frequencies are very different from the expected values, then we will conclude that the two events are contingent: We will reject the null hypothesis.

BOX 9.1 Numbers in the News

Because the strength of the military is dependent on the health of its soldiers, the U.S. military collects extensive data to measure that health. One variable that it tracks is the suicide rate. After a two-year decline, suicides among military personnel spiked in 2012, surpassing the number killed in action. The military responded by stepping up programs to help soldiers deal with potential causes such as stress. But although the military is committed to minimizing the occurrence of suicide, it could well be that it is not responsible for the 2012 spike. Actually, if you compare the number of suicides among service members to the number you would expect to see if there was no difference between military personnel and the civilian public, you find that the observed number of military suicides is lower than you would expect randomly. This difference is supported by the fact that many of the soldiers committing suicide were never deployed.[8] Perhaps factors on the home front are culpable for the 2012 spike.

What We Expect to See

In the first part of this chapter, we learned that if two events are independent, the probability that both will occur equals the probability of the first times the probability of the second. But the expected value is the number of cases we expect to see in the cell, not its probability. To find how many times we expect two events to occur jointly, we simply multiply the joint probability by the total number of cases. Table 9.4 showed the relationship between war and competition and included the marginals needed to calculate the *expected value* for each cell—which is the number of cases we would expect in each cell if the two variables are independent.

There are two ways to calculate the expected value of each cell. Using the equation from the first section of this chapter for the probability of the intersection of two events, we would multiply the row proportion by the column proportion to get the probability of being in a given cell. This is the same as the proportion of cases that should be in the cell if the two variables are independent. If we multiply that by the total number of cases, we get the expected value of the number of cases, or E(X).

$$E(X) = P(\text{domestic war} \cap \text{repressed}) \times n$$

$$E(X) = (\text{row proportion}) \times (\text{column proportion}) \times n$$

So assuming war is independent of competition, we calculate the expected number of countries that are repressed and involved in a domestic war in the following way:

$$E(X) = (0.158) \times (0.092) \times 152$$

$$= 2.2$$

This is how we found the values in Table 9.6.

The second way to calculate the expected value is to think about how we analyzed contingency tables in Chapter 8. When we looked at the column marginals, we knew that if the null hypothesis was correct, the probability of being in any row would stay constant across the columns. In that chapter, we compared the row marginal with the column percent for each category of the independent variable. If the null hypothesis is correct, we would expect that the proportion of cases in that row would be the same regardless of the column at which you look. For this table, if political competition does not affect the probability of domestic war, we would expect to see 15.8 percent of the cases involved in domestic war regardless of the level of political competition. To find the number of cases we would expect in any cell, we would simply multiply the marginal proportion in the given row by the number of cases in the column. With that approach, to get the expected value of a country being both repressed and involved in a domestic war, you would use the equation:

$$E(X) = (\text{row proportion}) \times (\text{column n})$$

$$= (0.158) \times 14$$

$$= 2.2$$

Table 9.8 shows the expected frequencies for each of the cells if we calculate it this way. It is comforting to see that the expected frequencies in Table 9.8 are identical to those found in Table 9.6. This is not a surprising finding because this equation is mathematically identical to the previous one: The column proportion times the total n is the same as the number of cases in the column. In what follows, we will calculate the expected frequency in this way because it is easier to multiply two numbers than three and so we are less likely to make a mistake.

In this example, I've created a second contingency table to calculate the expected value. But in calculating chi-square, we are going to need to use the expected values in a series of other mathematical operations. Theoretically, I could create a new table for each subsequent mathematical step. But that would take up a lot of space and be time consuming. Instead, it will be easier to set up a work table similar to the ones we've used previously. Once I describe what those subsequent mathematical operations will be, I'll show you how to set up the table.

Comparing What We Expect with What We Actually Observe

After calculating the expected value for each of the cells in the table, the next task is to compare that expectation with what actually is found in each cell. The observed frequencies are in Table 9.7. When we eyeballed Table 9.6 and Table 9.7, we saw that factional countries are

Table 9.8 Domestic War by Political Competition, Expected Values

Involved in Domestic War?	Political Competition					Total
	Repressed	Suppressed	Factional	Transitional	Competitive	
No	11.8	13.5	29.5	42.9	30.3	128 (84.2%)
Yes	(.158)14 = 2.2	2.5	5.5	8.1	5.7	24 (15.8%)
n =	14	16	35	51	36	152
Percent of Total	(9.2%)	(10.5%)	(23.0%)	(33.6%)	(23.7%)	(100.0%)

Sources: Polity Project IV, "PARCOMP," *Political Regime Characteristics and Transitions, 1800-2010*, Integrated Network for Societal Conflict Research, 2010. www.systemicpeace.org/inscr/inscr.htm. Gleditsch, Nils Petter; Peter Wallensteen, Mikael Eriksson, Margareta Sollenberg and Håvard Strand "Armed Conflict 1946–2001: A New Dataset," *Journal of Peace Research* 39 (2002): 615–637. www.pcr.uu.se/research/ucdp/datasets/ucdp_prio_armed_conflict_dataset/.

more bellicose than we would expect randomly. Conversely, there are a lot fewer politically competitive countries involved in domestic war than we would expect. It does appear that political competition affects involvement in war.

But we want a measure of how different the two tables are. To compare the observed with the expected, we subtract the expected frequency, E(X), from the observed frequency, f_x. If the two variables are contingent, some of the observations will be higher than expected and some will be lower. As a result, the difference between the expected and observed frequencies in some cells will be positive and others will be negative. We are more concerned with how different those two values are than with whether they are positive or negative. So we square the difference to make all of the differences positive.

Of course, with more cases, that squared difference will necessarily be more. A difference of two cases with a small sample should mean more than a difference of two with a large sample. To weight the difference to take into account the sample size, we take the squared difference and divide by the expected number of cases in the cell. If we total each of those values for all of the cells, we get what is called a chi-square value. Similar to z-scores and t-scores, the chi-square has a probability distribution. Most statistics textbooks will have a chi-square table in the appendix next to the z table and t table. Find it at the end of the book now on page 447. It looks like the t table because it gives a one-tailed probability connected to the degrees of freedom for the table. Unlike the t table, the degrees of freedom (d.f.) are dependent on the dimensions of the table rather than the number of cases. To find the degrees of freedom, you multiply one less than the number of columns by one less than the number of rows:

$$d.f. = (c - 1) \times (r - 1)$$

Table 9.9 Identifying the Rows for the Work Table for Calculating the Chi-Square

Cells
No, Repressed
No, Suppressed
No, Factional
No, Transitional
No, Competitive
Domestic War, Repressed
Domestic War, Suppressed
Domestic War, Factional
Domestic War, Transitional
Domestic War, Competitive

where c is the number of columns and r is the number of rows. Like a t table, in the chi-square table, the cells contain the minimum chi-square value you would need to be statistically significant at the level given at the top of the column.

Completing the foregoing math is easiest if you don't try to do it within the confines of the original table. Instead, you should create a work table where each of the original cells becomes its own row. You need to be very careful to make sure that each cell is in precisely one row. Begin with the first row and work your way across until you have included each of the categories of the independent variable with that first category of the dependent variable. Do the same thing for each of the rows. I do this for the possible combinations of our example in Table 9.9. When you are finished, double-check by counting all the cells in your contingency table. Then count the number of rows you've created in your work table to make sure you have the same number. In this case, I have ten cells in Table 9.7 and ten rows in Table 9.9, so it looks like I did it correctly.

In Table 9.10, I have added in the rest of the columns I need to calculate chi-square for a total of eight columns. After creating a row for each of the cells, enter the observed frequency for each cell in the next column. In the third column, copy the row proportion from the marginals. In the fourth, copy the total number of cases in the column. Notice the patterns that emerge in the third and fourth columns. In the third column, each row proportion repeats itself the same number of times as you have categories for the independent variable. In the fourth column, you keep running through the series of column ns the same number of times as you have categories for the dependent variable. As long as that pattern emerges, you can be sure that you've set up the table correctly.

Using the data from the contingency table that you've entered into the first four columns, you are now in a position to complete the computations for the remaining cells in the table. In the fifth column, you compute the expected value for each cell by multiplying the row proportion (column 3) by the number of cases in the column (column 4). In the sixth, you then subtract the expected value (column 5) from the observed value (column 2). To make these values positive, you square that difference (column 6) in the seventh column. This should feel familiar from calculating standard deviations. But the eighth column has a twist. In it, you divide the squared difference (column 7) by the expected value (column 5).

Once you have computed all the cell values, you are finally in a position to compute the chi-square value to determine whether the relationship is statistically significant. You add up the values of the last column to compute the chi-square value. In this case, chi-square equals 9.370. Looking at the chi-square table in Appendix 5, you notice at the top that it is

Table 9.10 Calculating the Chi-Square for Domestic War by Political Competition

Cells	f_x	Row Proportion	Column n	E(X)	$f_x - E(X)$	$[f_x - E(X)]^2$	$\dfrac{[f_x - E(X)]^2}{E(X)}$
No, Rep.	13	0.842	14	11.788	1.212	1.469	0.125
No, Sup.	14	0.842	16	13.472	0.528	0.279	0.021
No, Fact.	26	0.842	35	29.470	−3.470	12.041	0.409
No, Trans.	40	0.842	51	42.942	−2.942	8.655	0.202
No, Comp.	35	0.842	36	30.312	4.688	21.977	0.725
Dom., Rep.	1	0.158	14	2.212	−1.212	1.469	0.664
Dom., Sup.	2	0.158	16	2.528	−0.528	0.279	0.110
Dom., Fact.	9	0.158	35	5.530	3.470	12.041	2.177
Dom., Trans.	11	0.158	51	8.058	2.942	8.655	1.074
Dom., Comp.	1	0.158	36	5.688	−4.688	21.977	3.864

$X^2 = \Sigma = 9.370$

d.f. = (c − 1)(r − 1)
= (4)(1) = 4

Prob. < 0.10

a one-tailed distribution with the tail shaded. From that, you know that the probability contained in the table is the likelihood of observing these values if the two variables are actually independent of each other. That means that the probability given in the table is equivalent to the level of statistical significance.

You should notice that in order to find the probability of getting this value, you first need to find the degrees of freedom. This is one less than the number of columns times one less than the number of rows. In this case, we have five columns and two rows, so d.f. = (5 − 1)(2 − 1) = 4 × 1 = 4. Now, in the chi-square table, go down to the row for four degrees of freedom. In that row, the first column is headed 0.10 and the cell contains the number 7.78. So if the chi-square value for a table with four degrees of freedom is greater than 7.78, the relationship in it is statistically significant at the 0.10 level. In the next column, you see that a chi-square of 9.49 or greater would be statistically significant at the 0.05 level. Unfortunately, we have a chi-square of 9.370. We conclude that this relationship is not statistically significant at the 0.05 level and so we cannot reject the null hypothesis. The process of calculating chi-square is summarized in Box 9.2.

BOX 9.2 How to Calculate Chi-Square

1. Set up a work table with eight columns where each cell from the contingency table gets its own row. Identify each cell in the first column.

2. Copy the observed cell frequencies into the second column of your work table. Copy the marginal row proportion into the third column and the column n into the fourth.

3. In the fifth column, calculate the expected value by multiplying the row proportion (column 3) by the column n (column 4): E(X) = (row proportion) × (column n).

4. In the sixth column, subtract the expected value (column 5) from the observed frequency (column 2): $f_x - E(X)$.

5. In the seventh column, square the difference between the observed and expected values (column 6): $[f_x - E(X)]^2$.

6. In the last column, divide the squared difference (column 7) by the expected value (column 5): $[f_x - E(X)]^2 / E(X)$.

7. Calculate the chi-square by adding up the values in the final column: $\chi^2 = \Sigma[f_x - E(X)]^2 / E(X)$.

8. Find the statistical significance by looking this up on a chi-square table. Find the row for the correct degrees of freedom: d.f. = (c − 1)(r − 1). If your chi-square is larger than the value in the table, the relationship between the two variables is statistically significant and you can reject the null hypothesis.

CRAMER'S V

Chi-square calculates the probability of observing a relationship between two variables as a result of random error—if, in reality, there is no relationship. The next question we always ask is how strong the relationship appears to be. There are many measures of association, but, in this chapter we will look at Cramer's V, which is a measure of association related to chi-square. The limitation of chi-square is that it is strongly related to the number of cases and the size of the table. If you have a large enough sample size, the sum of the difference squared is almost inevitably going to get big. So it is really important to factor that in. But also the more cells a table contains, the more likely you are to find patterns simply due to chance. So in addition to compensating for sample size, Cramer's V also compensates for the size of the table. To compensate for sample size and table size, Cramer's V divides chi-square by the sample size multiplied by whichever is smallest—the number of rows minus one or the number of columns minus one. Finally, it takes the square root to find a value that approximates the average difference between the expected and the observed.

$$\text{Cramer's V} = \sqrt{\chi^2 / mn}$$

where χ^2 = the chi-square value for the table, m is the lesser of (c − 1) and (r − 1), and n is the total number of cases in the table. This process is summarized in Box 9.3. To compute Cramer's V for the relationship between war and competition, we would calculate

$$\text{Cramer's V} = \sqrt{9.370/(1 \times 152)}$$

$$= \sqrt{0.0616}$$

$$= 0.248$$

BOX 9.3   How to Calculate Cramer's V

1. Find the chi-square value for the table.

2. Identify the size of the table in terms of the number of rows (r) and columns (c). Find m, which is the lesser of (c − 1) and (r − 1).

3. Identify the sample size, n.

4. Calculate Cramer's V by substituting those values: $V = \sqrt{\chi^2/mn}$.

You'll recall that in Chapter 7, we gave a heuristic for interpreting measures of association. Table 9.11 repeats the general guidelines from Table 7.3. The table indicates how ranges of values can be translated into various levels of relationship strengths. A measure of just under 0.30 translates into a "moderate" relationship. So although we are not 95 percent confident that there is a relationship, if there were a relationship between the level of competition of a country and involvement in a domestic war, our best guess, based on a Cramer's V of 0.248, is that it would be moderate.

Table 9.11   How to Interpret Measures of Association

Measure of Association (X)	Qualitative Interpretation
0 ≤ X < 0.10	Very Weak
0.10 ≤ X < 0.20	Weak
0.20 ≤ X < 0.30	Moderate
X ≥ 0.30	Strong

SUMMARIZING THE MATH: CHI-SQUARE AND CRAMER'S V

With chi-square and Cramer's V, we compare the observed frequencies in the cells of a contingency table with what we would expect to see if the two variables are independent. Chi-square is a measure of statistical significance. It answers the question, Is there a relationship between our dependent variable and our independent variable? Cramer's V is a measure of substantive significance. It answers the question, How strong does the relationship appear to be?

Chi-Square

Chi-square is a measure of statistical significance. To calculate chi-square, set up a work table where each cell of the contingency table has its own row and a total of eight columns.

1. Set up a work table where each cell from the contingency table gets its own row. Begin with the upper left-hand cell and work your way across that row. Then do the same for each subsequent row. Count the number of cells in the original contingency table to make sure you have the same number of rows in your work table. Keep in mind that in the end, a chi-square work table will have eight columns.

2. From the contingency table, copy the observed cell frequencies (f_x) into the second column of your work table. Copy the marginal row proportion into the third column and the column n into the fourth.

3. In the fifth column, calculate the expected value by multiplying the row proportion (column 3) by the column n (column 4): E(X) = (row proportion) × (column n).

4. In the sixth column, subtract the expected value (from column 5) from the observed frequency (found in column 2): f_x − E(X).

5. In the seventh column, square the difference between the observed and expected values (in column 6): $[f_x - E(X)]^2$.

6. In the last column, divide the squared difference (from column 7) by the expected value (from column 5): $[f_x - E(X)]^2/E(X)$.

7. Calculate the chi-square by adding up the values in the final column:
$\chi^2 = \Sigma[f_x - E(X)]^2/E(X)$.

8. Find the statistical significance by looking this up on a chi-square table. Find the row for the correct degrees of freedom: one less than the number of columns multiplied by one less than the number of rows: d.f. = (c − 1)(r − 1). If your chi-square is larger than the value in the table, the relationship between the two variables is statistically significant and you can reject the null hypothesis.

Cramer's V

Cramer's V is a measure of substantive significance. Follow these steps to determine the measure.

1. Find the chi-square value for the table.
2. Identify the size of the table in terms of the number of rows (r) and columns (c). Find m, which is the lesser of (c − 1) and (r − 1).
3. Identify the sample size, n.
4. Cramer's V takes the chi-square and weights it by the number of cases and the size of the table. Calculate Cramer's V by substituting each of these values in the equation $V = \sqrt{\chi^2 / mn}$.

A Political Example: Race and Ideology. Because we are aware that blacks usually vote for the Democratic Party, we sometimes assume that blacks must be much more liberal than whites. Let's investigate that assumption. Table 9.12 shows the relationship between ideology and race. To analyze this table, we need to go through the three-step process of analysis. First, we describe the pattern; second, we identify the statistical significance to see the probability of observing that pattern simply due to chance; and third, we evaluate the strength of the relationship by reporting the appropriate measure of association.

1. Describe the pattern.
   To describe the pattern, I choose a row and compare the percentages across it. Because I framed the topic in terms of who is more likely to be liberal, I choose to compare the column percent for the row of liberals. The marginal shows that 30.0

Table 9.12 Political Ideology by Race

Ideology	White	Black	Hispanic	Other	Total
Liberal	27.9% (253)	35.0% (108)	28.2% (96)	43.8% (28)	30.0% (485)
Moderate	26.5% (240)	37.9% (117)	40.3% (137)	26.6% (17)	31.6% (511)
Conservative	45.6% (413)	27.2% (84)	31.5% (107)	29.7% (19)	38.5% (623)
Total	100.0%	100.1%	100.0%	100.1%	100.1%
n =	906	309	340	64	1619

Source: American National Election Study, ANES 2008 Time Series Study. www.electionstudies.org/studypages/2008prepost/2008prepost.htm).

percent of Americans consider themselves liberals. If the null hypothesis is correct, I will observe that each of the categories of race will be around 30 percent liberal. Although whites are close to that base, with 27.9 percent of them being liberal, 35 percent of blacks consider themselves to be liberal. It looks like blacks are more likely to be liberal than whites by 7 percentage points.

Table 9.13   Calculating the Chi-Square for Ideology by Race

Cells	f_x	Row Proportion	Column n	E(X)	$f_x - E(X)$	$[f_x - E(X)]^2$	$\dfrac{[f_x - E(X)]^2}{E(X)}$
Liberal, White	253	0.3	906	271.800	−18.800	353.440	1.300
Liberal, Black	108	0.3	309	92.700	15.300	234.090	2.525
Liberal, Hisp.	96	0.3	340	102.000	−6.000	36.000	0.353
Liberal, Other	28	0.3	64	19.200	8.800	77.440	4.033
Mod., White	240	0.316	906	286.296	−46.296	2143.320	7.486
Mod., Black	117	0.316	309	97.644	19.356	374.655	3.837
Mod., Hisp.	137	0.316	340	107.440	29.560	873.794	8.133
Mod., Other	17	0.316	64	20.224	−3.224	10.394	0.514
Cons., White	413	0.385	906	348.810	64.190	4120.356	11.813
Cons., Black	84	0.385	309	118.965	−34.965	1222.551	10.277
Cons., Hisp.	107	0.385	340	130.900	−23.900	571.210	4.364
Cons., Other	19	0.385	64	24.640	−5.640	31.810	1.291

$\chi^2 = \Sigma = 55.926$

d.f. = (4 − 1)(3 − 1) = 6

prob. < 0.001

2. To identify the statistical significance, calculate the chi-square in Table 9.13:
   a. Set up a table where each cell from the contingency table gets its own row. In the original table, there are twelve cells. I have twelve rows, so it looks like I didn't miss anything.
   b. Copy the observed cell frequencies into the second column, the marginal row proportion into the third column, and the column n into the fourth.

I notice that each of the column percentages repeat four times, which is the number of categories of my independent variable. There are four column ns in the fourth column (the same as the number of categories of my independent variable), which repeat three times (the same as the number of categories of my dependent variable). I am confident that I have properly created a row for each of the cells in the original table.

    c. In the fifth column, calculate the expected value by multiplying the row proportion by the column n.

This is shown in Table 9.13.

    d. In the sixth column, subtract the expected value from the observed frequency.

    e. In the seventh column, square the difference between the observed and expected values.

    f. In the last column, divide the squared difference by the expected value.

    g. Calculate the chi-square by adding up the values in the final column.
        The chi-square for the relationship between ideology and race is 55.926.

    h. Find the statistical significance by looking this up on a chi-square table.
        The original contingency table has four columns and three rows, so I find its degrees of freedom by multiplying 3 (one less than 4) times 2 (one less than 3). This table has six degrees of freedom. As I look across the "6" row in the chi-square work table, I find that my chi-square of 55.926 is larger than all the values given. I can conclude that the probability of getting this pattern if the null hypothesis is correct is less than 0.001. My results are statistically significant at the 0.01 level; I can reject the null hypothesis.

3. To evaluate the substantive significance, calculate Cramer's V for the relationship between ideology and race.

    a. Find the chi-square value for the table.
        The chi-square from Table 9.13 is 55.926.

    b. Identify the size of the table in terms of the number of rows (r) and columns (c). Find m, which is the lesser of (c − 1) and (r − 1).
        This table has four columns and three rows. Because there are fewer rows, m = 3 − 1 − 2.

    c. Identify the sample size, n.
        The sample size in Table 9.13 is given as n = 1619.

    d. Cramer's V takes the chi-square and weights it by the number of cases and the size of the table. Calculate Cramer's V by substituting each of these values in the equation $V = \sqrt{\chi^2/mn}$.

$$V = \sqrt{\chi^2/mn} = \sqrt{\frac{55.926}{(2)(1619)}} = \sqrt{0.017} = 0.13.$$ Looking this up in Table 9.13, I find that this indicates a weak relationship.

USE SPSS TO ANSWER A QUESTION WITH CHI-SQUARE AND CRAMER'S V

The cross tabulation command in SPSS allows you to request a contingency table. As usual, you will begin the process of analyzing data by opening your dataset in SPSS and making sure that your data are clean. As in the previous chapter, you get a cross tabulation by clicking on "Analyze," "Descriptive Statistics," and "Crosstabs." As usual, you will set these up so that the independent variable (the cause) is in the columns and the dependent variable (the effect) is in the rows. Once again, after specifying the variables, you will click on the "Cells" button and indicate that you want the "Observed" count and the "Column" percent. You will then return to the main crosstab command window. The addition for this chapter is that you also want SPSS to give you some statistics. This process is summarized in Box 9.4.

BOX 9.4 How to Get a Cross Tabulation in SPSS

After opening and cleaning your data in SPSS:

>Analyze
>>Descriptive Statistics
>>>Crosstabs
>>>>*"Dependent Variable"* → Row Variable
>>>>*"Independent Variable"* → Column Variable
>>>>\>Cells
>>>>>\>Observed
>>>>>\>Column
>>>>>\>Continue
>>>>\>Statistics
>>>>>\>Chi-square
>>>>>\>Phi and Cramer's V
>>>>>*>Any other statistics that you might want*
>>>>\>OK

You'll find the various measures of statistical significance and substantive significance used for contingency tables in the "Statistics" window. You'll notice that above the button to change the content of the "Cells," there is another button labeled "Statistics." When you click on this, you'll be given options to choose different statistical measures depending on the level of measurement of your variables. Chi-square is at the top and Cramer's V is in the box for nominal variables. Click on these so that SPSS will include these statistics with the output. Click on "Continue" and "OK" to get SPSS to process your request. Box 9.4 extends the contents of Box 8.4 to include the request for statistics.

Follow the same process for creating a professional-looking table that you used in Chapter 8. You will, however, want to add an additional row at the bottom to insert your measure of association, Cramer's V. Next to the value of Cramer's V, you will put stars corresponding to the level of statistical significance for the relationship. Be sure to put a key to the number of stars in a smaller font below the table. Normally, you will use one star if it is statistically significant at the 0.05 level and two stars if it is statistically significant at the 0.01 level. Box 9.5 appends Box 8.4 to include these last steps.

> **BOX 9.5 How to Create a Contingency Table in Word**
>
> 1. Give the table a title in the form of the name of the *"dependent variable"* "by" the *"independent variable."*
> 2. Insert a table with as many rows as it has categories in the dependent variable plus four. It has as many columns as it has categories in the independent variable plus two.
> 3. In the first row, type the name of the dependent variable in the first column, each of the categories of the independent variable in the next columns, and the word "Total" in the last column.
> 4. Below the name of the dependent variable in the first column, place each of the categories of the dependent variable. In the row third from the bottom, type "Total"; second from the bottom, type "n"; and leave the first column of the bottom row empty.
> 5. Calculate the column percents for each cell by dividing the cell count by the column count and multiplying by 100.
> 6. Include marginals for both the rows and the columns—in both percent and number of cases.
> 7. In the last column of the last line of the table, place the measure of association along with the appropriate number of stars for the statistical significance.
> 8. Place lines after the title, after the column headings and at the end using the "Borders and Shadings" command in Word.
> 9. Give any relevant information in a smaller font below the bottom line including a key to the number of stars for statistical significance.

An SPSS Application: Cross Tabulation of Religious Freedom

It is 2012, and you are interning at Human Rights Watch, a non-governmental organization that monitors the countries of the world in order to advocate for freedom and human rights. After the Arab Spring, Egypt is still in the process of drafting a new constitution. Your boss, Sarah Leah Whitson, the executive director for the Middle East and North Africa Division, is concerned that the current draft does not include a provision for religious freedom. In fact, it proposes maintaining its current status of having Sunni Islam as the state religion. She asks you to collect data analyzing whether countries with state religions are less likely to be free than other countries. You collect data for the countries of the world from Freedom House using a variable that distinguishes between free, partly free, and not free countries.[9] You then find an article in Wikipedia that lists different religions and which countries have adopted them as state religions.[10] Because of Whitson's instructions, you code this as a dichotomous variable (either "no" or "yes" there is a state religion), but you also wonder whether it might make a difference whether the country is Christian (with the attendant western culture) or non-Christian, and so you code a second variable that makes that distinction just in case you might need it.

1. Open your data in SPSS and clean them.
   Although there are no missing cases in the data for either variable, I notice that I need to add labels to the values of the variables. For "Freedom," I set "0" to "not free," "1" to "part free," and "2" to "free." I give StateRel the variable label "Is there a state religion?" and then give value labels "0" = "no" and "1" = "yes." Both variables have a limited number of categories, so I do not need to collapse them.

2. Request your contingency table with the dependent variable in the rows and the independent variable in the columns. Using the "Cells" button, request the cell frequency and the column percent. Using the "Statistics" button, request chi-square and Cramer's V.
   My dependent variable is freedom (so I place it in the rows) and my independent variable is the dichotomous version of state religion (so I place it in the columns). From the "Cells" button, I request both the observed count and the column percentage. From the "Statistics button, I request chi-square and Cramer's V. Figure 9.1 shows how I requested the appropriate statistics.

Figure 9.1 Using SPSS to Request Chi-Square and Cramer's V

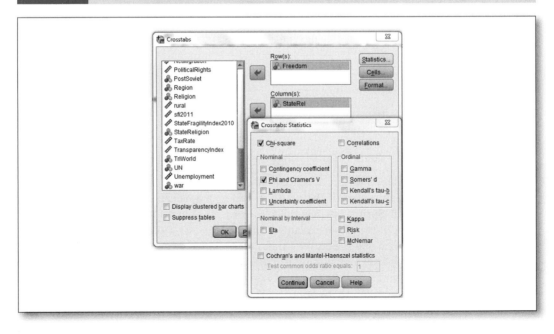

3. Translate the SPSS crosstab into a professionally constructed contingency table. Figure 9.2 shows the output from the above command. The first table gives the contingency table. The table titled "Chi-Square Tests" gives the chi-square value as well as its statistical significance. The table titled "Symmetric Measures" gives the value of Cramer's V.

I translate this into a professional-looking table by following the instructions in Box 9.4, adding the measure of association at the bottom, along with stars indicating the statistical significance. Below the table, I include a key to interpreting the number of stars.

Figure 9.2 Freedom by State Religion

Freedom * StateRel? Crosstabulation

			StateRel? no	StateRel? yes	Total
Freedom	not free	Count	31	17	48
		% within StateRel?	19.5%	48.6%	24.7%
	part free	Count	50	9	59
		% within StateRel?	31.4%	25.7%	30.4%
	free	Count	78	9	87
		% within StateRel?	49.1%	25.7%	44.8%
Total		Count	159	35	194
		% within StateRel?	100.0%	100.0%	100.0%

Chi-Square Tests

	Value	df	Asymp. Sig. (2-sided)
Pearson Chi-Square	13.596[a]	2	.001
Likelihood Ratio	12.476	2	.002
Linear-by-Linear Association	11.962	1	.001
N of Valid Cases	194		

a. 0 cells (0.0%) have expected count less than 5. The minimum expected count is 8.66.

Symmetric Measures

		Value	Approx. Sig.
Nominal by Nominal	Phi	.265	.001
	Cramer's V	.265	.001
N of Valid Cases		194	

4. Write a memo.

Memo

To: Sarah Leah Whitson, Human Rights Watch
From: T. Marchant-Shapiro, intern
Subject: Freedom and State Religions
Date: October 11, 2012

You asked me to compare the level of freedom for countries with and without an official state religion. To measure freedom, I used Freedom House's measure of whether states are free, part free, or not free. Comparing this variable for all states that have an official state religion to those that do not, I found that those without an official state religion are more likely to be free.

The table below shows the relationship between a country's level of freedom and whether it has a state religion. You can see that 49.1 percent of those countries without a state religion are categorized as free, whereas only 25.7 percent of those with an official religion are free. Thus, those without a state religion are more likely to be free than those with one by a difference of 23.4 percentage points. There is less than a 1 percent probability that this pattern is due to chance. The Cramer's V of 0.265 indicates that this is a moderate relationship.

You are on solid ground arguing that countries are more likely to be free if they do not have a state religion. But you should probably extend the analysis before you take a public position on the issue. It is possible that the effect you are seeing here has a cultural, rather than institutional, cause. Many of the countries with state religions are not European and do not have a democratic tradition. It would be worth separating the countries with state religions into Christian (presumably a western culture) and non-Christian. If we were to find that both groups of countries with state religions have the same level of freedom, we could conclude that it is the institution of state religion that is the problem. If not, then culture may be the cause, in which case eliminating the state religion from the constitution may not solve the problem.

Freedom by State Religion

Level of Freedom	State Religion? No	State Religion? Yes	Total
Not Free	19.5%	48.6%	24.7% (48)
Part-Free	31.4%	25.7%	30.4% (59)
Free	49.1%	25.7%	44.8% (87)
Total	100.0%	100.0%	100.0%
n =	159	35	194
		Cramer's V = 0.265**	

Source: Freedom House, Freedom in the World 2012. http://freedomhouse.org/report/freedom-world/freedom-world-2012.

*prob. < 0.05

**prob. < 0.01

Your Turn: Chi-Square and Cramer's V

YT 9.1

One possible explanation for why blacks are more likely to be liberal than whites is that blacks tend to be pro-choice. Although we usually talk about views on abortion in terms of the dichotomous categorization of pro-choice versus pro-life, the issue is actually much more complicated than that. Many Americans think abortion should be legal in some circumstances but not in others. To get a more nuanced measure of public opinion on the legalization of abortion, the standard is to ask a series of questions about whether the respondent believes abortion should be legal in various situations. The respondents are then placed on a scale depending on how many situations they feel that abortion should be legal. Six of these questions are asked, leading to a scale from 0 (the respondents don't feel that abortion should be legal in any of the situations) to 6 (they feel it should be legal in all the situations). Table 9.14 shows the relationship between views on abortion and race.

Table 9.14 Views on Abortion by Race

Number of Instances Abortion Should Be Legal	Race		Total
	White	Black	
None	9.2%	9.3%	9.2%
	(78)	(15)	(193)
1	6.5%	7.4%	6.6%
	(55)	(12)	(67)
2	11.3%	14.2%	11.7%
	(96)	(23)	(119)
3	17.9%	15.4%	17.5%
	(152)	(25)	(177)
4	7.9%	8.6%	8.0%
	(67)	(14)	(81)
5	5.6%	13.6%	6.9%
	(48)	(22)	(70)
All 6 instances	41.7%	31.5%	40.1%
	(355)	(51)	(406)
Total	100.1%	100.0%	100.0
n =	851	162	1013

Source: General Social Survey 2010, National Opinion Research Center, 2012. www3.norc.org/GSS+Website/Download/SPSS+Format/.

1. Calculate the chi-square by filling in the following table. Find the probability by looking up the chi-square value for those degrees of freedom on a chi-square table. What can you conclude?

2. Now calculate Cramer's V. What can you conclude?

Table 9.15 Work Table: Views on Abortion by Race

Cells	f_x	Row Proportion	Column n	$E(X)$	$f_x - E(X)$	$[f_x - E(X)]^2$	$\dfrac{[f_x - E(X)]^2}{E(X)}$
White, 0							
Black, 0							
White, 1							
Black, 1							
White, 2							
Black, 2							
White, 3							
Black, 3							
White, 4							
Black, 4							
White, 5							
Black, 5							
White, 6							
Black, 6							
						$X^2 =$	$\Sigma =$
						d.f. = $(c-1)(r-1)=$	
						Look up:	prob.

YT 9.2 Is the level of significance the same if you collapse the dependent variable? Table 9.16 shows the same data collapsed so that those thinking abortion should be legal in no more than two instances are given as "pro-life," three to five instances are "moderate" and all six instances are "pro-choice." Once again, calculate chi-square. What can you conclude? Calculate Cramer's V. What can you conclude?

Table 9.16 Collapsed Views on Abortion by Race

Support for Legal Abortion	White	Black	Total
Pro-Life	26.9% (229)	30.9% (50)	27.5% (279)
Sometimes	31.4% (267)	37.7% (61)	32.5% (328)
Pro-Choice	41.7% (355)	31.5% (51)	40.1% (406)
Total	100.0%	100.1%	100.1%
n =	851	162	1013

Source: General Social Survey 2010, National Opinion Research Center, 2012. www3.norc.org/GSS+Website/Download/SPSS+Format/.

1. Calculate the chi-square by filling in Table 9.17. Find the probability by looking up the chi-square value for those degrees of freedom on a chi-square table. What can you conclude?
2. Now calculate Cramer's V. What can you conclude?

Table 9.17 Work Table: Collapsed Views on Abortion by Race

Cells	f_x	Row Proportion	Column n	$E(X)$	$f_x - E(X)$	$[f_x - E(X)]^2$	$\dfrac{[f_x - E(X)]^2}{E(X)}$
White, life							
Black, life							
White, mod.							
Black, mod.							
White, choice							
Black, choice							
						$X^2 =$	$\Sigma =$
						d.f. = (c − 1)(r − 1) =	
						Look up:	prob.

YT 9.3 — Why did you get different results in YT 9.1 and YT 9.2 even though you began with the same data? Look at the math, focusing on the last two columns. What does this tell you about chi-square?

Apply It Yourself: Analyzing Data by Type

Sarah Leah Whitson has agreed that your initial analysis of state religions might be inchoate. She has requested that you reanalyze the freedom data, separating the types of state religions into Christian and non-Christian. Is it the institution of a state religion that leads to the lower level of freedom? Or are the earlier results a by-product of the fact that those countries with state religions tend not to have a western culture? In the "WordData" dataset, your dependent variable will be "Freedom" and your independent variable will be "StateReligion."

1. Open your data in SPSS and clean it using "Define Variable Properties." Add variable and value labels. For Freedom, set "0" to "not free," "1" to "part free," and "2" to "free"; for State Religion ("StateReligion"), designate the variable label "Is there a state religion?" and then give value labels "0" for "none," "1" for "Christian," and "2" for "non-Christian." (As you assign the value labels, remember to start at the bottom and work your way up.)

2. Use the "Analyze," "Descriptive Statistics," "Crosstabs" command to get the data for your contingency table. Remember that the dependent variable (Freedom) should be in the rows and the independent variable (StateReligion) should be in the columns. Go into the "Cells" window to request both the cell count and the column percentage and into the "Statistics" window to request chi-square and Cramer's V.

3. Translate the SPSS crosstab into a professionally constructed contingency table.

4. Write a memo. Please attach the SPSS output of the crosstab and statistics tables for grading purposes.

Key Terms

Chi-square statistic (p. 245)
Contingent (p. 247)
Cramer's V (p. 245)
Expected value (p. 253)

Independent (p. 247)
Intersection (p. 250)
Marginals (p. 248)

CHAPTER 10

Measures of Association

Making Connections

In the spring of 2012, policymakers had a problem that they couldn't explicate using the normal rationale. Gas prices were once again approaching $4.00 a gallon, but none of the normal causal mechanisms seemed to explain why. Normally, there is one of two culprits behind high gas prices: demand or supply. Sometimes, prices increase because demand goes up. Other times, prices increase because supply goes down. Figures 10.1 and 10.2 show these two economic mechanisms.

The puzzle of 2012 was that U.S. consumption of gasoline was falling and U.S. domestic production was up. Because there is a positive relationship between prices and demand, we would expect that as consumption declines, the price would mirror that decline. Conversely, because there is a negative relationship between prices and supply, we would expect that as production increases, prices would fall. With demand down and supply up, prices should have been declining. Apparently, neither demand nor supply could explain the high prices. The solution to the puzzle came when analysts changed their perspective from being focused entirely on the United States to looking at it in the context of the international oil market: The United States was producing enough oil to make international sales possible. And U.S. prices were so much lower than the rest of the world that European countries started importing our product. Thus, even though U.S. consumption was decreasing, total demand was increasing. In addition, because of

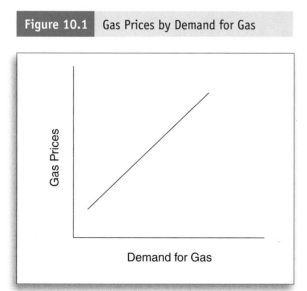

Figure 10.1 Gas Prices by Demand for Gas

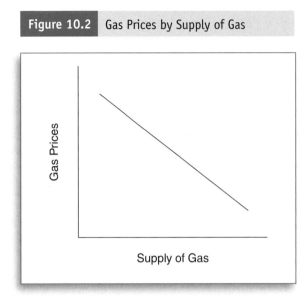

Figure 10.2 Gas Prices by Supply of Gas

foreign sales, even with the increased production, supply within the United States shrank. The increase in worldwide demand and the resulting decrease in U.S. supply combined to increase the U.S. price of gasoline despite the fact that U.S. consumption was down and U.S. production was up.

As we commence with our investigation of measures of association, we will begin by describing positive and negative relationships like the ones shown in Figures 10.1 and 10.2. To measure the association between two interval-level variables, we'll use Pearson's r—the statistical measure normally referred to as the correlation. But not all variables are interval level; we need to find other measures of association that assume less about the variables. To measure the association between two ordinal-level variables, we will learn about gamma. Although other measures of association can be used with ordinal-level variables, they tend to take an approach similar to gamma. As a result, by learning how to calculate gamma, you will also develop intuitions about how other ordinal measures of association work. With nominal-level variables, we already described how Cramer's V works, but in this chapter, we'll look at lambda, which takes a different approach. Having mastered Pearson's r, gamma, and lambda, you will have the tools necessary to measure the association between two variables regardless of their levels of measurement.

BASICS PRINCIPLES OF MEASURES OF ASSOCIATION

Before describing these three measures of association, let's do a quick review of the levels of measurement so that you can be sure to pick the correct measure of association for your variables. You'll recall from Chapter 2 that there are four levels of measurement that statisticians differentiate. Although in statistical analysis, we always assign a number to each possible value of the variable, how much those numbers actually mean can vary. When we measure a variable, we look at four characteristics: Measures can categorize; they can order; they can place on a scale; and they can allow you to locate absolute zero. Those measures that only categorize are called nominal-level variables. Those measures that can both categorize and order are called ordinal-level measures. Those variables that categorize, order, and place on a scale are called interval-level variables. Those variables that categorize, order, place on a scale, and have an absolute zero are called ratio-level variables.

In order to identify the level of measurement of any variable I suggested you follow the decision tree originally shown in Figure 2.2 and repeated in Figure 10.3. First ask the

Figure 10.3 Decision Tree to Determine Level of Measurement

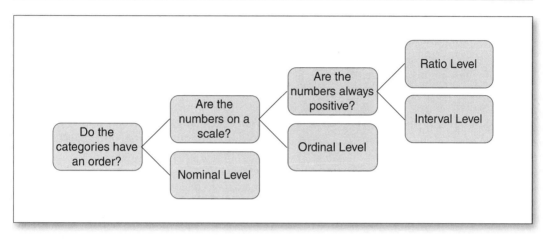

question, Do the categories have an order? If your answer is no, the variable has a nominal level of measurement. If the answer is yes, then you ask a second question: Are the numbers on a scale? If the answer is no, the variable has an ordinal level of measurement. If yes, ask, Do the numbers have to be positive? If the answer is no, the variable has an interval level of measurement. If the answer is yes, it has a ratio level of measurement.

The distinction between interval- and ratio-level variables is not particularly important for our purposes. Knowing a variable has a ratio level of measurement can be useful if we want to compare cases. Because zero is connected with the total absence of the attribute being measured, we know that a case with a value that is twice as big as another has twice as much of that attribute than the other. But that distinction just isn't relevant to how we do the math for measures of central tendency, measures of dispersion, or measures of association. So in choosing the appropriate statistical measures, I have lumped ratio-level variables with interval-level variables.

Choosing the correct measure of association depends on the level of measurement of the two variables, because mathematically each measure of association makes assumptions about those four characteristics (classify, order, scale, and absolute zero). In Chapter 9, when we calculated chi-square, we didn't assume anything about the directionality of the relationship—it wouldn't have mattered what order the categories were in, we would have gotten the same results. So both chi-square and Cramer's V can be used for nominal-level variables. They are also appropriate for ordinal- and interval-level variables when we don't think that there is a linear relationship between them. Similarly, in this chapter, lambda will not assume that the categories of the two variables have any order.

But when both of your variables have an ordinal level of measurement, you can both distinguish between the categories of the variables and place them in order from less to more. The ability to order the categories allows ordinal levels of measurement to compare whether the pattern of the joint distribution of the two variables is more supportive of a direct relationship or of an inverse relationship. In this chapter, we'll work through the math of gamma as a measure of association that is appropriate for ordinal-level variables, although I'll also mention tau as another ordinal measure of association that takes a similar approach.

With interval- (and ratio-) level variables, you add the knowledge that the two variables are measured on a scale where increasing the value by one unit is connected to a uniform increase in the attribute that is being measured. We call it a scale because units are equal regardless of where on the scale the increase occurs. With a uniform scale, the measure of association can not only assume that the relationship has a direction, but also begin to describe it as linear. In this chapter, we'll look at Pearson's r as the most common measure of association for interval variables. Study Table 10.1 and try to master when it is appropriate to use each measure of association because that is the most important job facing you as you use statistics to answer political questions. If you choose the wrong measure of association, you might well use perfectly good data to make the wrong policy decision.

You'll notice that as I described how to determine the correct measure of association, I made the assumption that both variables have the same level of measurement. That is not always the case. So keep in mind not just the levels of measurement, but also what the various measures of association assume about the relationship. Pearson's r assumes a linear relationship. As a result, even with two interval-level variables, Pearson's r would not be an appropriate measure of association if the relationship is not linear. Similarly, if one of the variables does not have an interval level of measurement, the relationship cannot be linear. How to deal with that depends on which variable is not interval level. If the independent variable is categorical (whether nominal or ordinal), we already indicated in Chapter 7 that the appropriate way to analyze it statistically is with ANOVA, and eta is the appropriate measure of association. If the dependent variable is categorical, your best bet is to collapse the interval-level independent variable into an ordinal-level variable and then choose the correct measure of association from there. If you have one nominal- and

Table 10.1 The Appropriate Measure of Association for Different Levels of Measurement

Level of Measurement	Measure of Association
Nominal	Cramer's V
	Lambda
Ordinal	Gamma
	Tau
Nominal/Ordinal Independent Variable, Interval Dependent Variable	ANOVA
Interval/Ratio	Pearson's r
	Regression

one ordinal-level variable, then you can no longer assume any direction to the relationship so you choose Cramer's V or lambda as the appropriate measure of association. You should have seen a pattern in this discussion: If one of the variables has a lower level of measurement, the data do not fulfill the assumptions made for the higher measure of association. The rule is, then, that you choose the measure of association appropriate for the variable with the lowest level of measurement.

The exception to the rule is that dichotomous variables, or those with only two possible values, can count as any level of measurement. Even a blatantly nominal variable like gender (although you may have more or less sex, you certainly cannot be more or less sex) can fit the attributes of a higher level measurement if it has only two categories. If you think about it, because there is only one unit on the scale, it must be a uniform size. So usually, you can pretend that variables with only two categories are ordinal or interval for mathematical purposes. As a result, if one variable is dichotomous, you would normally choose the measure of association appropriate for the other variable.

Most measures of association vary between zero and one. The closer they are to zero, the weaker the relationship; the closer to one, the stronger the relationship. If the measure of association is able to assume that the values of the variables have order, though, then it is also possible to distinguish between a positive and a negative relationship. As a result, ordinal measures of association can range from −1 to +1. In that case, being close to zero is still indicative of a very weak relationship. But −1 means a perfect negative relationship, whereas a +1 means a perfect positive relationship. Like the ordinal measures of association, Pearson's r ranges from −1 to +1 with the same interpretation. The exception to this rule is that when we get to regression in Chapters 12 and 13, we'll find that because the unstandardized regression coefficient is the same as the slope of the line that best fits the data, it no longer has to be less than one. The measures of association in this chapter, though, share the standard of having an absolute value that is less than one. This means that the magnitude of these measures can be interpreted in the same way as eta and Cramer's V. In Table 10.2, the contents of Table 7.3 are repeated, indicating how to interpret the magnitude of measures of association. Because Pearson's r measures the association between two interval-level variables, the relationship can be either positive or negative. If you have a negative correlation, you ignore the negative sign and interpret the magnitude in the same way as its inverse. Similarly, because gamma can measure a direct or inverse

Table 10.2 How to Interpret Measures of Association

Measure of Association (X)	Qualitative Interpretation
$0 \leq X < 0.10$	Very Weak
$0.10 \leq X < 0.20$	Weak
$0.20 \leq X < 0.30$	Moderate
$X \geq 0.30$	Strong

relationship, it can also yield a negative value. Again, to interpret its magnitude, ignore the negative sign. In contrast, because lambda, like Cramer's V, is measuring the relationship between two nominal-level variables, it can only yield a positive association. Interpret the magnitude of lambda in the same way as you did Cramer's V.

> **BOX 10.1 Numbers in the News**
>
> When the British granted India its independence in 1947, they knew that there would be conflict between the Hindus and Muslims in the new nation. As a result, the territory was divided into two countries, India and Pakistan, with the expectation that Hindus would migrate into the new India and Muslims into Pakistan. That migration was fraught with problems, and today, there are still many Muslims in India. Recently, Gallup did a study of India to see the impact that religious preference has on its citizens. The results indicate that although Muslims in India are equally satisfied with most aspects of the government, they struggle more economically.[1] Although we might assume on historical grounds that the religion of Indian citizens would have political consequences, measures of association allow us to identify those effects that are more pronounced.

PEARSON'S R

We use the language of positive and negative relationships because it reflects how we picture relationships graphically. By graphing a set of points, if we see that the values of our dependent variable (measured on the y-axis) increase as the independent variable (on the x-axis) increases, then we know that the best fitting line for those data points would have a positive slope like Figure 10.1. Conversely, if the values of the dependent variable decrease as you move to the right, the best fitting line will have a negative slope like Figure 10.2. For Pearson's r, though, we do not actually want the slope (although we'll calculate it in Chapter 12 when we address regression analysis) because the slope can take on any value depending on the units of the variables. Instead, we want our measures of association to be constrained to always be less than one. This allows us to compare the strength of relationships across different variables regardless of the scale we use to measure them.

Getting a Visual Intuition for Correlation

Intuitively, then, it would make sense to think of our variables in terms of z-scores instead of the absolute scale of our particular measures. You'll remember from Chapter 5 that for a z-distribution, the mean value of a variable is set to zero. If the variable is normally distributed, half the cases are above the mean and the other half are below it. Furthermore, two-thirds of the cases are within one standard deviation of the mean, although a few cases are spread beyond that point in both directions. To visualize the notion behind Pearson's r, you want to take what you know about a univariate z-distribution and change the y-axis from the frequency of the independent variable to the value of the dependent variable. If we talk about the joint distribution of two standardized variables, the data points will be centered at the origin of the table. Two-thirds of the cases should have values that are within one standard deviation of the origin (0, 0) as shown in Figure 10.4. The remaining third will be scattered somewhere outside the shaded region.

Think about this joint standardized distribution in terms of where you expect to see the data points for a positive relationship in contrast to where they would be for a negative relationship. Because the line describing the points has to include the mean for both variables, it will, by definition, go through the origin. In comparison to the origin, the only data points that support a positive relationship would be in the shaded areas found in Figure 10.5. All the points that include positive z-scores for both variables support a positive relationship. But all the points with negative z-scores for both variables also support a positive relationship. These two regions have one important characteristic in common. Remember from your math classes that a positive number multiplied by another positive number will always give you a positive product. Also remember that a negative multiplied by a negative will always give you a positive value. This means that for all cases that fall in these two quadrants, if we were to multiply the values of the z-scores for the two variables, we would get a positive value.

A negative relationship would look the opposite of Figure 10.5. Again, the line has to go through the origin. This means that the only points that combine with the origin to support a negative relationship are in the other two quadrants as shown in Figure 10.6. In one of these quadrants, the independent variable has a negative z-score and the

Figure 10.4 Joint Distribution for Two Standardized Variables

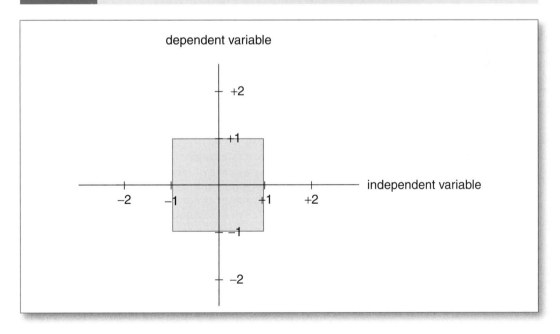

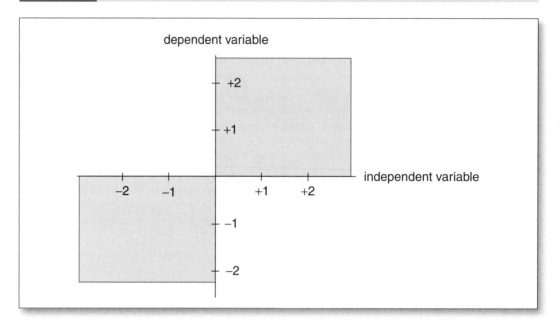

Figure 10.5 Points Supporting a Positive Relationship

dependent variable has a positive z-score. In the other quadrant, the independent is positive and the dependent is negative. These two regions share an important characteristic that distinguishes them from the other two quadrants. Remember from math class that a negative number multiplied by a positive number will always give you a negative value. This means that if we were to multiply the z-scores for our dependent and independent variables, we would get a negative value for all cases that fall in the two quadrants supporting a negative relationship.

We can begin to find the correlation between two interval-level variables by multiplying the z-scores for the two variables. This will give a positive value for all cases supporting a positive relationship and a negative value for all cases supporting a negative relationship. If we were to then add up all those products, we would discover whether, on balance, we have more points supporting a positive or a negative relationship. If we were to then divide by the number of cases, we would have identified the average product between the two z-scores.

The interesting thing about that final value is that its absolute value will always be less than one. There will be some individual cases that have a product of more than one. In a particularly strong positive relationship, you would expect that those cases with a z-score of two on the independent variable would have a z-score of two on the dependent variable as well. For that case, you would multiply two by two and get a value of four to add into the sum. In some particularly widely distributed variables, you might even have z-scores of three for each variable, yielding a product of nine to add to the sum. It might seem that with values above one, it might be possible to end up with a correlation above one. But keep in mind that we are going to be averaging all the products. (That's why we added them up and

Figure 10.6 Points Supporting a Negative Relationship

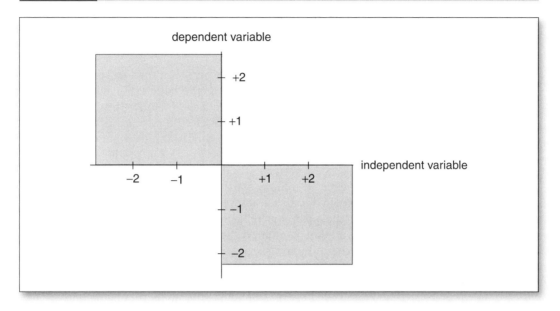

divided by the number of cases.) Less than 5 percent of the cases will have z-scores of greater than two. Remember also from Figure 10.4 that two-thirds of the cases have z-scores that are less than one. Another memory you should have from math is that if you multiply two decimals that are less than one, you get an even smaller decimal. Once you average the two-thirds of the cases that have products less than one with the one-third of cases greater than one, you will end up with a correlation that is less than one.

Setting up the Work Table for Pearson's r

As usual, the math is always easier to do in a work table than simply using an equation. For Pearson's r, though, we are also going to simplify the equation before we set up the work table. Computers can make short work of calculating z-scores for a bunch of cases on multiple variables, whereas doing so by hand gets messy. In addition, the steps I described above would require you to do the same thing multiple times, and in the end, some of the steps would cancel out mathematically. So although the previous section gives you an intuition for how correlation is measured, in this section, I will show you how to do the math in a simpler (although less intuitive) way.

The equation I'll have you use divides the covariance of the two variables by the product of their respective standard deviations. I'll make one change to it, though. In both the numerator and the denominator, you would normally need to divide by the number of cases. But if it is in both parts of the fraction, the number of cases in the top would cancel out the number of cases in the bottom. So there is no need to actually include it. As a result, the simplified equation to use is

$$\text{Pearson's } r = \frac{\Sigma(X-\bar{X})(Y-\bar{Y})}{\sqrt{\Sigma(X-\bar{X})^2 \; \Sigma(Y-\bar{Y})^2}}$$

As we move into measures of association, we need to adjust our work tables to take account of the values of two variables instead of one. As usual, the first column will contain the case identifier; the second column will contain the value of the independent variable for that case. We'll skip a couple of columns to leave room for the calculations we need to do with the independent variable, and then place the values of the dependent variable in the fifth column. At the bottom of the second and fifth columns, total all the values of the two variables and divide by the number of cases to get the means for both the independent and the dependent variables. In the third column, subtract the mean of X from each value of X. In the fourth column, square the difference and calculate the sum of squares for X at the bottom. In the sixth and seventh column, do the same thing for the dependent variable: subtract the mean, square the difference, and find the sum of squares for Y. In the eighth column, you want to multiply the unsquared differences for both X and Y. So take the values from the third and sixth columns, multiply them, and place that product in the eighth column. At the bottom of the eighth column, sum all the values found in it. Note that although you are accustomed to dividing the sum of squares by the number of cases, in the work table for Pearson's r, you do not bother doing that because the ns would have cancelled out in the end.

BOX 10.2 How to Calculate Pearson's r

1. Create a work table with eight columns and a row for each case.

2. In the first column, place the case identifier.

3. In the second and fifth columns, place, respectively, the values of the independent and dependent variables. Calculate the mean for each by summing the values and dividing by n.

4. In the third and sixth columns, subtract the respective mean from the value of the variable in the previous column.

5. In the fourth and seventh columns, square the difference found in the previous column. At the bottom of each, calculate the sum of squares.

6. In the eighth column, multiply the (unsquared) differences from the third and sixth columns and sum the products at the bottom.

7. Calculate Pearson's r by taking the sum of the eighth column and dividing it by the square root of the product you get when you multiply the sum of squares for X (from column 4) by the sum of squares for Y (from column 7).

Once you have the sum of squares at the bottom of the fourth, seventh, and eighth columns, you are ready to calculate Pearson's r. The entire process is summarized in Box 10.2. Remember that the equation is

$$\text{Pearson's } r = \frac{\Sigma(X-\bar{X})(Y-\bar{Y})}{\sqrt{\Sigma(X-\bar{X})^2 \, \Sigma(Y-\bar{Y})^2}}$$

This looks simpler if you translate it into the sums of squares you calculated in your work table.

$$= \frac{SS_{XY}}{\sqrt{SS_X \, SS_Y}}$$

The numerator is what you calculated in the eighth column. The denominator is the square root of the product of the sums of squares at the bottom of the fourth and seventh columns. The numerator will determine whether it is a positive or negative correlation, and the denominator will standardize it so that it varies between −1 and +1. You can use the same rule of thumb we've used previously to describe the strength of the relationship—go back to Table 10.2 if you need a refresher on that.

Working through an Example: Political Rights and Civil Liberties

We've discussed that Freedom House has two major components in its measure of democracy: political rights and civil liberties. In the United States, we take for granted that both aspects are necessary, but it is possible that some countries do better on one than another. How strong is the relationship between the two variables? Table 10.3 shows the raw values for these two variables for the countries in the Middle East. You'll recall that the maximum score for political rights is 40, and the score for civil liberties goes up to 60.

This example highlights one interesting feature of Pearson's r. In general, we have discussed measures of association as being useful to measure causal relationships with one variable dependent on another. For Pearson's r, we mirrored that by labeling one variable "X" and the other "Y." I even discussed the setup of the work table in terms of "dependent" and "independent" variables. But in this example, I am not actually hypothesizing a causal relationship. I am actually just thinking that the two variables might be correlated because I believe that the same countries that value one would value the other as well. For many measures of association, the causal direction I hypothesize will affect the math and so also the results. But for Pearson's r, that is not the case. The equation is set up so that regardless of the variables I call "X" and "Y," I will get the same results. In statistical terms, we call Pearson's r a **symmetrical measure of association** because you get the same results regardless of what you identify as the dependent variable. So although I placed Political Rights in the "X" column, I could just as easily have placed "Civil Liberties" there.

Table 10.4 is the work table to calculate the correlation between political rights and civil liberties. As I work through the table, I see that all the Middle Eastern countries have fairly

Table 10.3 Political Rights and Civil Liberties in Middle Eastern Countries

Country	Political Rights X	Civil Liberties Y
Bahrain	8	12
Iran	6	10
Iraq	12	13
Israel	36	45
Jordan	10	25
Kuwait	19	25
Lebanon	17	34
Oman	9	18
Qatar	10	18
Saudi Arabia	3	7
Syria	0	6
UAE	8	16
Yemen	8	15

Source: Freedom House, *Freedom in the World 2012*. Accessed October 11, 2012. http://freedomhouse.org/report/freedom-world/freedom-world-2012.

low scores for both political rights and civil liberties. The average political rights score for these countries is 11.2 out of 40, and the average civil liberties score is 18.8 out of 60. After comparing each case with the mean, I get my sum of squares for Political Rights and for Civil Liberties that I can use to find the bottom part of the Pearson's r fraction. In the far right-hand column, I multiply the difference for the two variables and see that most of the cases support a positive relationship. The values in this column need to be totaled too. You can see at the bottom of this column that I call its summation "SS_{XY}." Technically, this isn't a sum of squares because I am not squaring anything. But calling it that makes the equation for the correlation easier to understand and remember.

After totaling the values in the eighth column, I place the sum in the top part of the fraction.

$$\text{Pearson's } r = \frac{\Sigma(X-\bar{X})(Y-\bar{Y})}{\sqrt{\Sigma(X-\bar{X})^2 \ \Sigma(Y-\bar{Y})^2}}$$

$$= \frac{SS_{XY}}{\sqrt{SS_X \ SS_Y}}$$

Table 10.4 Work Table for Calculating Pearson's r for Political Rights and Civil Liberties, Middle Eastern Countries

Country	Political Rights X	Column 3 $X - \bar{X}$	Column 4 $(X - \bar{X})^2$	Civil Liberties Y	Column 6 $Y - \bar{Y}$	Column 7 $(Y - \bar{Y})^2$	Column 8 $(X - \bar{X})(Y - \bar{Y})$
Bahrain	8	–3.2	10.24	12	–6.8	46.24	21.76
Iran	6	–5.2	27.04	10	–8.8	77.44	45.76
Iraq	12	0.8	0.64	13	–5.8	33.64	–4.64
Israel	36	24.8	615.04	45	26.2	686.44	649.76
Jordan	10	–1.2	1.44	25	6.2	38.44	–7.44
Kuwait	19	7.8	60.84	25	6.2	38.44	48.36
Lebanon	17	5.8	33.64	34	15.2	231.04	88.16
Oman	9	–2.2	4.84	18	–0.8	0.64	1.76
Qatar	10	–1.2	1.44	18	–0.8	0.64	0.96
Saudi Arabia	3	–8.2	67.24	7	–11.8	139.24	96.76
Syria	0	–11.2	125.44	6	–12.8	163.84	143.36
UAE	8	–3.2	10.24	16	–2.8	7.84	8.96
Yemen	8	–3.2	10.24	15	–3.8	14.44	12.16
	$\Sigma X = 146$	$SS_X =$	968.32	$\Sigma Y = 244$	$SS_Y =$	1478.32	$SS_{XY} = 1105.68$
$\bar{X} =$	146/13 = 11.2			$\bar{Y} =$	244/13 = 18.8		

$$= \frac{1105.68}{\sqrt{(968.32)(1478.32)}}$$

$$= 0.92$$

The Pearson's r for political freedom and civil rights equals 0.92. This suggests that I was correct in my assumption that the two facets of democracy tend to go together: There is a very strong relationship between them.

GAMMA

Although our intuition about positive and negative relationships is strongly connected to the graphic presentation of the relationship between interval-level variables, we will sometimes apply that language to ordinal-level variables as well. Unfortunately, because ordinal-level variables are better presented in tables than in graphs, the language of positive and negative relationships doesn't translate well into intuitions about what the tables should look like. Because tables are set up differently from graphs, we can no longer say that positive relationships go up and to the right. We'll begin by describing the patterns that different relationships take in contingency tables. Once we have a feel for how positive and negative relationships look in contingency tables, we can describe how to measure those relationships with the statistic known as gamma.

Positive and Negative Relationships in Contingency Tables

When you are plotting a relationship on a graph, the origin is located in the lower left-hand corner. As your independent variable increases, you move to the right. As your dependent variable increases, you move up. But a contingency table is set up differently. In it, the origin is placed in the upper left-hand corner. Once again, as your independent variable increases, you move to the right across the columns. But as your dependent variable increases, you move down through the rows. This means that the patterns you expect for a positive and a negative relationship look the opposite of what you expect in a graph. Table 10.5 shows the pattern that emerges when you see a positive relationship in a contingency table, and Table 10.6 shows what you will see for a negative relationship. To measure the association between two ordinal-level variables, any statistics you use will have to distinguish between these two patterns, giving patterns that look like Table 10.5 a positive value and those that look like Table 10.6 a negative value.

Table 10.5 Positive Relationship

	Independent Variable		
Dependent Variable	*Category 1*	*Category 2*	*Category 3*
Category 1	X		
Category 2		X	
Category 3			X

Table 10.6 Negative Relationship

	Independent Variable		
Dependent Variable	*Category 1*	*Category 2*	*Category 3*
Category 1			X
Category 2		X	
Category 3	X		

Gamma fulfills our expectation for an ordinal measure of association by looking at each case in the table and pairing it with each of the other cases. Because a positive relationship goes down and to the right, each pair going down and to the right gives evidence of a positive relationship. These are called **concordant pairs**. Because a negative relationship goes down and to the left, each pair going down and to the left gives evidence of a negative relationship. These are called **discordant pairs**. Gamma compares the number of concordant and discordant pairs to compute a number indicating which pattern dominates in a particular table.

Gamma for Dichotomous Variables

To illustrate this, let me set up some hypothetical 2 × 2 tables that indicate various levels of association. Table 10.7 shows what a perfect positive relationship would look like. In this instance, all of the cases are found in the diagonal going down and to the right. For each of the fifty cases in the upper left cell, there are fifty cases in the lower right with which it can be paired to support a positive relationship. So if you multiply 50 by 50, you get 2500 different possible pairs of cases that support a positive relationship. Gamma calls these "Concordant Pairs" or "C" for short: C = 50 × 50 = 2500.

Table 10.8 shows what a perfect negative relationship would look like—with all of the cases in the diagonal going down and to the left. Again, each of the fifty cases in the upper

Table 10.7 Perfect Positive (Direct) Relationship

	Independent Variable		
Dependent Variable	Category 1	Category 2	Total
Category 1	100.0% (50)	0.0% (0)	50.0% (50)
Category 2	0.0% (0)	100.0% (50)	50.0% (50)
Total	100.0%	100.0%	100.0%
n =	50	50	100

Table 10.8 Perfect Negative (Inverse) Relationship

	Independent Variable		
Dependent Variable	Category 1	Category 2	Total
Category 1	0.0% (0)	100.0% (50)	50.0% (50)
Category 2	100.0% (50)	0.0% (0)	50.0% (50)
Total	100.0%	100.0%	100.0%
n =	50	50	100

right can combine with the fifty cases in the lower left to support a negative relationship. Gamma calls these possible combinations "Discordant Pairs" or "D" for short. In this table, there are 50 × 50 = 2500 discordant pairs: D = 2500.

Finally, Table 10.9 shows what the null hypothesis would look like—with equal proportions in the cells found in each row. You should notice that in this table, there are both "Concordant Pairs" and "Discordant Pairs." To calculate the number of concordant pairs, you would multiply the number of cases in the upper left cell by the number of cases in the lower right cell. In this instance, you would multiply 25 × 25 = 625 concordant pairs, or C = 625. Conversely, to calculate the number of discordant pairs, you would multiply the number of cases in the upper right cell by the number of cases in the lower left cell. In this instance, you would multiply 25 × 25 = 625 discordant pairs, or D = 625. With the null hypothesis, you would expect to see the same number of concordant and discordant pairs, and that is exactly what we have in Table 10.9.

Gamma is given by the difference between the number of concordant (C) and discordant (D) pairs, proportional to the total number of pairs. The equation for gamma (given by the lowercase Greek letter "γ"—which looks like an upside-down cursive e) is

$$\gamma = (C - D)/(C + D)$$

where the number of concordant pairs is the sum of the number of cases in each cell multiplied by the number of cases down and to the right of it; the number of discordant pairs is the sum of the number of cases in each cell multiplied by the number of cases down and to the left of it. For Table 10.7,

$$C = \text{Concordant Pairs} = 50 \times 50 = 2500$$

$$D = \text{Discordant Pairs} = 0 \times 0 = 0$$

$$\gamma = (2500 - 0)/(2500 + 0)$$

$$= 1$$

Table 10.9 No Relationship—Null Hypothesis

Dependent Variable	Independent Variable		
	Category 1	Category 2	Total
Category 1	50.0% (25)	50.0% (25)	50.0% (50)
Category 2	50.0% (25)	50.0% (25)	50.0% (50)
Total	100.0%	100.0%	100.0%
n =	50	50	100

As desired, the table with a perfect positive relationship gives an association of a positive 1. Solving for gamma for Table 10.8,

$$C = 0 \times 0 = 0$$

$$D = 50 \times 50 = 2500$$

$$\gamma = (0 - 2500)/(0 + 2500)$$

$$= -1$$

So the perfectly negative data have an association of −1. Finally, for Table 10.9,

$$C = 25 \times 25 = 625$$

$$D = 25 \times 25 = 625$$

$$\gamma = (625 - 625)/(625 + 625)$$

$$= 0$$

The null hypothesis has an association of zero.

Working through an Example: Gun Control and Gun Ownership. Of course, real data will always be somewhere between these three extremes. Take the case of the relationship between support for gun laws and gun ownership. I hypothesize that there is an inverse relationship between support for gun control and owning a gun. Table 10.10 shows this relationship. I've included the number of cases in each cell as well as the column percent because we need the number of cases to calculate gamma.

Table 10.10 Support for Gun Control by Gun Ownership

There Should be More Restrictions on Handguns	Gun in House?		Total
	No	Yes	
Disagree	15.6% (86)	46.2% (144)	26.7% (230)
Agree	84.4% (464)	53.8% (168)	73.3% (632)
Total	100.0%	100.0%	100.0%
n =	550	312	862

Source: General Social Survey 2004, National Opinion Research Center, 2012. Accessed October 9, 2012. www3.norc.org/GSS+Website/Download/SPSS+Format/.

We first need to calculate the number of concordant pairs. The (no, disagree) cell is the only cell with another cell down and to the right. Its 86 cases in combination with the 168 cases in the (yes, agree) cell form the concordant pairs:

$$\text{Concordant Pairs} = 86 \times 168$$

$$C = 14{,}448$$

Support for the negative relationship is found with the discordant pairs by combining the 144 cases in the (yes, disagree) cell with the only cell down and to the left—(no, agree) with 464 cases.

$$\text{Discordant Pairs} = 144 \times 464$$

$$D = 66{,}816$$

With those two numbers, we can calculate gamma:

$$\gamma = (C - D)/(C + D)$$

$$\gamma = (14{,}448 - 66{,}816)/(14{,}448 + 66{,}816)$$

$$= -52{,}368/81{,}264$$

$$= -0.644$$

As hypothesized, there is a strong negative relationship between support for gun control and gun ownership.

Gamma for More than Two Categories

Calculating gamma for a 2 × 2 table is fairly straightforward. Doing it for a larger table requires a little care. Keep in mind that concordant pairs are always down and to the right, whereas discordant pairs are always down and to the left. (In order to connect concordant pairs with a positive or direct relationship and discordant pairs with a negative or inverse relationship, you need to make sure that your variables are coded so that the categories for both the dependent and the independent variables go from small to large.) I recommend setting up a work table similar to the one used for calculating chi-square—where each cell in the original contingency table becomes a row in which you can calculate the number of concordant pairs and discordant pairs for the cases within it.

The identity of these cells will become the first column of my work table, shown in Table 10.11. The second column will contain the number of cases in each cell. In the third column, I will identify the number of cases down and to the right of this cell (which, in combination, will support a positive relationship). In the fourth column, I will calculate the number of concordant pairs by multiplying the number of cases in the cell (column 2) by the number of concordant cases (column 3). In the fifth column, I will identify the number of cases down and to the left of this cell (which, in combination, will support a negative relationship). In the sixth column, I will calculate the number of discordant pairs by multiplying the number of cases in the cell (column 2) by the number of discordant cases (column 5). By totaling the fourth column, I get C, or the total number of concordant pairs; by totaling the sixth, I get D, or the discordant pairs. Finally, I plug these two numbers into the equation to calculate gamma: $\gamma = (C - D)/(C + D)$.

The key to identifying concordant and discordant pairs is remembering the basic patterns we mapped out at the beginning of the section for positive and negative relationships. Everything down and to the right supports a positive relationship and so goes into concordant pairs. The concordant pairs for the first cell are pictured in Table 10.12. Everything down and to the left supports a negative relationship and so goes into discordant pairs. The discordant pairs for the third cell are pictured in Table 10.13. Box 10.3 summarizes the process of calculating gamma.

Table 10.11 Work Table for Calculating Gamma

Cells	f_x	Concordant Cases	Concordant Pairs	Discordant Cases	Discordant Pairs
R1, C1					
R1, C2					
...					
R2, C1					
R2, C2					
...					
R3, C1					
R3, C2					
...					
			$\Sigma = C$		$\Sigma = D$

Table 10.12 Finding Concordant Pairs

	C1	C2	C3
R1			
R2			
R3			

| Table 10.13 | Finding Discordant Pairs |

	C1	C2	C3
R1			
R2			
R3			

BOX 10.3 How to Calculate Gamma

1. Set up a work table where each cell gets its own row. (Double check to make sure you have the right number.) There will be six columns in the table. In the first column, identify the cell.

2. In the second column, place the number of cases in the cell.

3. In the third column, find the number of concordant cases by identifying all the cells that are down and to the right and then totaling the number of cases in all of them.

4. In the fifth column, find the number of discordant cases by identifying all the cells that are down and to the left and then totaling the number of cases in all of them.

5. In the fourth column, calculate the number of concordant pairs for each cell by multiplying the cell frequency (in column 2) by the concordant cases (column 3). At the bottom, total the number of concordant pairs in the column to find C.

6. In the sixth column, calculate the number of discordant pairs for each cell by multiplying the cell frequency (in column 2) by the discordant cases (column 5). At the bottom, total the number of discordant pairs in the column to find D.

7. Calculate gamma by substituting the values of C and D into the equation $\gamma = (C - D)/(C + D)$.

Working through an Example: Gun Control and Ideology. Table 10.14 is a 3 × 3 table that shows the relationship between support for gun control and political ideology. This time, I hypothesize that there is a negative relationship between support for gun control and political conservativism. I have set up the table so that political ideology begins with the least conservative category (liberal) and increases to the most conservative category (conservative). Support for gun control begins with the least support (oppose background check) and increases to the most support (support background check).

The first step of analyzing data is always to describe the pattern. If I take the marginal row percent and look for discrepancies, I see that there are more conservatives who oppose background checks than you would expect randomly, and more liberals who support it than you would expect. The cells where the observed frequencies exceed the expected frequencies are found along the diagonal going down and to the left. This is the same pattern we saw when we mapped out what a negative relationship looks like. Apparently, support for background checks is negatively related to how conservative an individual is. Specifically, conservatives are less likely to support background checks than liberals by a difference of 9 percentage points.

Table 10.14 Support for Gun Control by Political Ideology

Support Background Check?	Political Ideology			Total
	Liberal	Moderate	Conservative	
Oppose	10.6% (37)	8.7% (41)	16.4% (79)	12.0% (157)
Neither	5.7% (20)	9.1% (43)	8.7% (42)	8.1% (105)
Support	83.7% (293)	82.2% (387)	74.9% (362)	79.9% (1042)
Total	100.0%	100.0%	100.0%	100.0%
n =	350	471	483	1304

Source: General Social Survey 2004, National Opinion Research Center, 2012. Accessed October 9, 2012. www3.norc.org/GSS+Website/Download/SPSS+Format/.

Questions: *In most states, a gun owner may legally sell his or her gun without proof that the buyer has passed a criminal history check. How strongly do you favor or oppose a law that required private gun sales to be subject to the same background check requirements as sales by licensed dealers? Is the respondent liberal, moderate, or conservative?*

My job in calculating gamma is to systematically make sure I get all possible combinations precisely once. I set up my work table to contain one row for each of the nine cells in the table. As I did in Chapter 9 for chi-square, I begin with the leftmost column of the first row and work my way to the right. I systematically do this for each of the subsequent rows. This is shown in Table 10.15. In the end, I count the number of rows and make sure I have the same number as I had cells in the original table. I have nine rows—the same as the number of cells in the contingency table, so I don't think I made any mistakes. I have set up the other five columns necessary to calculate gamma in Table 10.16.

After entering the labels for each of the cells in the first column, I enter the frequencies in the second. I then begin gathering the evidence supporting a positive relationship. Systematically going through each cell on the table, I find the number of cases down and to the right of it. I begin in the top left-hand corner: There are four cells down and to the right of Oppose, Liberal. They contain 43, 42, 387, and 362 cases each. I add them into the third column of the first row of my work table. My next cell is Oppose, Moderate. There are only two cells

Table 10.15 Cells in Gun Control by Political Ideology

Cell
Oppose, Liberal
Oppose, Moderate
Oppose, Conservative
Neither, Liberal
Neither, Moderate
Neither, Conservative
Support, Liberal
Support, Moderate
Support, Conservative

down and to the right, which contain 42 and 362 cases. There are no cells down and to the right of Oppose, Conservative. When I look at Neither, Liberal, I see two cells down and to the right, containing 387 and 362 cases. Neither, Moderate has one cell down and to the right, containing 362 cases. None of the remaining cells has any cells down and to the right, so I enter a zero for each of them.

For now, I'm going to skip the calculating of concordant pairs and jump over to the fifth column to find the evidence for a negative relationship. For this column, I want to enter all the cases down and to the left of each cell. There are no cases down and to the left of Oppose, Liberal so I enter 0. There are two cells down and to the left of Oppose, Moderate, containing 20 and 293 cases. There are four cells down and to the left of Oppose, Conservative, containing 20, 43, 293, and 387 cases. There are no cases down and to the left of Neither, Liberal. There is one cell down and to the left of Neither, Moderate, containing 293 cases. Neither, Conservative connects to two cells to support a negative relationship, and they contain 293 and 387 cases. None of the remaining cells has any cells down and to the left of it, so I enter zeros for them. All the foregoing work is shown in Table 10.17.

Table 10.16 Work Table for Calculating Gamma, Blank

Cells	f_x	Concordant Cases	Concordant Pairs	Discordant Cases	Discordant Pairs
Oppose, Lib.					
Oppose, Mod.					
Oppose, Cons.					
Neither, Lib.					
Neither, Mod.					
Neither, Cons.					
Support, Lib.					
Support, Mod.					
Support, Cons.					

Table 10.17 Work Table for Calculating Gamma, Populated

Cells	f_x	Concordant Cases	Concordant Pairs	Discordant Cases	Discordant Pairs
Oppose, Lib.	37	43 + 42 + 387 + 362		0	
Oppose, Mod.	41	42 + 362		20 + 293	
Oppose, Cons.	79	0		20 + 43 + 293 + 387	
Neither, Lib.	20	387 + 362		0	
Neither, Mod.	43	362		293	
Neither, Cons.	42	0		293 + 387	
Support, Lib.	293	0		0	
Support, Mod.	387	0		0	
Support, Cons.	362	0		0	

Table 10.18 Work Table for Calculating Gamma, Complete

Cells	f_x	Concordant Cases	Concordant Pairs	Discordant Cases	Discordant Pairs
Oppose, Lib.	37	43 + 42 + 387 + 362 = 834	37 × 834 = 30,858	0	0
Oppose, Mod.	41	42 + 362 = 404	16,564	20 + 293 = 313	41 × 313 = 12,833
Oppose, Cons.	79	0	0	20 + 43 + 293 + 387 = 743	58,697
Neither, Lib.	20	387 + 362 = 749	14,980	0	0
Neither, Mod.	43	362	15,566	293	12,599
Neither, Cons.	42	0	0	293 + 387 = 680	28,560
Support, Lib.	293	0	0	0	0
Support, Mod.	387	0	0	0	0
Support, Cons.	362	0	0	0	0
			$\Sigma = C =$ 77,968		$\Sigma = D =$ 112,689

I now complete the math in Table 10.18. First, I sum the concordant and discordant cases for each cell. I find the number of concordant pairs for each cell by multiplying the number of cases in the cell (column 2) by the number of concordant cases (column 3) and placing that product in column 4. At the bottom of that column, I total the number of concordant pairs. I follow the same process for discordant pairs: In column 6, I multiply the frequency from column 2 by the number of discordant cases in column 5, and then I total the number of discordant pairs at the bottom.

When we described the pattern we saw in the original contingency table, we observed a negative relationship. The pattern of concordant and discordant pairs supports our initial finding: There are more discordant pairs than concordant pairs, so the evidence supports a negative relationship. To find out how strong that association is, we need to now compute gamma. To do so, we take the difference between concordant and discordant pairs and divide by the total number of pairs.

$$\gamma = (C - D)/(C + D)$$

$$\gamma = (77,968 - 112,689)/(77,968 + 112,689)$$

$$= -34,721/190,657$$

$$= -0.182$$

We can conclude that there is a weak inverse relationship between support for gun control and level of conservatism.

Kendall's Tau

Gamma compares the support for a positive relationship with the support for a negative relationship. Based on the preponderance of evidence for one versus the other, it calculates a measure of association between −1 and +1. There are other measures of association appropriate for ordinal-level data. For example, Kendall's tau also calculates the difference between the number of concordant pairs and the number of discordant pairs. The difference is how tau handles ties. Gamma ignores possible pairs that support neither a positive nor a negative relationship—cells that are either in the same row or in the same column. Tau includes those possible pairs in the denominator as part of the total number of possible pairs. As a result, tau is generally smaller than gamma. My personal bias is to use gamma rather than tau because the ties for concordant pairs will be included with the discordant pairs and vice versa—so the larger gamma is a perfectly reasonable measure of association to use for ordinal-level data. If you choose to use the more conservative tau, realize that there are actually two tau measures. You choose the appropriate one depending on the shape of your original contingency table. If the table is square (the number of rows equals the number of columns), you use tau-b. If it isn't square (you have either more or fewer rows than columns), you use tau-c.

LAMBDA

In Chapter 9, we looked at Cramer's V as a measure of association for nominal-level variables. Cramer's V is based on a comparison of what we observe with what we expect to see if the null hypothesis is true. But that is not the only criterion on which we can evaluate a potential relationship between two variables. Alternatively, we could ask the question of how well we can predict the dependent variable if we know the value of the independent variable. Remember when we discussed eta-squared? We talked about eta-squared in terms of proportional reduction in error: how well the value of the independent variable explained variance in the dependent variable. With nominal-level variables, we cannot calculate variance like we did with eta, but like eta, lambda is a PRE measure of how well we can predict the dependent variable.

Calculating Lambda

As with chi-square and Cramer's V, to calculate lambda (given by the lower case Greek letter λ), we need to have the data presented in a tabular form with cell counts and marginals. By looking at the row marginals, we can see how well we can predict the dependent variable without knowing the independent variable. The marginals tell what proportion of cases fall into each category of the dependent variable. With that information, our best guess is the category with the most cases (the mode). If we predict the most frequent category for all of the cases, we will, of course, be wrong for all the cases that do not fall into that category. But we will be right more times with the largest category than with any other category. (And we will, on average, be right more times than if we randomly chose.) For lambda, after identifying the largest category of the dependent variable and choosing it for all the cases, we find the total number of cases for which we are wrong. We call this number the "errors without" knowing the independent variable. This is this number against which we will compare how well the independent variable can predict the dependent variable.

To find out the predictive capability of the independent variable, we then find the best prediction for each category of the independent variable. For each of these categories, we go through the same process that we did for the row marginals. For each category of the independent variable, we find the modal category of the dependent variable and predict it for all the cases in that column. For each of the independent categories, we again find the number of errors we make and total them across all of the columns. We call this total the number of "errors with" knowledge of the independent variable. We will never make more errors by knowing the independent variable than without knowing it, but if we are correct that the independent variable is a good predictor of the dependent variable, then we should make fewer.

The question that is answered by lambda is, Proportionally, how many of the original errors we made without knowing the independent variable do we eliminate by knowing into which category of the independent variable a case falls? Mathematically, we calculate lambda with the following equation:

$$\lambda = (\text{errors without} - \text{errors with})/\text{errors without}$$

Think about the math of that equation. If by knowing the category of the independent variable into which a case falls you do not make any errors, the equation for lambda would simplify to the number of errors without divided by itself. Any number divided by itself equals one. So lambda would approach one, which is our standard value for a perfect association. If, on the other hand, you make just as many errors when you know the value of the independent variable as you did without knowing it, then the top of the fraction would approach zero. Because zero divided by any non-zero number is always zero, in this case, lambda would approach zero, which is our standard value for no association. Lambda is thus set up perfectly to measure the range of association from zero to one for two variables whose categories are not ordered in such a way as to imply a directional (either positive or negative) relationship.

Working through an Example: Generations of Partisanship

Let's walk through an example of calculating lambda. One of the phenomena of American politics is that partisanship follows generational patterns: Historically, the dominant party in the United States changed every forty years. Beginning in the 1970s, political scientists began looking for a new party system to replace the New Deal realignment of the 1930s. Some political scientists saw a shift in partisanship during the Reagan years—but not at the level of prior realignments. The question is whether those voters socialized during the Reagan years exhibit different partisanship than their predecessors of the New Deal realignment and the younger millennial generation. Because this relationship is cyclical, it cannot be described as either positive or negative. Depending on where you are in the cycle, the proportion of Democrats could be either increasing or decreasing.

Table 10.19 Partisanship by Age

Party	Age				
	Less than 30	30-49	50-64	65 and up	Total
Democrat	33.5% (122)	35.1% (248)	32.0% (168)	41.3% (158)	35.2% (696)
Independent	48.9% (178)	41.2% (291)	44.4% (233)	31.3% (120)	41.5% (822)
Republican	17.6% (64)	23.8% (168)	23.6% (124)	27.4% (105)	23.3% (461)
Total	99.9%	100.0%	100.0%	100.0%	100.0%
n =	364	707	525	383	1979

Source: General Social Survey 2008, National Opinion Research Center, 2012. www3.norc.org/GSS+Website/Download/SPSS+Format/.

The relationship between partisanship and generational age is found in Table 10.19. Because we are hypothesizing that partisanship is dependent on age, we place partisanship (the dependent variable) in the rows and age (the independent variable) in the columns. The first step in analyzing any data is to describe the pattern that we see. If we choose the row of Republicans, a pattern emerges. Among those respondents aged 65 and up (the Baby Boomer generation along with the remaining New Deal generation), 27.4 percent identify as Republicans. In the next generation (aged 50-64—sometimes called generation Jones), only 23.6 percent identify as Republicans. Generation X is similar to Generation Jones, with 23.8 percent identifying as Republicans. Generation Y (the Millennials) are even less likely to identify as Republicans with 17.6 percent. Overall, Millennials are less likely to identify as Republicans than Baby Boomers by a difference of 10 percentage points. The odd thing is that Millennials are also less likely to identify as Democrats than Baby Boomers—by a difference of 8 percentage points. It appears that those generations socialized during the Reagan years were not socialized into the Republican Party. Rather, each subsequent generation has become more likely to identify as Independents.

The question we are asking with lambda is whether knowing an individual's age helps us in predicting his or her partisanship. To answer this question, we compare the partisanship we would predict if we didn't know age with our prediction if we did know age. First, we look at the row marginals to see the distribution of partisanship throughout our sample. We see that 35.2 percent are Democrats, 41.5 percent are Independents, and 23.3 percent are Republicans. Because the largest category is Independent, our best guess on any respondent is that he or she is an Independent. We will be wrong 58.5 percent of the time (for all the Democrats and Republicans), but it is still our best guess—there is no way to do better unless we know more information about the respondent. Our hypothesis is that knowing the respondent's age will allow us to do a better job predicting partisanship.

The amount of information that age communicates is found in the cells of the table. For each category of our independent variable (age), we can make a new prediction of our dependent variable (partisanship). If we know that a particular respondent is under the age of 30, we would predict that he or she is an Independent because that is the most common partisanship in that age group. Similarly, for those between 30 and 49 and 50 to 59 years old, we would predict that they are Independents. Among respondents who are 65 years or older, we would predict that they are Democrats.

In making our predictions of partisanship based on our knowledge of age, we would still make mistakes. The question is whether we would make fewer mistakes. Our standard for measures of association is that a value of zero means a very weak association—in this case, we would be making the same number of errors. A value of one means that we have a perfect association—in this case, we would not make any mistakes in predicting partisanship. For lambda (λ), we get the proportion of errors that we eliminated by knowing our independent variable in comparison to the number we made originally using the equation

$$\lambda = (\text{errors without} - \text{errors with})/\text{errors without}$$

The errors without is the original number of errors we made when we predicted that any given respondent would be an Independent. We would be correct 41.5 percent of the time, for a total of 822 cases. But we would be wrong for the 696 Democrats and the 461 Republicans. Our number of errors without knowing our independent variable is 696 + 461 = 1157.

To find the number of errors we make with knowing the value of our independent variable, we look at each column individually. For those respondents under the age of 30, we decided to predict a partisanship of Independent. In this category, we are incorrect for the 122 Democrats as well as the 64 Republicans. In the category of 30- to 49-year-olds, we incorrectly predict that the 248 Democrats and the 168 Republicans will be Independents. Among 50- to 64-year-olds, we err in predicting that the 168 Democrats and 124 Republicans are Independents. Finally, among those 65 and older, we incorrectly predict that the 120 Independents and 105 Republicans are Democrats. Our total number of errors knowing the independent variable is 122 + 64 + 248 + 168 + 168 + 124 + 120 + 105 = 1119.

Plugging these two values into the equation for lambda, we get

$$\lambda = (\text{errors without} - \text{errors with})/\text{errors without}$$

$$= (1157 - 1119)/1157$$

$$= 0.033$$

The process of calculating lambda is summarized in Box 10.4. Looking up our lambda of 0.033 on our guideline Table 10.2, we would say that there is a very weak relationship between partisanship and generational age. We can also interpret lambda as a PRE measure: By knowing a respondent's age, we make 3.3 percent fewer mistakes in predicting their partisanship.

Limitations of Lambda

Although lambda is a useful measure of association for nominal-level data, it does have its limitations. Specifically, if the differences show up in categories that have few cases,

BOX 10.4 How to Calculate Lambda

1. Using the row marginals, identify the modal category of the dependent variable and calculate the number of errors without knowing the independent variable by adding up the number of cases that are *not* in the modal category.

2. Within each of the columns in the body of the table, find the modal category and add up the number of cases that are not in it. The number of errors with knowing the independent variable is the sum of errors across each of the columns.

3. Calculate lambda by getting the difference between the number of errors without and errors with and then dividing by the errors without: $\lambda = (\text{errors without} - \text{errors with})/\text{errors without}$. Interpret lambda as a PRE measure: By knowing the independent variable, we make that proportion fewer errors in predicting the dependent variable.

lambda is not going to do a good job measuring those differences. This can be seen to a certain degree in the partisanship/age example. Most of the age categories had more Independents than any other partisanship. The exception was among the Baby Boomers. This group showed a markedly different partisan preference. But because this age group amounted to less than one-fifth of the population, the difference did not translate into very many fewer errors. As a result, even though the percent differences in partisanship between the categories of age are large enough to make a difference in elections, the differences just don't show up well in lambda.

If you look at race as a predictor of partisanship, this problem is even more obvious. This relationship is shown in Table 10.20. The difference in partisanship is pretty stark: Blacks are more likely to be Democrats than whites by a difference of 35.6 percentage points. This relationship is statistically significant at the 0.01 level, so we can reject the null hypothesis. But because there are so few minorities, knowing the race of a respondent just doesn't eliminate very many errors in predicting partisanship. As a result, the lambda for this table is only 0.094—knowing race eliminates only 9.4 percent of our errors in predicting partisanship. Whenever the distribution of the independent variable is skewed in this way, lambda will fail to reflect the magnitude of the relationship between the dependent and independent variables. If the distribution of the independent variable is skewed, Cramer's V will probably be a better choice as a measure of association.

Similarly, a skewed distribution of the dependent variable will cause lambda to underestimate the strength of the relationship. Take the relationship between voter turnout and age. Younger citizens are less likely to vote than older citizens. Table 10.21 shows this

Table 10.20 Partisanship by Race

Party	White	Black	Other	Total
Democrat	29.0% (435)	64.6% (199)	35.8% (62)	35.2% (696)
Independent	43.1% (645)	29.2% (90)	50.3% (87)	41.5% (822)
Republican	27.9% (418)	6.2% (19)	13.9% (24)	23.3% (461)
Total	100.0%	100.0%	100.0%	100.0%
n =	1498	308	173	1979
			Lambda =	0.094**
			Cramer's V =	0.205**

Source: General Social Survey 2010, National Opinion Research Center, 2012. www3.norc.org/GSS+Website/Download/SPSS+Format/.

*prob. < 0.05

**prob. < 0.01

relationship. Those respondents who are at least 65 years old are more likely to vote than those under 30 by a difference of 34.5 percentage points. According to chi-square, the relationship is statistically significant at the 0.01 level—we can reject the null hypothesis. If we use Cramer's V, the association is 0.18. But lambda is zero—knowing the age of the respondent does not decrease the number of errors in predicting whether they voted or not. You can see why λ equals zero by looking at the data. From the marginals, we would predict that any given individual would vote. But in each of the age categories, we would also predict that all the members would vote. Even among the youngest group, which is least likely to vote, more respondents vote than do not vote. Because we would predict voting regardless of the age, knowing age does not decrease the number of errors we make in predicting voter turnout. If the dependent variable has a skewed distribution, Cramer's V would once again be a better measure of association.

SUMMARIZING THE MATH: MEASURES OF ASSOCIATION

Choosing the correct measure of association depends on the level of measurement of the two variables. The association between two interval-level variables can be measured with Pearson's r. If you have two ordinal variables, you can measure their association using gamma (or tau). For nominal variables, you can use lambda (or Cramer's V). Most measures

Table 10.21 Voter Turnout by Age

	Age				Total
	Less than 30	30-49	50-64	65+	
Voted	48.7% (181)	66.3% (477)	75.1% (405)	83.2% (327)	68.7% (1390)
Didn't Vote	39.8% (148)	28.1% (202)	21.9% (118)	15.8% (62)	26.2% (530)
Ineligible	11.6% (43)	5.6% (40)	3.0% (16)	1.0% (4)	5.1% (103)
Total	100.0%	100.0%	100.0%	100.0%	100.0%
n =	372	719	539	393	2023
				Lambda =	0.000**
				Cramer's V =	0.180**

Source: General Social Survey 2010, National Opinion Research Center, 2012. www3.norc.org/GSS+Website/Download/SPSS+Format/.

*prob. < 0.05

**prob. < 0.01

of association are set up so they give a value of one for a perfect relationship and a value of zero for no relationship. But when you are describing the relationship between two variables that have order (either ordinal or interval), you have the ability to distinguish between positive and negative relationships. In that case, you would want your measure to find an association of −1 for a perfect negative relationship. Both Pearson's r and gamma do that. If the two variables have different levels of measurement, you would generally use the measure of association for the lower-level variable. But if one variable is dichotomous, you generally use the measure of association appropriate for the other variable.

Pearson's r for Two Interval-Level Variables

Pearson's r is the appropriate measure of association for two interval-level variables. It takes the values of the two variables and sees how they co-vary, standardized by the magnitude of the standard deviations of both variables.

1. Create a work table with eight columns and a row for each case.
2. In the first column, place the case identifier.
3. In the second and fifth columns, place, respectively, the values of the independent and dependent variables. Calculate the mean for each by summing the values and dividing by n.
4. In the third and sixth columns, subtract the respective mean from the value of the variable in the previous column.
5. In the fourth and seventh columns, square the difference found in the previous column. At the bottom of each, calculate the sum of squares.
6. In the eighth column, multiply the (unsquared) differences from the third and sixth columns and sum the products at the bottom.
7. Calculate Pearson's r by taking the sum of the eighth column and dividing it by the square root of the product you get when you multiply the sum of squares for X (from column 4) by the sum of squares for Y (from column 7).

A Political Example: The Arab Spring and Internet Availability

In 2011, political turmoil rippled across the Arab League, leading to the replacement of four governments and turmoil in most of the other member states. Those protesting their governments communicated with each other digitally, getting ideas and making plans through Twitter and Facebook. Did the availability of the Internet lead these countries to be less stable? Table 10.22 shows the level of Internet usage along with the stability for these twenty-one member states. Internet usage is given as the number of Internet users per thousand population, and stability is the State Fragility Index we've used before, where a 25 indicates the most fragile a state can be. Does the communication that Internet usage makes available lead to greater instability in these countries? Calculate the Pearson's r.

Table 10.22 — Arab League Internet Usage and State Fragility

Country	Internet Users (per 1000)	State Fragility Index, 2010
Algeria	105	15
Bahrain	353	4
Comoros	34	13
Djibouti	14	13
Egypt	107	13
Iraq	2	17
Jordan	186	6
Kuwait	359	3
Lebanon	234	6
Libya	44	7
Mauritania	10	16
Morocco	216	6
Oman	106	5
Qatar	387	6
Saudi Arabia	225	10
Somalia	11	25
Sudan	38	24
Syria	180	10
Tunisia	168	8
United Arab Emirates	518	3
Yemen, Rep.	14	16

Sources: "Internet Statistics: Users (Per Capita)," *Nation Master*, 2012. www.nationmaster.com/graph/int_use_percap-internet-users-per-capita. Marshall, Monty G. and Benjamin R. Cole, "State Fragility Index and Matrix 2011." Center for Systemic Peace, 2012. www.systemicpeace.org/SFImatrix2011c.pdf.

1. Complete the work table to find the various sums of squares.
   I do this by completing the work table in Table 10.23.

Table 10.23 Work Table for Calculating Pearson's r for Arab League Internet Usage and State Fragility

Country	Internet	$X - \bar{X}$	$(X - \bar{X})^2$	SFI	$Y - \bar{Y}$	$(Y - \bar{Y})^2$	$(X - \bar{X})(Y - \bar{Y})$
Algeria	105	−52.6	2766.76	15	4.2	17.64	−220.92
Bahrain	353	195.2	38111.63	4	−6.8	46.24	−1327.51
Comoros	34	−123.4	15221.64	13	2.2	4.84	−271.43
Djibouti	14	−143.9	20720.74	13	2.2	4.84	−316.68
Egypt	107	−50.3	2529.99	13	2.2	4.84	−110.66
Iraq	2	−155.6	24222.56	17	6.2	38.44	−964.94
Jordan	186	28.6	816.99	6	−4.8	23.04	−137.20
Kuwait	359	201.6	40642.96	3	−7.8	60.84	−1572.49
Lebanon	234	76.7	5879.67	6	−4.8	23.04	−368.06
Libya	44	−114.0	12999.19	7	−3.8	14.44	433.25
Mauritania	10	−148.1	21930.06	16	5.2	27.04	−770.06
Morocco	216	58.7	3439.82	6	−4.8	23.04	−281.52
Oman	106	−51.5	2653.49	5	−5.8	33.64	298.77
Qatar	387	229.3	52574.82	6	−4.8	23.04	−1100.60
Saudi Arabia	225	67.0	4492.89	10	−0.8	0.64	−53.62
Somalia	11	−146.9	21565.80	25	14.2	201.64	−2085.31
Sudan	38	−119.5	14282.40	24	13.2	174.24	−1577.52
Syria	180	22.1	486.42	10	−0.8	0.64	−17.64
Tunisia	168	10.0	99.44	8	−2.8	7.84	−27.92
UAE	518	360.0	129564.00	3	−7.8	60.84	−2807.61
Yemen	14	−143.2	20507.67	16	5.2	27.04	−744.67
n = 21	ΣX = 3310		SS$_X$ = 435508.95	ΣY = 226		SS$_Y$ = 817.84	SS$_{XY}$ = −14024.3
	$\bar{X}$ = 157.6			$\bar{Y}$ = 10.8			

2. Calculate Pearson's r by taking the sum of the eighth column and dividing it by the square root of the product you get when you multiply the sum of squares for X (from column 4) by the sum of squares for Y (from column 7).

$$\text{Pearson's } r = \frac{\Sigma(X-\bar{X})(Y-\bar{Y})}{\sqrt{\Sigma(X-\bar{X})^2 \ \Sigma(Y-\bar{Y})^2}}$$

$$= \frac{SS_{XY}}{\sqrt{SS_X \ SS_Y}}$$

$$= \frac{-14{,}024.30}{\sqrt{(435{,}508.95)(817.84)}}$$

$$= -0.743$$

We can conclude that there is a very strong negative relationship between state fragility and Internet usage. Those countries with higher Internet usage per thousand population are less fragile politically.

Gamma for Two Ordinal-Level Variables

You use gamma to measure the level of association between two ordinal-level variables. Gamma varies from a negative one to a positive one. If there is a positive relationship, gamma reports a positive number. If the relationship is negative, gamma is negative. As with all measures of association, zero indicates that there isn't a relationship. Gamma compares the number of pairs of data points that support a positive relationship (called concordant pairs) to the number of pairs of data points that support a negative relationship (called discordant pairs).

1. Set up a table with six columns and two more rows than you have cells in the original contingency table.

2. In the first column, identify each of the cells in your table, placing its frequency in the second column.

3. In the third column, calculate the number of concordant cases for each cell by totaling the number of cases in all the cells down and to the right of the cell in question.

4. In the fifth column, calculate the number of discordant cases for each cell by totaling the number of cases in all the cells down and to the left of the cell in question.

5. In the fourth column, calculate the number of concordant pairs by multiplying the frequency in column 2 by the number of concordant cases in column 3. Total this column to get the total number of concordant pairs (C).

6. In the sixth column, calculate the number of discordant pairs by multiplying the frequency in column 2 by the number of discordant cases in column 5. Total this column to get the total number of discordant pairs (D).

7. Calculate gamma by taking the difference between the number of concordant and discordant pairs and dividing by the sum of the number of concordant and discordant pairs: $\gamma = (C - D)/(C + D)$.

A Political Example: The Arab Spring and Internet Availability Revisited. Perhaps rather than being the cause of the Arab Spring, availability of the Internet simply served as a catalyst to incite agitators to be more incendiary in their behavior. In that case, those Arab League countries with more Internet access might have had more extreme results. Table 10.24 shows the relationship between the level of activity during the Arab Spring and the level of Internet access. Calculate gamma to see whether Arab League countries with high levels of Internet access (more than 200 per thousand population) had higher levels of involvement in the Arab Spring.

Table 10.24 Impact of Arab Spring by Level of Internet Access

Arab Spring	Low	High	Total
Minor Protests	35.7% (5)	57.1% (4)	42.9% (9)
Major Protests	28.6% (4)	28.6% (2)	28.6% (4)
Civil Upheaval	35.7% (5)	14.3% (1)	28.6% (6)
Total	100.0%	100.0%	100.1%
n =	14	7	21

Sources: "Internet Statistics: Users (Per Capita)," *Nation Master*, 2012. www.nationmaster.com/graph/int_use_percap-internet-users-per-capita. "Arab Spring," *Wikipedia, The Free Encyclopedia*, October 16, 2012. http://en.wikipedia.org/wiki/Arab_Spring.

1. Set up the appropriate work table to find gamma.
   This is found in Table 10.25.

2. Total the product of cell frequencies and concordant cases in the fourth column to get the total number of concordant pairs (C).
   I get a total of 19 concordant pairs.

3. Total the product of cell frequencies and discordant cases in the sixth column to get the total number of discordant pairs (D).
   I get a total of 55 discordant pairs.

Table 10.25	Work Table for Calculating Gamma				
Cells	f_x	Concordant Cases	Concordant Pairs	Discordant Cases	Discordant Pairs
Minor, Low	5	3	15	0	0
Minor, High	4	0	0	9	36
Major, Low	4	1	4	0	0
Major, High	2	0	0	5	19
Upheaval, Low	5	0	0	0	0
Upheaval, High	1	0	0	0	0
			$\Sigma C = 19$		$\Sigma D = 55$

4. Calculate gamma by taking the difference between the number of concordant and discordant pairs and dividing by the sum of the number of concordant and discordant pairs: $\gamma = (C - D)/(C + D)$.

$$\gamma = (C - D)/(C + D)$$
$$= (19 - 55)/(19 + 55)$$
$$= -0.486$$

There is a strong inverse relationship between Arab Spring involvement and Internet availability. Those countries with higher Internet usage had lower levels of violence during the Arab Spring.

Lambda for Nominal-Level Variables

Lambda can be a very useful measure of association for nominal-level variables. Like most measures of association, it ranges between zero and one. And also like most measures of association, it can be interpreted as an indication of the strength of a relationship. As a PRE measure, it is directly interpreted as the proportion of errors in predicting the dependent variable explained by the independent variable. Unfortunately, it is very sensitive to variables with skewed data. If either the independent or dependent variables are dominated by one category, lambda will underestimate the strength of the relationship.

1. Find the number of "errors without" knowing the independent variable: Predict the most frequent category of the dependent variable and total the number of cases in all the other categories.

2. Calculate the number of "errors with" knowing the independent variable: For each column, predict the most frequent category of the dependent variable and total the number of cases in the other categories.

3. Calculate lambda by taking the difference between the errors without and the errors with and dividing by the errors without: λ = (errors without − errors with)/errors without. Interpret lambda as a PRE measure: the proportions of errors you eliminate by knowing the independent variable.

A Political Example: Freedom and Religion. How does the religious composition of a country affect its level of freedom? Are Christian-dominated countries any freer than Muslim-dominated countries? How about countries that are religiously diverse? Table 10.26 shows the relationship between the Freedom House score we've used before and the religious makeup of the country. Calculate lambda. Does knowing what religion dominates a country help us to predict how free it is?

1. Find the number of "errors without" knowing the independent variable: Predict the most frequent category of the dependent variable and total the number of cases in all the other categories.

   By looking at the marginals, I see that the mode for my Freedom variable is the category "Free," so I predict that all countries fall into this category. In doing so, I incorrectly predict the 48 "Not Free" and 59 "Part Free" countries. My "errors without" = 48 + 59 = 107.

Table 10.26 Freedom by Dominant Religion

	No Dominant Religion	Christian	Muslim	Other Dominant Religion	Total
Not Free	42.1% (8)	10.3% (12)	54.3% (25)	25.0% (3)	24.7% 48
Part Free	26.3% (5)	27.4% (32)	39.1% (18)	33.3% (4)	30.4% 59
Free	31.6% (6)	62.4% (73)	6.5% (3)	41.7% (5)	44.8% 87
Total	100.0%	100.0%	99.9%	100.0%	100.0%
n =	19	117	46	12	194
				Cramer's V =	0.373**

Source: Freedom House, Freedom in the World 2012. http://freedomhouse.org/report/freedom-world/freedom-world-2012; "Religions," *CIA World Factbook*, 2012. www.cia.gov/library/publications/the-world-factbook/fields/print_2122.html.

*prob. < 0.05

**prob. < 0.01

2. Calculate the number of "errors with" knowing the independent variable: For each column, predict the most frequent category of the dependent variable and total the number of cases in the other categories.
Among the countries with no dominant religion, I predict that they will all be "Not Free." In this category, because there are five countries that are "Part Free" and six countries that are "Free," I make eleven errors in this column. Among the Christian countries, I predict that they will be "Free" because this is the most frequent row in this column. In doing so, I incorrectly predict the twelve "Not Free" and thirty-two "Part Free" countries. Among the Muslim countries, "Not Free" is the most frequent category, so that is what I predict for these countries. I err for the eighteen "Part Free" and three "Free" countries. Among the countries with another dominant religion, I predict that they will be "Free." My errors in this column include the three "Not Free" and four "Part Free" countries. I total these errors to calculate my "errors with": 5 + 6 + 12 + 32 + 18 + 3 + 3 + 4 = 83.

3. Calculate lambda by taking the difference between the errors without and the errors with and dividing by the errors without: λ = (errors without − errors with)/errors without. Interpret lambda as a PRE measure: the proportions of errors you eliminate by knowing the independent variable.
Lambda = λ = (errors without − errors with)/errors without = (107 − 83)/107 = .224. This suggests that there is a moderate relationship between freedom and religious composition. We can reduce our error in predicting the level of freedom by 22.4 percent if we know which religion is dominant in a country.

USING SPSS TO ANSWER A QUESTION WITH MEASURES OF ASSOCIATION

We have already learned how to get contingency tables in SPSS. In order to properly describe the pattern, you need to be sure that you have the dependent variable in the rows and the independent variable in the columns. In addition, you learned how to use the "Cells" command to make sure that the cells contain the column percent so that you can compare it across a specific row to see if a pattern emerges. In order to identify the statistical significance, you already learned to use the "Statistics" command to request the chi-square for the table. In order to evaluate the strength of the association, you need to augment the request you made in the "Statistics" window to include the appropriate measure of association. Your choice of measure of association depends on two things: the level of measurement of your variables and the nature of their relationship. If at least one of the variables is nominal, the appropriate measures of association are Cramer's V and lambda. If both variables are ordinal and the relationship is directional (either direct or inverse), the appropriate measures of association are gamma or tau (tau-b if there is an equal number of columns and rows, tau-c if not). All of these measures of association are found in the "Statistics" window of the Crosstabs command in SPSS.

If, however, both variables have either an interval or a ratio level of measurement and the relationship is linear, the appropriate measure of association is Pearson's r.

This measure of association is not found within the Crosstabs command because you do not normally create a contingency table for two interval-level variables because the size of the table would be unwieldy. Sometimes, you do want to present this kind of a relationship in a table. If so, you would first collapse the two variables into ordinal-level variables. But if you decide to do that, you should still report the Pearson's r for the uncollapsed variables.

BOX 10.5 How to Get a Correlation Matrix with SPSS

>Analyze
>>Correlate
>>>Bivariate
>>>>*First Variable*→Variables
>>>>*Second Variable*→Variables
>>>>*Third Variable*→Variables
>>>>...
>>>>>OK

To get Pearson's r, you use the "Correlate" command under "Analyze." You'll recall that normally when you see a statistical analysis that concludes that two variables are correlated, the analysis has measured Pearson's r. Because you are measuring the relationship between two variables, once you indicate you want a correlation, you will designate that it be "Bivariate." The window that opens up will have one box that contains all your variables, and an empty "Variables" box into which you can put as many interval-level variables as you want. This process is summarized in Box 10.5. The table that SPSS produces in the output window is called a correlation matrix. It will have each of the variables you requested listed at the top of each column and to the left of each row. Each cell contains the bivariate correlation between the variable identified at the top of the column and the variable identified at the left of each row.

Because the same variables are in the rows and the columns, correlation matrices have an interesting structure. When you look at one, you'll notice that along the diagonal from the upper left down to the lower right, all the correlations are a perfect 1.0. That makes sense because that diagonal is the correlation of each variable with itself. Because there is a perfect correlation between a variable and itself, the Pearson's r always equals one. You should also notice that the two triangles on each side of the diagonal are mirror images of each other. This is because Pearson's r is a symmetrical measure of association: It calculates the same value regardless of which variable is the dependent and which the independent variable. Finally, be sure to look within each cell of the matrix; it contains not only the Pearson's r, but also the statistical significance. If you get confused about what each of the numbers are, check the row heading to the left—it gives a key to the contents of the cell. As you analyze your results, keep in mind that even though the correlation matrix can include more than two variables, each of the correlations is only for the relationship between those two variables; it does not control for the other variables in the matrix. We'll be able to do that when we get to multiple regression in Chapter 13.

An SPSS Application: The Internet and Political Instability

It is June 2013. The National Security Agency (NSA) has come under media scrutiny for the revelation of two different programs, the first requiring Verizon to release all its phone records and the second (called PRISM) monitoring Internet usage in real time. You are interning at Free Press, an organization advocating for universal and affordable Internet. Your supervisor, Timothy Karr, the senior director of strategy, is concerned that a Pew survey indicates widespread support for surveillance on the basis of protecting national security.[2] He wants to respond by talking about Internet usage as a source of national stability. He asks you to look at how strongly correlated Internet usage is with stability in comparison with other factors that are known to be highly correlated: GDP, freedom, and corruption.

1. Open your data and clean them.
   I collect the following data for all the countries of the world. For my dependent variable, I will use the State Fragility Index (which is actually the inverse of stability—the countries with the lowest values are the most stable). For my independent variables, I get Internet usage from Nation Master on the number of Internet users per thousand population. GDP per capita comes from GapMinder; freedom comes from Freedom House; and corruption comes from Transparency International (for this variable, the most corrupt countries have the lowest values). Once I have the data in SPSS, I check to make sure all my "−9" values have been set to missing.

2. Get a correlation matrix between your variables using the "Correlate," "Bivariate" command.
   Figure 10.7 shows the location of the commands. As shown in Figure 10.8, I enter my five variables into the Variables box. Figure 10.9 shows the resulting correlation matrix. I notice that all of the correlations are negative. This is because my dependent variable is measuring fragility, not stability. As Internet usage, GDP, freedom, and transparency increase, fragility decreases. I find it interesting that Internet usage has a stronger correlation than GDP.

3. Write a memo. Be sure to describe the pattern, identify the statistical significance, and evaluate the strength of the association.

Memo

To: Timothy Karr, Free Press
From: T. Marchant-Shapiro, intern
Subject: Internet Usage and National Stability
Date: June 14, 2013

CHAPTER 10 Measures of Association: Making Connections   313

Figure 10.7   Commands to Get Pearson's r in SPSS

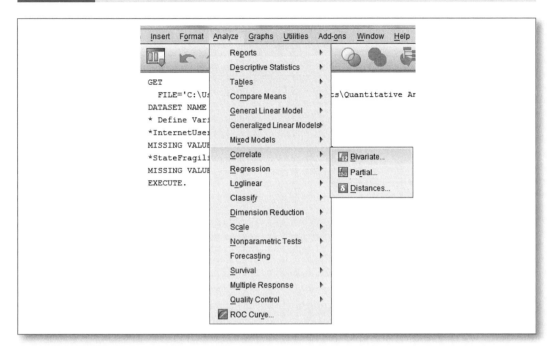

Figure 10.8   How to Get a Correlation Matrix in SPSS

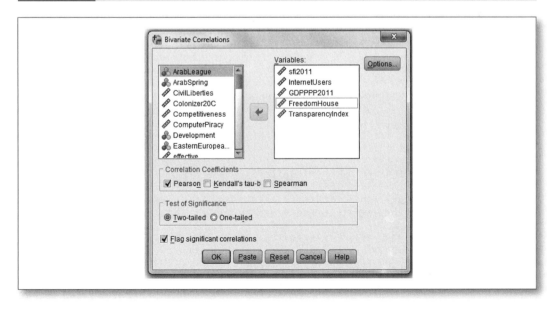

Figure 10.9 SPSS Correlation Matrix for State Fragility Index, 2011, Internet Users, GDP, and Freedom

Correlations

		sfi2011	InternetUsers	GDPPPP2011	FreedomHouse	TransparencyIndex
sfi2011	Pearson Correlation	1	-.766**	-.651**	-.702**	-.710**
	Sig. (2-tailed)		.000	.000	.000	.000
	N	164	163	164	164	163
InternetUsers	Pearson Correlation	-.766**	1	.759**	.581**	.824**
	Sig. (2-tailed)	.000		.000	.000	.000
	N	163	189	183	189	178
GDPPPP2011	Pearson Correlation	-.651**	.759**	1	.395**	.781**
	Sig. (2-tailed)	.000	.000		.000	.000
	N	164	183	184	184	179
FreedomHouse	Pearson Correlation	-.702**	.581**	.395**	1	.649**
	Sig. (2-tailed)	.000	.000	.000		.000
	N	164	189	184	194	179
TransparencyIndex	Pearson Correlation	-.710**	.824**	.781**	.649**	1
	Sig. (2-tailed)	.000	.000	.000	.000	
	N	163	178	179	179	179

**. Correlation is significant at the 0.01 level (2-tailed).

You asked me to assess how Internet usage affects the stability of countries in comparison to other factors. I was able to find a measure of Internet usage on Nation Master and decided to use the State Fragility Index as the inverse of stability. The correlation of Internet usage with stability is on a par with the other factors.

Increases in Internet usage, GDP per capita, freedom, and governmental transparency are all associated with increasing levels of stability. The correlation between Internet usage and the State Fragility Index is −0.766; for GDP, −0.651; for freedom, −0.702; and for transparency, −0.710. For all relationships, there is less than a 1 percent chance that we are observing the correlation simply due to chance.

I find it interesting that Internet usage actually has the highest correlation for all the variables. This would suggest that the impact of the Internet on the stability of a country is not simply an artifact of the Internet being more readily available in wealthy, free countries. Keep in mind, though, that these are bivariate relationships. I would advise you to do further analysis in order to determine whether the impact of the Internet on stability remains equally strong once you have controlled for these other factors.

Your Turn: Measures of Association

YT 10.1 Identify the levels of measurement (nominal, ordinal, interval/ratio) of the following variables:

1. Political Knowledge (number of correct answers to ten political questions)
2. Religious Affiliation (To which church do you belong?)

CHAPTER 10 Measures of Association: Making Connections **315**

3. Political Activism (the number of specified political activities in which the respondent has participated)
4. Frequency of Voting (never, sometimes, usually, always)
5. Current Elected Office (school board, city council, mayor, state legislature)
6. Region (East, South, Midwest, West)
7. Campaign Contacts (never, once, 2-5 times, 5-10 times, more than 10 times)

YT 10.2 For each of the following variables, identify the level of measurement. For each pair, choose the appropriate measure of association (Pearson's r, gamma, lambda) for the relationship:

1. IV = Military Service (Army, Navy, Air Force); DV = Ideology (Liberal, Moderate, Conservative)
2. IV = Convention Viewing (number of speeches watched); DV = Support for Political Party (100-point feeling thermometer)
3. IV = Freedom (Least Free, Some Free, Most Free); DV = Economy (Pre-emerging, Emerging, Advanced)

YT 10.3 If Internet usage did not lead to the instability among Arab League members that led to the Arab Spring, is it possible that political corruption did? (Remember that the Transparency Index is a measure of corruption where 0 is the most and 10 is the least corrupt.)

1. Complete the worktable in Table 10.27.
2. Calculate Pearson's r.
3. What can you conclude?

YT 10.4 In the chapter, Tables 10.10 through 10.12 are three sample contingency tables showing a perfect positive relationship, a perfect negative relationship, and the null hypothesis. In a similar way, set up three sample contingency tables, except instead of making them 2 × 2 tables like those in the chapter, make them 3 × 3 tables. This will be easiest if you have a total of three hundred cases distributed evenly between the three columns. Adjust the cell counts in the three tables to show relationships that you think would approximate the following strengths:

1. A strong negative relationship (not quite as strong as Table 10.9)
2. A weak negative relationship (a little more negative than Table 10.10)
3. A moderate positive relationship (halfway between Table 10.8 and Table 10.10)

YT 10.5 Table 10.28 shows the relationship between views on abortion and political ideology.

1. Eyeball the data and describe the pattern you see.
2. Calculate gamma.
3. What can you conclude?

YT 10.6 Table 10.29 returns to the gender gap.

1. Calculate lambda.
2. As a PRE measure, how do you interpret it?

Table 10.27 Work Table to Find Correlation between State Fragility and Corruption

Country	TI	$X - \bar{X}$	$(X - \bar{X})^2$	SFI	$Y - \bar{Y}$	$(Y - \bar{Y})^2$	$(X - \bar{X})(Y - \bar{Y})$
Algeria	2.9			15			
Bahrain	5.1			4			
Comoros	2.4			13			
Djibouti	3.0			13			
Egypt	2.9			13			
Iraq	1.8			17			
Jordan	4.5			6			
Kuwait	4.6			3			
Lebanon	2.5			6			
Libya	2.0			7			
Mauritania	2.4			16			
Morocco	3.4			6			
Oman	4.8			5			
Qatar	7.2			6			
Saudi Arabia	4.4			10			
Somalia	1.0			25			
Sudan	1.6			24			
Syria	2.6			10			
Tunisia	3.8			8			
UAE	6.8			3			
Yemen	2.1			16			
n = 21	$\Sigma X =$		$SS_X =$	$\Sigma Y =$		$SS_Y =$	$SS_{XY} =$
	$\bar{X} =$			$\bar{Y} =$			

Sources: Marshall, Monty G. and Benjamin R. Cole, "State Fragility Index and Matrix 2011." *Center for Systemic Peace,* 2012. www.systemicpeace.org/SFImatrix2011c.pdf; Transparency International, *Corruption Perceptions Index 2011.* http://www.transparency.org/cpi2011/in_detail.

Table 10.28 Support for Legal Abortions by Political Ideology

		Ideology		Total
	Liberal	Moderate	Conservative	
Never	8.7% (31)	19.3% (21)	18.8% (96)	15.1% (148)
If rape or incest	17.9% (64)	22.9% (25)	37.5% (182)	28.7% (281)
Other times	15.1% (54)	22.0% (24)	15.6% (80)	16.1% (158)
Always	58.4% (209)	35.8% (39)	28.1% (144)	40.0% (392)
Total	100.1%	100.0%	100.0%	99.9%
n =	358	109	512	979

Source: American National Election Study, *ANES 2008 Time Series Study*. www.electionstudies.org/studypages/2008prepost/2008prepost.htm.

Table 10.29 Partisanship by Gender, 2010

Party	Men	Women	Total
Democrat	28.4% (244)	40.3% (452)	35.2% (696)
Independent	45.2% (388)	38.7% (434)	41.5% (822)
Republican	26.3% (226)	21.0% (235)	23.3% (461)
Total	99.9%	100.0%	100.0%
n =	858	1121	1979

Source: General Social Survey 2010, National Opinion Research Center, 2012. www3.norc.org/GSS+Website/Download/SPSS+Format/.

Apply It Yourself: Measure Poor Student Graduation Rates

It is 2013. You are working for the summer as a research assistant at the Urban Institute that has just published its study, "The Moynihan Report Revisited."[3] The original Moynihan Report had called for change because in 1965, race and poverty were so intertwined in America that life for many seemed hopeless. The new report found that, although high school graduation for minorities now is very close to the average for whites, black men in particular are still less likely to graduate from college. While you were helping with the report, you noticed that there is a great deal of variation between states in their high school graduation rates for the poor. You ask Robert I.

Lerman, an Institute Fellow who is interested in education policy, whether it might be a good idea to study why some states are able to get more poor students to graduate than others. He suggests that you begin by finding how various factors (spending on education, pre-K attendance, median income, and percent minority) correlate with that rate.

1. Open up the "StateData.sav" dataset in SPSS. In the data window, go to the "Variable View." Make sure that SPSS is set to show the variable names alphabetized.

    >Data Set Window
    >>Variable View
    >>>Edit
    >>>>Options
    >>>>>Display Names
    >>>>>Alphabetical

2. Go into the "Define Variable Properties" window to make sure that the variables you are going to use (GradPoor, EdSpending, PreK, MedianIncome, Minority) are clean with all the "–9" values set to missing.

3. Get a correlation matrix.

    >Analyze
    >>Correlate
    >>>Bivariate
    >>>>GradPoor→Variables
    >>>>EdSpending→Variables
    >>>>PreK→Variables
    >>>>MedianIncome→Variables
    >>>>Minority→Variables
    >>>>OK

4. Write your memo answering the question. In the introduction, you need to describe the question, identify your data source and measures, and briefly answer the question. In the body, remember that you always need to describe the pattern, decide whether there is a relationship (statistical significance), and indicate how strong it is (measure of association). Conclude by summarizing your results and indicate what you would like to do next with the research. For grading purposes, attach the SPSS correlation matrix to your memo.

Key Terms

Concordant pair (p. 287)
Dichotomous variable (p. 277)
Discordant pair (p. 287)
Gamma (p. 274)

Kendall's tau (p. 296)
Lambda (p. 274)
Pearson's r (p. 274)
Symmetrical measure of association (p. 283)

CHAPTER 11

Multivariate Relationships

Taking Control

When Serena Williams prepares for a game, she carefully brings her shower sandals to the court. She ties her shoelaces in a particular way. Then, before her first serve, she bounces the ball precisely five times. For all subsequent serves, she limits herself to precisely two bounces. She credits her wins to faithfully following this routine; she credits her losses to her failure to comply with it.[1] Williams is not alone among athletes who credit their successes (and failures) to superstitious behaviors. But observers of these quirks can't help but remember that correlation does not prove causation.

In *How to Lie with Statistics,* Darrell Huff gives three reasons that a correlation might not imply causation. First, the correlation could be due to chance. Second, you might have reversed the order of the causal relationship. Third, both variables could be the effects of a third cause.[2] All three of these problems can occur.

First, the correlation could be produced by chance. Recently, the *Journal of Personality and Social Psychology* published a study by a respected psychologist, Daryl J. Bem,[3] purporting to find evidence of extrasensory perception (ESP—mindreading!). Critics of the study say that Bem's findings are simply a result of chance—if you keep studying something long enough, eventually you'll randomly select a sample in which the data happen to support your hypothesis. The failure of at least three studies to replicate Bem's findings seems to support the critics.[4] In discussing the scientific method, we emphasize the importance of reliability because the ability to replicate an interesting finding indicates that it is not a random event. If you find a correlation, you always need to check to see if it could be the result of chance.

Second, it could be a real relationship, but you might have gotten it backwards when you identified which variable is the effect and which the cause. For example, a new study in the *Journal of the American Medical Association* indicates that the faster older people walk, the longer they can expect to live.[5] The suggestion is that walking fast will help you live longer. But isn't it equally possible that healthy people can walk faster? It is always worth asking whether the effect you want to explain could actually be influencing what you thought of as the cause.

Third, the correlation could indicate a spurious relationship that shows up because both variables are correlated with something else. For example, if you correlated the price of wheat with the price of gold over the past hundred years, you would probably get a very high correlation. But beyond their color, the two commodities have nothing in common. The correlation can be attributed to the fact that the cost of everything has increased for the past century. Because you know that correlation does not prove causation, when you find a correlation, you should always ask whether there could be a third variable that is influencing both your "cause" and your "effect." If there is such a variable, you need to be sure to control for it in your analysis.

A political example of this happened in the midst of congressional budget hearings in March 2011. During these hearings, Republican representative Dennis Ross of Florida went on the warpath against the salaries paid to federal employees. He pointed out that federal workers earn, on average, $101,628 in wages and benefits, whereas private employees average only $60,000.[6] His argument suggested that at least some of our budget woes are attributable to overpaid federal employees. We could conclude that balancing the budget should come, at least in part, by fixing that discrepancy.

What Representative Ross failed to do, however, was control for the types of jobs that the workers perform. If he had done so, he would have found that the average government lawyer is paid $127,500, whereas the average lawyer in the private sector is paid $137,540. The reason that the overall average federal employee is paid so much can be attributed to the fact that the government has outsourced many of its lower paying jobs, leaving the balance of government jobs in the higher paying professional and administrative categories. If Representative Ross had compared the public and private salaries for specific jobs, he would have found that, controlling for type of job, public employees are paid less than private employees. In this chapter, we will learn how to control for a third variable by using three-way contingency tables.

SPURIOUS RELATIONSHIPS

Frequently, when we want to answer a specific question about the relationship between our dependent and independent variables, we need to consider the possibility that there might be another factor connected to both variables for which we need to control. Social and political behavior is very messy—there are many variables that can affect any one dependent variable. Sometimes (frequently!) there is a third variable that is associated with both our independent and dependent variables. If we do not control for this third variable, then our independent variable will pick up its effect on our dependent variable and so will appear to be correlated with the dependent variable much more strongly than it actually is. For example, if you were to go to a high school parking lot and identify the value of each of the cars, you would probably find that it is highly correlated with the driver's GPA. But that doesn't mean that giving fancy cars to teens will increase their GPAs. The causal relationship would actually be connected to a third variable: family income. Children from wealthier families have nicer cars and children from wealthier families do better in school. When two variables appear to be correlated but aren't, we call it a **spurious relationship**.

Once you control for the third variable, the correlation vanishes. The third variable can be related in two different ways: It can be an antecedent variable or an intervening variable.

Antecedent Variables

Sometimes, we need to control for a third variable that has a causal effect on both our independent and dependent variables. When a variable comes before both our variables in the causal chain, we call that an antecedent variable. Figure 11.1 shows a model for that kind of relationship. It appears that there is a relationship between the independent variable and the dependent variable, but the only real relationships are between the antecedent and independent variables and the antecedent and dependent variables. For example, for many years, we took for granted that Catholics in the United States would vote for the Democratic Party. But it wasn't being Catholic that affected the vote, it was being an immigrant. Irish and Italian immigrants came initially to big cities in the United States, where the Democratic Party machines were well established. These machines helped new immigrants find housing and jobs, in return for which the new immigrants were socialized into the Democratic Party. So for many years, being Catholic was highly correlated with being a Democrat, but that correlation reflected a spurious relationship. Immigrants tended to be Catholic and they tended to join the Democratic Party.

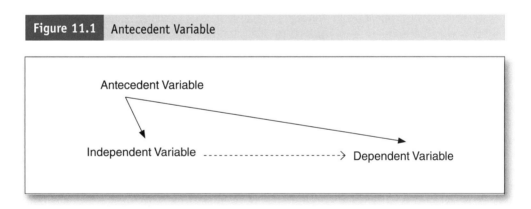

Figure 11.1 Antecedent Variable

Intervening Variables

Sometimes, our independent variable affects our dependent variable only indirectly, through an intervening variable. Figure 11.2 shows this kind of relationship. In this scenario, the independent variable affects the intervening variable, which in turn affects the dependent variable. In the past three decades, as the political socialization experienced by new immigrants a century ago has faded, Catholics have begun to vote for the Republican Party in greater numbers. Much of that shift has been fueled by an increasing focus on abortion as a political issue. To the degree that Catholics are pro-life and the Republican Party has, since 1980, increasingly taken a pro-life position on abortion, Catholics have moved to the Republican Party. In this case, an individual's view on abortion has become an intervening variable that pushes Catholics toward the Republican Party.

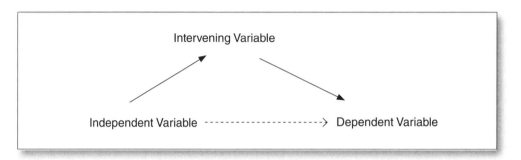

Figure 11.2 Intervening Variable

BOX 11.1 Numbers in the News

In 1990, the World Bank issued goals for improving the lives of the people of the world, called the Millennium Development Goals. One of these goals is the elimination of extreme poverty. One objective was to cut extreme poverty in half by 2015. In 2010, that objective had already been met. Unfortunately, that pattern was not consistent across the world. Much of the decline was attributable to increasing prosperity in one country: China.[7] Although poverty had declined in most of the countries of the world, once you control for region, you can see that the magnitude of the decline varies. Poverty has not declined in Africa and Southeast Asia as much as it has elsewhere in the world.[8]

SPURIOUS NON-RELATIONSHIPS

We've seen that with spurious relationships, correlation does not prove causation. It is important to include control variables that we know are associated with both our independent and dependent variables in order to be confident that the original correlation reflects reality. But sometimes, we'll be faced with the opposite of a spurious relationship. On occasion, we need to include a control variable because its absence will mask a real relationship. In that case, including a control variable will allow a real relationship that was previously invisible to appear. This happens when one of two highly correlated variables has a positive relationship with the dependent variable and the other has a negative relationship. If only the variable with the positive relationship is included, it will pick up the negative relationship of the excluded variable.

If either variable is excluded, the correlation between the included variable and the control variable will compound the underlying positive and negative relationships. The two opposing pressures cancel each other out and there appears to be no relationship. Only if you include both the independent and control variables do you see the separate positive and negative pressures on the dependent variable. This is shown in Figure 11.3. You can find an example of this kind of countervailing relationship once again with Catholics. To the degree that the political views of Catholics are highly influenced by their views on the sanctity of life, those views may, in some instances, push them toward the Republican Party (as in the case of abortion) and in others push them toward the Democratic Party (as in the case of the death penalty). In order to unmask the influence of each, you would want to include views on both these issues as independent variables.

Figure 11.3 Countervailing Relationship

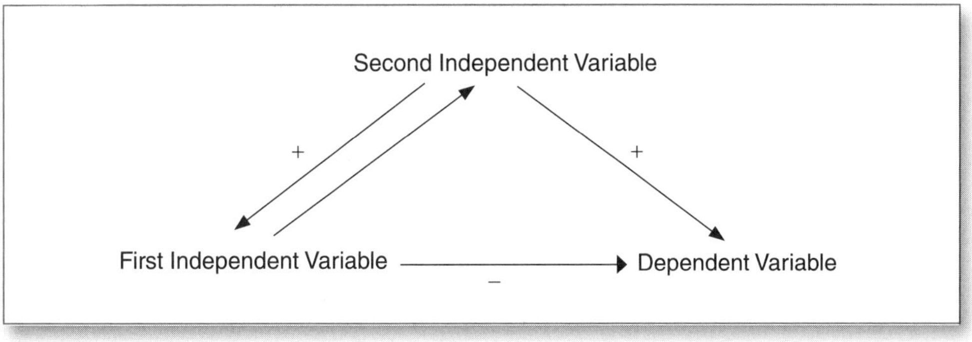

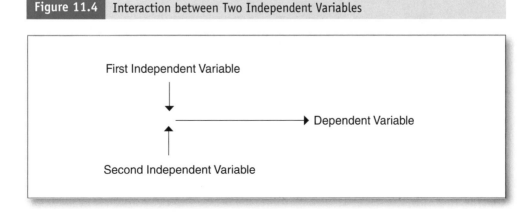

INTERACTION EFFECTS

Finally, on occasion, two variables interact in affecting the dependent variable. Figure 11.4 models how this can occur. For example, in some instances, foreign aid can help a country, and in others, it can hurt. In well-established democracies, foreign aid is normally used to build infrastructure. In contrast, in countries with weak governments, it tends to be used to

Figure 11.4 Interaction between Two Independent Variables

enrich office holders. Thus, there is an interaction between foreign aid and government type that influences how well the aid is used. Again, the nature of the relationship can be seen only when you include the proper controls.

THREE-WAY CONTINGENCY TABLES

In the social sciences, we describe messy processes. The decisions that people make and the ways in which they interact with others are affected by many variables. This is not to say that we need to control for every factor that can influence our dependent variable. Thus far in this chapter, we've seen several ways that a third variable can affect the relationship between our independent and dependent variables. What is important to remember is that in each of them, the control variable is connected to both the independent and dependent variables. If there is a variable that is associated with both of our variables, then we need to control for it in order to see the direct effects of our independent variable on our dependent variable.

Table 11.1 Views on Abortion by Race

Views on Abortion	White	Black	Total
Pro-Life	26.9%	30.9%	27.5% (279)
Sometimes	31.4%	37.7%	32.5% (328)
Pro-Choice	41.7%	31.5%	40.1% (406)
Total	100.0%	100.1%	100.1%
n =	851	162	1013
		gamma = −0.147*	

Source: General Social Survey 2010, National Opinion Research Center, 2012. www3.norc.org/GSS+Website/Download/SPSS+Format/.

*prob. < 0.05

**prob. < 0.01

Take, for example, our finding in Chapter 9 that blacks are more likely to be pro-life than whites. Table 11.1 shows that relationship. The pattern that you see in this table is that blacks are more likely to be pro-life than whites by a difference of 4 percentage points. This relationship is statistically significant at the 0.05 level, so we can reject the null hypothesis. The gamma of −0.147 indicates that it is a weak relationship.

The question for this chapter is, Why are blacks more pro-life than whites? Because black churches tend to be more conservative than white churches, I hypothesize that it is not race, per se, that leads to the difference in views on abortion, but rather the level of religious fundamentalism. If I am correct, I would expect that controlling for religious fundamentalism should eliminate the relationship between views on abortion and race.

Table 11.2 Views on Abortion by Race, for Fundamentalists

Views on Abortion	White	Black	Total
Pro-Life	47.4%	32.5%	42.6% (109)
Moderate	32.9%	41.0%	35.5% (91)
Pro-Choice	19.7%	26.5%	21.0% (56)
Total	100.0%	100.0%	100.1%
n=	173	83	256
		gamma = 0.236*	

Source: General Social Survey 2006, National Opinion Research Center, 2012. Accessed October 9, 2012. www3.norc.org/GSS+Website/Download/SPSS+Format/.

*prob. < 0.05

**prob. < 0.01

In testing my hypothesis, one option available to me is to select only those respondents who consider themselves religious fundamentalists. In that case, I would end up with Table 11.2. In it, we can see that the relationship we observed in Table 11.1 reverses itself. Black fundamentalists are less likely to be pro-life than white fundamentalists by a difference of 15 percentage points. This direct relationship is statistically significant at the 0.05 level, so we can reject the null hypothesis. The gamma indicates that this is a moderate relationship. In a similar fashion, I could select for each of the categories of the control variable, set up a contingency table, and analyze it.

But it would be more efficient to create a single table that includes all the smaller tables for the subgroups of the control variable. We call the resulting table a three-way contingency table. In combining the tables for the subgroups, we end up with a table that has a subsection for each of the values of our control variable. If Representative Ross had analyzed the relationship between salary and public/private employment correctly, he would have done the comparison for each type of job. If he had done so, he would have found that within the jobs of lawyers, managers, engineers, and scientists, government employees have salaries that are about 20 percent lower than those of their private counterparts. He would also have found that among cooks and mailroom clerks, government employees make more than private employees—hence the outsourcing.[9]

Table 11.3 Views on Abortion by Race, Controlling for Religious Fundamentalism

Religious Fundamentalism	Views on Abortion	Race		Total
		White	Black	
Fundamentalists	Pro-Life	47.4%	32.5%	42.6% (109)
	Moderate	32.9%	41.0%	35.5% (91)
	Pro-Choice	19.7%	26.5%	21.0% (56)
		(173)	(83)	n = 256
				gamma = 0.236*
Religious Moderates	Pro-Life	31.6%	33.3%	31.8% (117)
	Moderate	34.0%	41.7%	34.8% (128)
	Pro-Choice	34.3%	25.0%	33.4% (123)
		(332)	(36)	n = 368
				gamma = −0.113
Religious Liberals	Pro-Life	11.8%	26.5%	13.3% (46)
	Moderate	28.4%	29.4%	28.5% (99)
	Pro-Choice	59.7%	44.1%	58.2% (202)
		(313)	(38)	n = 347
				gamma = −0.317*

Source: General Social Survey 2010, National Opinion Research Center, 2012. Accessed October 9, 2012. www3.norc.org/GSS+Website/Download/SPSS+Format/.

*prob. < 0.05

**prob. < 0.01

Question: *The number of situations in which the respondent said abortion should be legal: Pro-Life (Abortion should be legal in 0-2 instances); Moderate (3-4); Pro-Choice (5-7).*

Table 11.3 is the three-way contingency table for views on abortion by race, controlling for religious fundamentalism. In it, you'll notice that each category of the control variable has its own section divided by a line. The categories of the dependent variable repeat themselves for each section, and each section has its own measure of association. But the subsections share a common heading at the top of the table identifying the categories of the independent variable. To make it easier to compare the sections of the table, I've minimized the column marginal so that at the bottom of each section, I only give the column ns in parentheses.

It is much easier to analyze the data when they are set up in a three-way contingency table rather than separate tables for each category of the control variable. First, you can analyze each section of the table as if it were an individual table: describe the pattern, identify the statistical significance, and evaluate the strength of the relationship. If we were to do this for this table, we would analyze the first section of the table in the same way we analyzed Table 11.2. We would then move down to the second section and see that among religious moderates, blacks are more likely to be pro-life than whites by a difference of 2 percentage points, but this relationship is not statistically significant, so we cannot reject the null hypothesis. Among religious liberals, we see that blacks are more likely to be pro-life than whites by a difference of 15 percentage points. This relationship is statistically significant at the 0.05 level. The gamma of −0.317 indicates that there is a strong relationship between views on abortion and race among religious liberals.

After analyzing each section that is of interest, the second step is to compare the relationships in the different sections. This table highlights one of the advantages of a three-way contingency table: It allows you to see how the relationship can change with changes in the control variable. In Table 11.3, you observe three different relationships in the three sections. Among religious fundamentalists, blacks appear to be more pro-choice than whites; among religious moderates, blacks have views similar to whites; among religious liberals, blacks are more likely to be pro-life than whites. This table suggests that religious fundamentalism is an important intervening variable in the relationship between views on abortion and race.

If you look across the sections, you'll see that religious fundamentalism does not affect the views of abortion among blacks in the same way it does whites. In the column of whites, the proportion of respondents who are pro-life varies dramatically across the three categories of fundamentalism. In contrast, in the column of blacks, each section shows a very similar pattern. In addition, two-thirds of the blacks surveyed identified as fundamentalists as opposed to one-fourth of whites. Because the respondents in this group are uniformly more likely to be pro-life, the greater proportion of blacks who identify as religious fundamentalists leads them to appear in the aggregate to be more pro-life than whites even though black fundamentalists are less pro-life than white fundamentalists.

Box 11.2 summarizes how to set up a three-way contingency table. The title at the top of the table will be the same as for a regular contingency table, except that it will indicate the control variable: "*Dependent Variable* by *Independent Variable* Controlling for *Control Variable*." You will essentially have a separate table for each of the categories of your control variable. You will merge these together into a single table with a line dividing the subsections for each of the values of the control variable. The category of the control variable is at the top of each section in the far left column. In the second column, the categories of the dependent variable will be repeated in each section. The last column, headed "Total," has the row marginals for each subsection. The categories of the independent variable are given as headings at the top of the table above each of the remaining columns. Because your goal is to set it up in such a way that comparing the patterns of each of the subsections is easy to do, do not give the full column marginals. Instead, for each section, just give the column n at the bottom in parentheses. But do be sure to include separate statistics for each subsection at the bottom on the far right. Include both the appropriate measure of association and the appropriate number of stars for statistical significance. After the full table, include a single key for how many stars are connected to what level of statistical significance.

BOX 11.2 How to Create a Three-Way Contingency Table

1. Give the table a title in the form of the name of the "*dependent variable*" "by" the "*independent variable*" "controlling for" the "*control variable*."
2. As with a regular contingency table, the independent variable is found in the columns; the dependent variable, in the rows. Insert a table with three more columns (c) than the number of categories in the independent variable (IV); that is, $c = IV + 3$.
3. In addition to two rows for column headings, each section will need enough rows for the number of categories of the dependent variable (DV) plus one for the number of cases and one for the measure of association. Think of the number of categories of the control variable as "CV." To get the number of rows (r), use this formula: $r = (DV + 2)CV + 2$.
4. The first two rows contain headings. The first column heading should identify the control variable; the second, the dependent variable. In the middle columns, place the categories of the independent variable in the second row and center the label of the independent variable in the first row above the categories. The last column of the second row should be headed "Total" for the row marginals.
5. Below the heading in the second column, give the categories of the dependent variable followed by two blank lines. Repeat this as many times as there are categories in the control variable.
6. Go to each time you repeated the first category of the dependent variable. Consecutively place each of the categories of the control variable to the left in the first column.
7. Within the cells, you will include the column percent within each of the sections of the table.
8. Include the row marginals for each section as you would for a regular contingency table, with both the row percent and the number of cases in parentheses.
9. Instead of the complete column marginal, in the second to the last row of each section, just include the column n in parentheses.
10. In the lower right-hand cell for each section, give the value of the measure of association for that section with the appropriate number of stars to indicate statistical significance.
11. Divide the table by lines: at the top and bottom, below the column headings, and between each subsection.
12. Below the table, in a smaller font, give the information necessary to understand it: the data source, the key for statistical significance, and the question wording.

You will notice that there is a separate section for each category of the control variable, separated by a line. Those categories are noted in the leftmost column. Indented from that, each section is set up in the same way as described in Table 8.4 for a simple contingency table with the exception that there is a single title at the top of the full table that identifies the dependent and independent variables and identifies the control variable. Each section has its own level of statistical significance and measure of association that need to be noted.

To analyze a three-way contingency table, you treat each of the sections as separate contingency tables and analyze them using the same three-step process we have used previously. Beginning with the first section of interest to you, you describe the pattern by choosing a row and comparing the column percentages across it. Then you identify that section's statistical significance to determine whether you can reject the null hypothesis. Finally, you evaluate the strength of the association using the appropriate measure of association. You follow this same process for each section.

After completing the three-step analysis of the subsections, you can address the impact of the control variable on the relationship. To do this, you will compare how the measure of association changes across each of the sections of the table. There are three possibilities here. First, if the association you originally saw was spurious, it will evaporate if you have controlled for the appropriate causal factor. Second, if the control variable is not affecting the relationship between the dependent and independent variables, the magnitude of the measure of association will stay fairly constant. Note here that because you are dividing up your sample size, the statistics may be somewhat inconsistent just due to having fewer cases in the subsections. Finally, if there is some kind of interaction between your independent variable and the control variable, the measures of association might be different for the different categories of the control variable. This is what happened in Table 11.3 when we controlled for religious fundamentalism. By comparing the measures of association of the different sections of the three-way contingency table, you can give a more refined evaluation of the relationship between your dependent and independent variables.

SUMMARIZING THE PROCESS: SETTING UP THREE-WAY CONTINGENCY TABLES

Sometimes, the relationship we originally see between two variables is spurious: It is actually the result of a third variable that is associated with both our dependent and independent variables. If the third variable influences both our independent and dependent variables, it is called an antecedent variable; if our independent variable influences the third variable, which in turn influences our dependent variable, it is called an intervening variable. We need to control for any antecedent or intervening variables if we want to get an accurate picture of the influence that our independent variable has on our dependent variable. Sometimes, controlling for the third variable will leave the relationship unchanged; sometimes, it will disappear; and sometimes, it will emerge. One way to control for a third variable is with a three-way contingency table.

1. Give the table a title in the form of the name of the dependent variable "by" the independent variable "controlling for" the control variable.

2. Insert a table. The number of columns (c) equals three plus the number of categories in the independent variable (IV), or c = IV + 3. The number of rows (r) equals the number of categories in the dependent variable (DV) plus two, multiplied by the number of categories in the control variable (CV), with an additional two rows added at the end for the column headings: r = (DV + 2)CV + 2.

3. The first column heading should identify the control variable; the second, the dependent variable; and the middle columns, the independent variable and its categories. The last column should be headed "Total."

4. As you work down the second column, give the categories of the dependent variable followed by two blank lines. Repeat this as many times as there are categories in the control variable.

5. To the left of each repetition of the first category of the dependent variable, consecutively place each of the categories of the control variable in the first column.

6. Within the cells, you will include the column percent for each of the sections of the table.

7. Include the row marginals for each section as you would for a regular contingency table with both the row percent and the number of cases in parentheses.

8. Instead of the complete column marginal, at the bottom of each section just include the column n in parentheses.

9. In the lower right-hand cell for each section, give the value of the measure of association for that section with the appropriate number of stars to indicate statistical significance.

10. Divide each section of the table by lines.

11. Below the table, in a smaller font, give the information necessary to understand it: the key for statistical significance, the data source, and the question wording if needed.

A Political Example: Abortion and Religious Fundamentalism

The previous example looked at the impact of race on abortion views controlling for religious fundamentalism. If we had come at the question from the direction of the impact of religious fundamentalism on abortion views, we would have wanted to control for race. Table 11.4 shows this relationship. In contrast to Table 11.3, each subsection of the table corresponds with a different race, designated in the first column. The second column once again contains the categories of the dependent variable: pro-life, moderate, and pro-choice. The columns in the table proper designate the categories of the new independent variable, level of religious fundamentalism. One interesting thing to note on this table is that its cells correspond with the cells in Table 11.3. For example, in both tables, 47.4 percent of white religious fundamentalists are categorized as pro-life. But because we've changed our focus to asking about the relationship between religious fundamentalism controlling for race, the order of the cells has changed.

Table 11.4 Views on Abortion by Religious Fundamentalism, Controlling for Race

Race	Views on Legalizing Abortion	Religious Fundamentalism			Total
		Fundamentalists	Moderates	Liberals	
White	Pro-Life	47.4%	31.6%	11.8%	27.4% (224)
	Moderate	32.9%	34.0%	28.4%	31.7% (259)
	Pro-Choice	19.7%	34.2%	59.7%	41.0% (335)
		(173)	(332)	(313)	n = 818
					gamma = 0.478**
Black	Pro-Life	32.5%	33.3%	26.5%	31.4% (48)
	Moderate	41.0%	41.7%	29.4%	38.6% (59)
	Pro-Choice	26.5%	25.0%	44.1%	30.1% (46)
		(332)	(36)	(34)	n = 368
					gamma = 0.135

Source: GSS 2010 General Social Survey 2010, National Opinion Research Center, 2012. www3.norc.org/GSS+Website/Download/SPSS+Format/.

*prob. < 0.05

**prob. < 0.01

Question: *The number of situations in which the respondent said abortion should be legal: Pro-Life (Abortion should be legal in 0-2 instances); Moderate (3-4); Pro-Choice (5-7).*

Among whites, religious fundamentalists are more likely to be pro-life than religious liberals by a difference of 36 percentage points. This relationship is statistically significant at the 0.01 level, so we can reject the null hypothesis. The gamma of 0.478 indicates the relationship between views on abortion and religious fundamentalists is strong for whites. Among blacks, though, religious fundamentalists are more likely to be pro-life than religious liberals by only 6 percentage points. This relationship is not statistically significant at the 0.05 level, so we cannot reject the null hypothesis. We can conclude that there is an interaction between religious fundamentalism and race so that fundamentalism has more of an impact on views about abortion among whites than it does among blacks. By setting up the table in this way, we can see more clearly than we did in Table 11.3 that we have not yet fully answered our question about why blacks are more pro-life than whites.

USE SPSS TO ANSWER A QUESTION WITH A THREE-WAY CONTINGENCY TABLE

You already have the skills to use SPSS to create a contingency table, along with the appropriate data for the contents of the cells and the statistics for evaluating it. To create a three-way contingency table, you use the "Analyze," "Descriptive Statistics," and "Crosstabs" commands exactly as you have previously, with one addition. After you put the dependent variable in the rows and the independent variable in the columns, you place the control variable in the box that has the heading "Layer 1 of 1." Once again, you will choose the appropriate measure of association, but in determining that, you ignore the level of measurement of the control variable. The measure of association will be the same one you would have chosen in your original two-variable crosstab: Cramer's V or lambda for nominal-level variables and gamma for ordinal variables. As usual, you make sure that the cells contain the observed count and the column percent. Box 11.3 modifies Table 9.4 to allow for a control variable.

BOX 11.3   How to Get a Three-Way Cross Tabulation in SPSS

After opening and cleaning your data in SPSS:

>Analyze
>>Descriptive Statistics
>>>Crosstabs
"*Dependent Variable*"→Row Variable
"*Independent Variable*"→Column Variable
"*Control Variable*"→Layer 1 of 1
>>>>Cells
>>>>>Observed
>>>>>Column
>>>>>Continue
>>>>Statistics
>>>>>Chi-square
>>>>>Phi and Cramer's V
>>>>>*any other statistics that you might want*
>>>>OK

In the output, you'll notice that the crosstab has multiple sections—one for each of the categories of the control variable. Also, at the end you'll see that there is an additional section for the original two-way crosstab. This means that although you will normally want to present the uncontrolled relationship, you do not actually need to request a separate crosstab for it.

After the contingency table, the output will include each of the tables you got earlier for the statistics you requested. The statistics tables are divided up into sections in the same way as the crosstab so that you can find the statistics connected with each of the sections of the table. As you prepare your professional-looking contingency table, you will insert the appropriate statistics into each section. The statistics in each section will have the number of stars corresponding to its statistical significance, and there will be a single key to the interpretation of those stars at the end of the full table.

An SPSS Application: Views on Legal Abortion for Rape among Catholics

It is fall 2012, and you are interning at Catholic Democrats, a nonprofit organization that expresses the views of Catholics within the Democratic Party. The past few months have witnessed many Republican leaders defending their opposition to abortion in all cases. In the process of explaining their opposition to abortion in the context of rape, several Republican leaders have improvised new adjectives to categorize different rape situations. These include "legitimate rape," "forcible rape," "honest rape," and "emergency rape." Catholic Democrats is in the process of updating its issue pages from the 2008 election. You have been assigned to update the "Abortion" page. You remember that the catechism condemns rape in any circumstances, so you wonder whether Catholics tend to support legalized abortion in the case of rape or follow the Republican opposition to abortion without exception. Your thought is that a more expansive definition of what it means to be pro-life (which allows for abortion in the case of rape) might appeal to moderate voters as well as liberals. When you run your idea by Victoria Kennedy, widow of Senator Edward Kennedy and member of the board of directors, she asks you to research how ideology affects support for legalized abortion in the case of rape and write her a memo with your results.

Because both political ideology and views on abortion tend to be correlated with partisanship, you decide to control for partisanship to eliminate any compounding effects. Using the 2000-2010 General Social Survey, analyze the relationship between ideology and support for legalized abortion among Catholics. Include both a contingency for the two-way relationship and a three-way contingency table controlling for partisanship.

1. Open GSS. Your dependent variable will be "Abrape," your independent variable will be "Polviews," your control variable will be "PartyID," and you will be selecting case by using "Relig" to identify Catholics. The data have already been limited to the surveys conducted in the twenty-first century. Use the "Define Variable Properties" command to see how the data need to be cleaned.
   All the variables need "8" to be set to missing. Polviews has too many categories. I take notes of to which categories each value corresponds: Values 1-3 consider themselves some sort of liberal; 4 includes moderates; and 5-7 are conservatives. PartyID also has too many categories: Values 0-1 are those respondents who

originally identified as Democrats; values 2-4, as Independents; and 5-6, as Republicans. In Relig, I see that a value of "2" designates someone who is Catholic.

2. Because PartyID has too many categories, collapse it into only three: Democrats, Independents, and Republicans.
Using the "Transform" and "Recode into New Variables" commands, I collapse the categories in the following ways: 0-1 become 1, 2-4 become 2, 5-6 become 3, and all other values become system-missing.

3. Because Polviews also has too many categories, collapse it into only three: Liberal, Moderate, and Conservative.
Before I start, I notice that SPSS has left the prior recode in place, so I click on "Reset" to empty out the prior commands. Using the "Transform" and "Recode into New Variables" commands, I collapse the categories in the following ways: 1-3 become 1, 4 becomes 2, 5-7 become 3, and all other values become system-missing.

4. Go into "Define Variable Properties" to give labels to the values of these new variables.
I assign the value labels of "Democrat," "Independent," and "Republican" to cPartyID and "Liberal," "Moderate," and "Conservative" to cPolviews.

Figure 11.5 Command for Selecting a Combination of Two Criteria for Cases

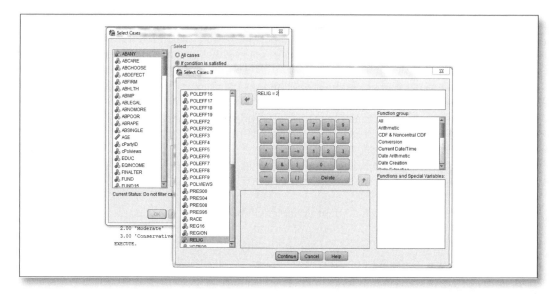

5. Use the "Select" command to limit your respondents to Catholics.
Under "Data," I "Select Cases" and enter the condition: "Relig=2." Figure 11.5 shows what the command window looks like.

6. Get a three-way crosstab for Abrape by cPolviews by cPartyID.
   Abrape is the dependent variable, so it goes in the rows; Polviews is the independent variable, so it goes in the columns; and PartyID is the control variable, so it goes into "Layers."

   Because the rape variable is dichotomous, I choose the measure of association appropriate for the independent variable, ideology. This has an ordinal level of measurement, so gamma should be a good measure of association. My request is shown in Figure 11.6.

Figure 11.6 Requesting a Three-Way Contingency Table

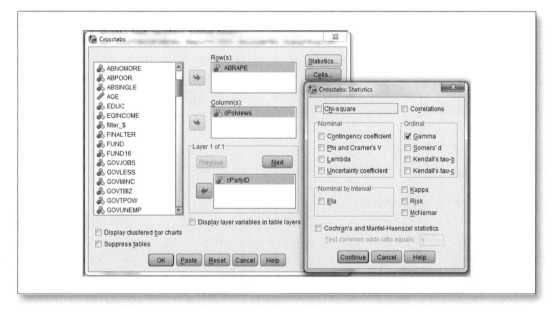

7. Formulate a professional-looking, two-way contingency table from the bottom of the crosstab and a separate, three-way contingency table from the main part.
   The bottom part of the contingency table (which includes the data of the uncontrolled relationship) is shown in Figure 11.7. The top part (with the controls) is in Figure 11.8. The statistics table is found in Figure 11.9. I translate these into professional-looking tables, inserting the appropriate measure of association and stars for each section. You can see these in Tables 1 and 2 of the memo.

8. Write a memo first evaluating the impact of ideology on abortion views and, second, what happens when you control for partisanship. Is this likely to be an issue affecting whether Catholic Democrats can attract supporters?

Figure 11.7 Portion of Three-Way Contingency Table Showing Uncontrolled Relationship

	PREGNANT AS RESULT OF RAPE						
Total	PREGNANT AS RESULT OF RAPE	YES	Count	374	569	385	1328
			% within collapsed Polviews	81.1%	76.8%	67.0%	74.7%
		NO	Count	87	172	190	449
			% within collapsed Polviews	18.9%	23.2%	33.0%	25.3%
	Total		Count	461	741	575	1777
			% within collapsed Polviews	100.0%	100.0%	100.0%	100.0%

Figure 11.8 Portion of Three-Way Contingency Table Showing Controlled Relationship

PREGNANT AS RESULT OF RAPE * collapsed Polviews * collapsed PartyID Crosstabulation

collapsed PartyID					collapsed Polviews			
					Liberal	Moderate	Conservative	Total
Democrat	PREGNANT AS RESULT OF RAPE	YES	Count		193	210	83	486
			% within collapsed Polviews		83.2%	78.1%	71.6%	78.8%
		NO	Count		39	59	33	131
			% within collapsed Polviews		16.8%	21.9%	28.4%	21.2%
	Total		Count		232	269	116	617
			% within collapsed Polviews		100.0%	100.0%	100.0%	100.0%
Independent	PREGNANT AS RESULT OF RAPE	YES	Count		148	235	138	521
			% within collapsed Polviews		80.4%	74.1%	67.6%	73.9%
		NO	Count		36	82	66	184
			% within collapsed Polviews		19.6%	25.9%	32.4%	26.1%
	Total		Count		184	317	204	705
			% within collapsed Polviews		100.0%	100.0%	100.0%	100.0%
Republican	PREGNANT AS RESULT OF RAPE	YES	Count		33	124	164	321
			% within collapsed Polviews		73.3%	80.0%	64.3%	70.5%
		NO	Count		12	31	91	134
			% within collapsed Polviews		26.7%	20.0%	35.7%	29.5%
	Total		Count		45	155	255	455
			% within collapsed Polviews		100.0%	100.0%	100.0%	100.0%

Memo

To: Victoria Kennedy, Board Member Catholic Democrats
From: T. Marchant-Shapiro, intern
Subject: Abortion and Rape
Date: October 25, 2012

Figure 11.9 Statistics Table for a Three-Way Contingency Table

Symmetric Measures

collapsed PartyID				Value	Asymp. Std. Error[a]	Approx. T[b]	Approx. Sig.
Democrat	Ordinal by Ordinal	Gamma	Zero-Order	.204	.080	2.476	.013
	N of Valid Cases			617			
Independent	Ordinal by Ordinal	Gamma	Zero-Order	.204	.069	2.894	.004
	N of Valid Cases			705			
Republican	Ordinal by Ordinal	Gamma	Zero-Order	.284	.093	3.047	.002
	N of Valid Cases			455			
Total	Ordinal by Ordinal	Gamma	Zero-Order	.240	.044	5.363	.000
			First-Order Partial	.219			
	N of Valid Cases			1777			

a. Not assuming the null hypothesis.
b. Using the asymptotic standard error assuming the null hypothesis.

You asked me to assess the relationship between views on abortion in the instance of rape with the political ideology among American Catholics. I conducted a statistical analysis using the General Social Survey for the years 2000 to 2010, including only those respondents who identified as Catholics. I found that political ideology does affect the views of Catholics about whether abortion should be legal in the case of rape.

Table 1 shows the relationship between views on abortion and ideology. You can see that liberals are more likely to support legalized abortion in the case of rape than conservatives by a difference of 14 percentage points. We can be 99 percent confident that this pattern is not just random. The gamma of 0.240 indicates that these two variables have a moderate relationship.

Table 2 shows the same relationship, except this time I controlled for the political party of the respondent. Among Democrats, liberals are more likely than conservatives to support legal abortion in the case of rape by a difference of 8 percentage points. Among independents, liberals are more likely than conservatives to support it by a difference of 13 percentage points. Finally, among Republicans, both liberals and moderates are more likely than conservatives to support legal abortion in the case of rape by a difference of 9 percentage points and 16 percentage points, respectively. For all three of these groups, we can be at least 95 percent confident that there is a relationship between ideology and view on abortion, and for all three groups, the relationship appears to be moderate. Two things stand out as you compare the three sections of the table. First, support for the legalization of abortion if the pregnancy is the result of rape is universally high among all categories of ideology and partisanship. Second, the level of support among moderates approaches (if not surpasses) that for liberals.

From these data, you can conclude that among Catholics, political ideology does affect an individual's support for legal abortions after a rape. Liberals are more likely to support it than conservatives. This suggests that we might be able to make inroads in gaining support from moderates. Perhaps more importantly, the data suggest that those Catholic groups that echo the Republican opposition to abortion regardless of the circumstances are out of step with the American Catholic mainstream. The majority of Catholics, regardless of ideology and regardless of partisanship, support legal abortion in the case of rape. This is an issue that we could potentially use to our advantage.

Memo Table 1 Views on Abortion in the Case of Rape, by Ideology

Should Abortion be Legal after Rape?	Ideology			
	Liberal	Moderate	Conservative	Total
Yes	81.1%	76.8%	67.0%	74.7% (1328)
No	18.9%	23.2%	33.0%	25.3% (449)
Total	100.0%	100.0%	100.0%	100.0%
n =	461	741	575	1777
				gamma = 0.240**

Source: General Social Survey 2000-2010, National Opinion Research Center, 2012. www3.norc.org/GSS+Website/Download/SPSS+Format/.

Memo Table 2 Views on Abortion by Ideology, Controlling for Party

Party	Should Abortion be Legal after Rape?	Ideology			
		Liberal	Moderate	Conservative	Total
Democrat	Yes	83.2%	78.1%	71.6%	78.8% (486)
	No	16.8%	21.9%	28.4%	21.2% (131)
		(232)	(269)	(116)	n = 617
					gamma = 0.204*
Independent	Yes	80.4%	74.1%	67.6%	73.9% (521)
	No	19.6%	25.9%	32.4%	26.1% (184)

Party	Should Abortion be Legal after Rape?	Ideology			
		Liberal	Moderate	Conservative	Total
		(184)	(317)	(204)	n = 705
					gamma = 0.204**
Republican	Yes	73.3%	80.0%	64.3%	70.5% (321)
	No	26.7%	20.0%	35.7%	29.5% (134)
		(45)	(155)	(255)	n = 455
					gamma = 0.284**

Source: General Social Survey 2000-2010, National Opinion Research Center, 2012. www3.norc.org/GSS+Website/Download/SPSS+Format/.

*prob. < 0.05
**prob. < 0.01

Your Turn: Multivariate Relationships

YT 11.1 — For the following spurious relationships, identify a possible control variable that explains it. Is that control variable an antecedent or intervening variable?

1. In Congress, women are more likely to be liberals than men.
2. Presidents are taller than other Americans.
3. Ohio and Florida have higher September and October hotel revenues in presidential election years than in other years.
4. In Congress, there is a negative relationship between an incumbent's share of the vote and how much he or she spends on the race. (Hint: What prompts representatives to raise and spend more money than usual?)

YT 11.2 — Analyze the following three-way contingency table.

1. Describe the pattern.
2. Identify the statistical significance.

Table 11.5	Partisanship by Ideology, Controlling for Year				
		Ideology			
Year	Party	Liberal	Moderate	Conservative	Total
1980	Democrat	40.6%	43.4%	31.5%	38.6% (548)
	Independent	42.2%	36.0%	35.3%	38.2% (541)
	Republican	17.2%	18.7%	33.2%	23.2% (329)
		(360)	(579)	(479)	n = 1418
				gamma = 0.175**	
2010	Democrat	58.0%	34.3%	16.3%	35.2% (673)
	Independent	35.1%	50.6%	35.3%	41.0% (785)
	Republican	6.9%	15.2%	48.4%	23.8% (456)
		(553)	(724)	(308)	n = 1914
				gamma = 0.569**	

Source: General Social Survey 1972-2010, National Opinion Research Center, 2012. www3.norc.org/GSS+Website/Download/SPSS+Format/.

*prob. < 0.05

**prob. < 0.01

3. Evaluate the strength of the relationship.
4. Draw conclusions.

Apply It Yourself: Analyze Data on Race for Partisanship and Income

You are interning at the Republican National Committee in the political office. Your boss, the political director, is planning a strategy of outreach to blacks. Historically, blacks have tended to vote for the Democratic Party. Your boss is wondering whether that might just be an artifact of blacks being poor. He wonders if it is possible that as the Republican plans for prosperity bring increased wealth to all Americans, minorities might just turn to the Republican Party as a natural outgrowth of their increased income. Your assignment is to analyze GSS data from 2006 to 2010 to see whether race (RACE) continues to be associated with partisanship (PARTYID) even when you control for income (INCOME06).

1. Open up the GSS dataset. Make sure that SPSS is set to show the variable names alphabetized.

    >Data Set Window
    >>Variable View
    >>Edit
    >>>Options
    >>>>Display Names
    >>>>Alphabetical

2. Your dependent variable will be "PartyID," your independent variable will be "Race," your control variable will be "Income06," and you will be selecting cases by using "Race" to exclude "Other" and "Year" to limit the surveys since 2006. Use the "Define Variable Properties" command to see how the data need to be cleaned. Take notes about the values connected to categories you want to recode. For Income06, you'll be making a dichotomous variable separating those below $50,000 from those above $50,000, so take note of the ranges of values that include those incomes.

3. Because PartyID has too many categories, collapse it into a new variable, "rPartyID," with only three. Use "Transform" and "Recode into New Variable," and then collapse the following ranges. For Democrats, recode the range of 0-1 into "1." For Independents, collapse the range of 2-4 into "2." For Republicans, collapse the range of 5-6 into "3." Set all other values to "System Missing."

4. Because $50,000 is the median income for whites, recode Income06 into a new variable, "cIncome06," with two categories, "less than $50,000" and "$50,000+." When you were in "Define Data," you should have noted that the values 1-18 included incomes under $50,000 and the values 19-25 included those above $50,000. Use "Transform" and "Recode into New Variable," and then collapse the following ranges. For "less than $50,000," recode the range of 1-18 into "1." For "$50,000+," collapse the range of 19-25 into "2." Set all other values to "System Missing."

5. Go into "Define Variable Properties" to give labels to the values of these new variables.

6. You want to limit your cases in two ways. First, you want to include only the years from 2006 on. Second, you want to contrast only whites and blacks, so you want to exclude those respondents who identified as an "Other" race. In the "Data" command, "Select Cases" combining both criteria with an "&."

    >"Data"
    >>"Select Cases"
    >>>If condition is satisfied
    >>>If

             Year>=2006 & Race<3
             >Continue
      >OK

7. Get frequencies of the three variables to make sure that the categories are set up the way you want them.

8. Get your three-way crosstab. Be sure that your dependent variable is in the rows and the independent variable is in the columns. The control variable goes in the "Layers" box. You will need to tell SPSS what information you want in your cells and what statistics you want. Copy the tables into a Word document so that you can use them later.

    >Analyze
        >Descriptives
            >Crosstabs
                Row=rPartyID
                Column=Race
                Layer 1 of 1 = cIncome06
                >Cells
                    >(Counts) Observed
                    >(Percentages) Column
                    >Continue
                >Statistics
                    >(choose the appropriate measure of association)
                    >Continue
                >OK

9. Formulate a professional-looking, two-way contingency table from the bottom of the crosstab and a separate, three-way contingency table from the main part.

10. Write your memo answering the question. Begin by describing the relationship between partisanship and race (Table 1). Then describe how the relationship changes when you control for income (Table 2). Remember that you always need to describe the pattern, decide whether there is a relationship (statistical significance), and evaluate how strong it is (measure of association). Include two properly set-up tables—do not cut and paste! Please attach the SPSS output for the crosstab and the statistics table for grading purposes.

Key Terms

Antecedent variables (p. 275)

Intervening variables (p. 275)

Spurious relationship (p. 274)

CHAPTER 12

Bivariate Regression

Putting Your Ducks in a Line

In *Freakonomics,* economist Steven Levitt and journalist Stephen Dubner[1] pointed out an event that seemed to have escaped the rest of us. On analyzing crime data, they found that, contrary to the portrayal of the news media, the crime rate in the United States has actually been on the decline since the early 1990s. Of course, they weren't the only ones to have noticed the event, and actually, several hypotheses had been proposed to explain it, including innovative policing strategies, increased reliance on prisons, changes in crack and other drug markets, aging of the population, tougher gun control laws, a stronger economy, and an increased number of police. But Levitt and Dubner were unique in two respects. First, they added one more hypothesis to the mix of explanations: They suggested that the *Roe v. Wade* decision in 1973, paving the way for legal abortions, resulted in fewer unwanted births and so led to a generation with a lower propensity toward crime twenty years later. Second, they actually tested all of the hypotheses to see which of them did the best job of explaining the pattern of the 1990s decline, concluding that the previous explanations were all specious. One way to test multiple hypotheses in this way is to use regression analysis.

In previous chapters, we found that the way we test our hypotheses depends on whether the data have nominal, ordinal, or interval levels of measurement. We set up the data visually using contingency tables. After describing the pattern we saw in the table, we then identified the statistical significance of the relationship using chi-square and other significance tests. Finally, we analyzed the strength of the relationship using the appropriate measure of association depending on the level of measurement. For nominal data, we discussed using Cramer's V or lambda; for ordinal, gamma or Kendall's tau; and for interval, Pearson's r. In the discussion of Pearson's r, I promised we would be learning a different technique to measure the relationship between interval-level variables: regression. In this chapter, we see how to do the three-step analysis of a relationship using regression.

GRAPH A RELATIONSHIP

Although regression is mathematically more complex than the other measures of association, it actually is much clearer to understand intuitively. When we think about correlations, it is almost second nature to think about the relationship between the variables having a linear relationship. If I say that there is a positive correlation between weight and height, we almost immediate picture a petite gymnast standing next to a buff football player. From there, it is not a huge mental jump to picturing a graph (like the one found in Figure 12.1) that relates weight to height. Similarly, if I say that there is a negative relationship between GPA and playing World of Warcraft, it is easy to picture two friends, one of whom stays up all night gaming, the other of whom makes it to class the next day. From there, it is easy to visualize Figure 12.2.

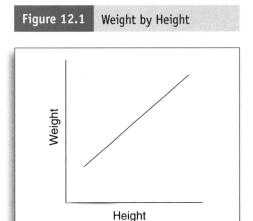

Figure 12.1 Weight by Height

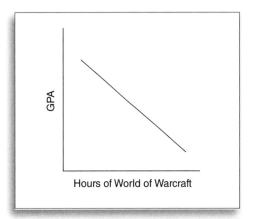

Figure 12.2 GPA by Hours Spent Playing World of Warcraft

Plot the Data

Of course, if we were to collect data, it would be messier than the line indicates. For either of these figures, some of the real data would fall above the line and some would fall below it. There are short people who weigh a lot and gamers who are able to keep their grades up. Let's look within the United States at the size of state budgets. I would expect that states with higher populations would spend more on government services than states with lower populations. Figure 12.3 shows a scatter plot of these two variables for all of the states.

There appears to be a linear relationship between expenditures and population. Just eyeballing the data points, I would begin the line by predicting that a state with no citizens would have no government expenditures. As the population increases, I would extend the line so that there are roughly the same number of cases above the line as there are below it. Although California appears in the upper right-hand corner because its population (and so also its spending) is so much higher than all the other states, it seems to be right in line with where you would expect it to be if you drew a line from the other states. Figure 12.4 shows the line that I would draw, just eyeballing the data.

Figure 12.3 State Government Expenditures by State Population, 2010

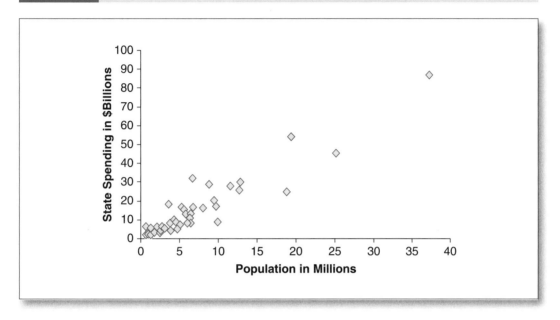

Figure 12.4 A Linear Relationship: State Expenditures by Population, 2010

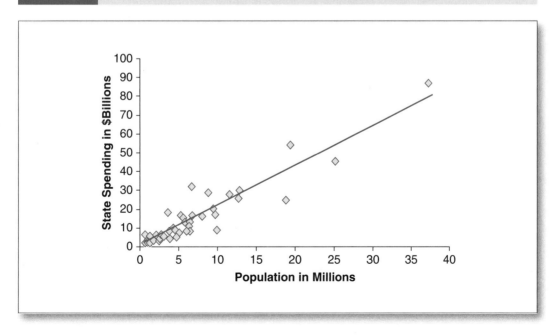

Find a Line

But we want to do more than just eyeball the line that we think best fits the data: We want an equation for the line. When you took algebra, your math teacher taught you that the equation for a line is given as

$$y = mx + b$$

where m is the slope of the line and b is the y-intercept. You learned to interpret the slope as the amount that y increases for every unit increase in x. You learned that the y-intercept is the value of y when x equals zero.

In statistics, we change that equation a bit:

$$Y = B_0 + B_1 X_1$$

X and Y are **variables** in this equation because they vary across cases. These are the two characteristics that we measure for different cases in order to see whether changes in one lead to consistent changes in the other. We call X the independent variable because we are thinking of it as the cause and Y the dependent variable because we are thinking of it as the effect. In statistics, we replace m with B_1 for the slope and b with B_0 for the intercept. Although X and Y vary across cases, we assume that there is an underlying relationship between the two that is constant. For any value of X, we will multiply it by the same slope and add the same intercept to get our best estimate of Y.

(Later on, we'll call this estimate of the dependent variable $\hat{Y}$ or Y hat). Because both the slope and the intercept stay constant for a particular dataset, we call them **coefficients**. We interpret the two coefficients in the second (statistics) equation in exactly the same way we did in the first (algebra) equation. As the intercept, B_0 is the value of the dependent variable when your independent variable has a value of zero. As the slope, B_1 is how much your dependent variable changes for every unit increase in your independent variable.

FIT THE DATA WITH THE ORDINARY LEAST SQUARES ESTIMATE OF THE LINE

When I eyeballed a line for the state-level data, I began it at the origin—where both population and expenditures are 0—but this is not the ideal approach to finding the line that best fits the data. The best approach to fitting a line is to begin with the point in the middle of the data—the point where X equals the mean value for our independent variable and Y equals the mean value for our dependent variable. For these data, the mean population is 6.16 million and the mean expenditure is $12.9 billion. The point (6.16 million, $12.9 billion) is shown in Figure 12.5 with a star. This is only the beginning, though, because there are an infinite number of lines we could draw through that point. How do you choose between them?

Figure 12.5 Fitting a Line: State Expenditures by Population

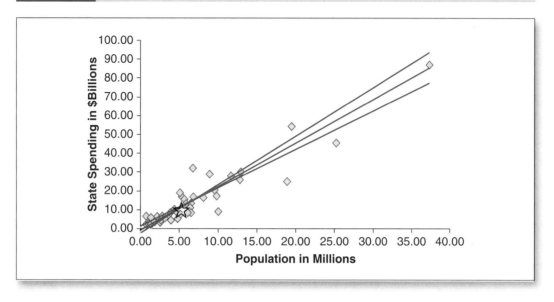

In regression, we choose the line that on average is closest to all of the points. For any line, you can find the distance from it to each of the points. This distance (the difference between the case's actual value of the dependent variable Y and its estimate $\hat{Y}$ based on the value of the independent variable X) is called the **residual**. Symbolically, we designate the residual as e because it is given as an error term in the relationship between the dependent variable and its estimate:

$$Y = \hat{Y} + e$$

Given the messiness of social and political behavior, it is inevitable that any equation for the line will make errors in predicting the dependent variable. What we want to do, though, is choose the line that minimizes overall the total amount of error. In order to find the total amount of error, we cannot just add up the errors for all of the cases because sometimes they are positive and sometimes negative. Just as we have done previously, we square the differences. We want the line for which the sum of the squared difference is minimized. Because we are minimizing the squared difference, the kind of regression we are talking about in this chapter is frequently called ordinary least squares, or OLS.

Statisticians have a procedure for finding the line that best fits the data. Its slope is given by the equation:

$$B_1 = \frac{\sum (X - \bar{X})(Y - \bar{Y})}{\sum (X - \bar{X})^2}$$

where X is the value of the independent variable for each case, $\bar{X}$ is its mean across all the cases, Y is the value of the dependent variable, and $\bar{Y}$ is its mean. This should look familiar because it is very similar to how we calculated Pearson's r. The difference here is that we want the actual slope, so we no longer want to constrain the magnitude of the relationship to be less than 1.0. For Pearson's r, we standardized the relationship by dividing by the sum of squares for the dependent variable. But here, because we want the actual slope, we eliminate that standardizing factor from the equation.

As usual, we will calculate this value by using a work table. Each case will have its own row where we find the difference between its values of X and Y and the average values of X and Y. At the bottom, we'll find the appropriate summations that we can then use to calculate the slope according to the equation. This work table will have eight columns and will be identical to the table we used to find Pearson's r. In the first column, identify each of the cases; in the second, record the value of the independent variable; and in the fifth, the dependent variable. Before you fill in the other columns, you need the mean values for both X and Y. You know the drill: Calculate the mean value of X by adding up all the values in the second column and dividing by the total number of cases. Repeat this for the fifth column to find the mean value of Y.

Once you have the means, you can complete the other columns. In the third column, take the difference between the case's value of X (from column 2) and the mean of X. Square this difference in the fourth column. In the sixth column, find the difference between the value of Y for this case (from column 5) and the mean of Y. In the seventh column, you will, as usual, square the difference for the dependent variable. As you did for Pearson's r, in column 8, multiply the difference between Y and its mean (from column 6) by the difference of X and its mean (from column 3). At the bottom of columns 4, 7, and 8, add up all of the values in that column. We'll use the sum of squares for Y later; for now, you will find the regression slope by plugging the totals from columns 4 and 8 into the equation above—the summation of column 8 (SS_{XY}) is the numerator, and the summation of column 4 (SS_X) is the denominator.

In Table 12.1, I regress state expenditure on state population for all of the New England states. In New England, the average population is 2.65 million and average state expenditure is $9.715 billion. I get the slope of the regression line by substituting the sums of squares:

$$B_1 = \frac{\sum(X-\bar{X})(Y-\bar{Y})}{\sum(X-\bar{X})^2}$$

$$= \frac{SS_{XY}}{SS_X}$$

$$= \frac{155.07}{31.03}$$

$$= 5.00$$

To interpret the slope, check out the units that we used in measuring the two variables. For every unit, increase in the independent variable, the dependent variable changes in size by

Table 12.1 Work Table for Regressing Expenditures on Population

State	Population (in millions) X	$X - \bar{X}$	$(X - \bar{X})^2$	Spending (in $billions) Y	$Y - \bar{Y}$	$(Y - \bar{Y})^2$	$(X - \bar{X})(Y - \bar{Y})$
CT	5.03	2.38	5.66	17.92	8.205	67.32	19.5279
ME	1.33	−1.32	1.74	2.86	−6.855	46.99	9.0486
MA	6.55	3.9	15.21	32.08	22.365	500.19	87.2235
NH	1.32	−1.33	1.7689	1.31	−8.405	70.64	11.17865
RI	1.05	−1.6	2.56	2.96	−6.755	45.63	10.808
VT	0.63	−2.02	4.0804	1.16	−8.555	73.19	17.2811
n = 6	ΣX = 15.91 $\bar{X}$ = 2.65		SS_X = 31.03	ΣY = 58.29 $\bar{Y}$ = 9.715		SS_Y = 803.97	SS_{XY} = 155.07

Source: National Association of State Budget Officers, *The Fiscal Survey of States 2012*. www.nasbo.org/sites/default/files/Spring%202012%20Fiscal%20Survey%20of%20States.pdf.

the magnitude of the slope. In this case, we have learned that for every million more people a New England state has in population, it makes $5 billion more in state expenditures.

In algebra, you learned that as long as you know the slope of a line and one point on it, you can determine the equation of the line. We just calculated the slope of the line as 5.00. We decided earlier that we wanted the line to go through the middle of the data as defined by the point containing the mean value of X and the mean value of Y. So to find the intercept, we will modify the equation of the line to include the two means:

$$Y = B_0 + B_1 X_1$$

$$\bar{Y} = B_0 + B_1 \bar{X}_1$$

In Table 12.1, we determined that the mean value of X was 2.65 (New England states have an average population of 2.65 million) and the mean value of Y was 9.715 (New England states spend an average of $9.715 billion). We can substitute these values along with our estimate of the slope (5.00) to estimate the intercept:

$$9.715 = B_0 + 5.00(2.65)$$

$$9.715 = B_0 + 13.25$$

$$9.715 - 13.25 = B_0 + 13.25 - 13.25$$

$$-3.54 = B_0$$

Interpret the intercept as the value of the dependent variable if the value of the independent variable is zero. In this case, if a state's population is zero, we would expect its spending to be a negative $3.54 billion. For the case at hand, the interpretation does not mean very much because you can't have a state without any population.

In calculating the intercept, we could go through the process of substituting the means into the equation and then solving for B_0 like we did earlier. But instead, let's go through that process without the substitutions so that we get a simpler equation for the intercept.

$$Y = B_0 + B_1 X_1$$
$$\bar{Y} = B_0 + B_1 \bar{X}_1$$
$$\bar{Y} - B_1 \bar{X}_1 = B_0$$

This gives us an equation for the intercept into which we can substitute the values of the means of the two variables and the slope of their line:

$$B_0 = \bar{Y} - B_1 \bar{X}_1$$

Once you have the values of the slope and the intercept, you can substitute them into the general equation for a line in order to find the specific equation for the line that best fits the data. For these data, we would give the equation as

$$Y = B_0 + B_1 X_1$$
$$Y = -3.54 + 5.00 X_1$$

FIND THE STATISTICAL SIGNIFICANCE

As with all relationships, after we describe the pattern that we see, we next need to address the possibility that what we see is due to random fluctuations. In regression, we find statistical significance with a statistic called the standard error of the slope. Like the standard error of the mean, this measure looks at the sample size and finds the probability of observing the slope due to random error—if, in reality, the slope is zero. The null hypothesis is that there is no relationship between X and Y. If that is the case, we would expect the slope determined by the regression would equal zero. But because random things happen, it is possible that on occasion, a sample will yield a slope that is not zero. We would, however, expect that multiple samples would be normally distributed around zero. And we would expect that as the sample size increased, the standard deviation for those sample slopes would get tighter around zero. So what we are looking for with our measure of statistical significance is a t-test of whether, given the sample size and the magnitude of the estimated slope, we can reject the null hypothesis.

In order to get to that point, we're going to need to set up another work table. Theoretically, we could expand our work table from finding the regression slope in the previous section, but there just isn't much more space in Table 12.1 to do that. So we'll copy the values of our independent and dependent variables into Table 12.2 and work from there. We have the case identifier in the first column and the independent variable in the second. In the third column, we will plug the value of X into our estimated equation to find our estimate of Y. This estimate is given by the symbol $\hat{Y}$, which we call "Y hat." For Connecticut, we would plug X = 5.03 into the equation

$$Y = -3.54 + 5.00X_1$$
$$\hat{Y} = -3.54 + 5.00(5.03)$$
$$= 21.60$$

This tells us the equation's estimate of Connecticut's spending based on its population is $21.60 billion. Similarly, we substitute the populations of each of the states to get its estimated spending level.

In the fourth column, place the value of the dependent variable. We then take the difference between the actual value of the dependent variable and our estimate in the fifth column. For Connecticut, the estimate is $3.68 billion lower than the actual spending by the state. Finally, in the sixth column, we square that difference and get the sum of squares at the bottom. I'm going to call the sum of those squares $SS_{Y|X}$, or the sum of squares of Y given X. Notice that this is not the same as SS_{XY}. The order of X and Y are reversed, and there is a line between them. That line indicates that the estimated values of Y are dependent on the values of X. This is different from what we did to find SS_{XY}. In that case, we

Table 12.2 How to Find the Statistical Significance of Expenditures Regressed on Population

State	Population (in millions) X	$\hat{Y}$	Spending (in $billions) Y	$Y - \hat{Y}$	$(Y - \hat{Y})^2$
CT	5.03	21.60	17.92	−3.68	13.55
ME	1.33	3.11	2.86	−0.25	0.06
MA	6.55	29.20	32.08	2.88	8.30
NH	1.32	3.06	1.31	−1.75	3.06
RI	1.05	1.71	2.96	1.25	1.56
VT	0.63	−0.39	1.16	1.55	2.40
n = 6	d.f. = 4				$SS_{Y|X}$ = 28.93

Source: National Association of State Budget Officers, *The Fiscal Survey of States* 2012. www.nasbo.org/sites/default/files/Spring%202012%20Fiscal%20Survey%20of%20States.pdf.

simply multiplied the differences between X and Y and their means to measure how X and Y covary. Instead, here we are getting at the notion of how far off the estimates are from their true values.

The standard error of the slope is given by the equation

$$s^2_{B1} = SS_{Y|X}/(d.f.)(SS_X)$$

In the end, we are going to take the square root of this equation to get the standard error of the estimate. To calculate this equation, substitute all the appropriate values. You calculated the sum of squares of Y given X ($SS_{Y|X}$) in the most recent work table, you calculated the sum of squares of X (SS_X) in the previous work table, and the degrees of freedom are the number of cases minus 2 (because you've used one degree of freedom to calculate the mean of Y and another one to calculate the mean of X). Because d.f. = n − 2, let's substitute that directly into the equation for the squared standard error of the slope:

$$s^2_{B1} = SS_{Y|X}/(n-2)(SS_X)$$

Now substitute all the values:

$$s^2_{B1} = (28.93)/(4)(31.03)$$
$$= 0.233$$
$$s_{B1} = \sqrt{0.233}$$
$$\text{s.e.} = 0.483$$

Once you have the standard error, you find the t-score in the same way you did when you used the standard error of the mean to do a means test. Remember from Chapter 6 the equation for a t-score:

$$t = (\bar{X} - \mu)/\text{s.e.}$$

The difference this time is that we are comparing slopes, not means. Because we are testing the null hypothesis, we are finding the probability of observing the sample slope if the slope is actually zero. So we are simply going to get the ratio of the sample slope to its standard error:

$$t = B_1/\text{s.e.}$$
$$= 5.00/0.483$$
$$= 10.35$$

Now it's a simple matter of looking up this value of t on a t table (found in Appendix 4 on page 447) in the row for the appropriate number of degrees of freedom, remembering that we are doing a two-tailed test. For 4 degrees of freedom, we need a t-score of 4.604 to be statistically significant at the 0.01 level. Because our t-score of 10.35 is greater than that, we know that there is less than a 1 percent probability that what we are observing is due to chance. We can reject the null hypothesis.

If you have a large sample size, keep in mind that the magic number for t-scores is 1.96—if the ratio between the estimate of the slope and its standard error is 1.96 or larger, you know that the estimate is statistically significant at the 0.05 level. The rule of thumb when you are eyeballing a regression is that if the slope is twice as big as its standard error, you know that the variable will be a statistically significant predictor of the dependent variable. This rule holds true for sample sizes over sixty. But even if your sample size is around twenty, looking to see if the slope is twice as big as the standard error is a good way of eyeballing whether the relationship is statistically significant.

FIND THE STRENGTH OF THE RELATIONSHIP

The last question we ask is how strong the relationship is. In regression, we conclude by reporting the R^2 or **R-square**. This is closely related to Pearson's r. For bivariate regression, you can literally square the value of Pearson's r to find the R^2 of a relationship. As a reminder, Pearson's r is given by the equation:

$$\text{Pearson's r} = \frac{\sum(X-\bar{X})(Y-\bar{Y})}{\sqrt{\sum(X-\bar{X})^2 \sum(Y-\bar{Y})^2}}$$

Because we are actually calculating R^2, we need to square this equation. This means that we will square both the top and the bottom of the fraction. For the denominator, we will just refrain from taking the square root. For the numerator, we will actually square this summation. You'll recall that although the expression in the numerator isn't actually a sum of squares, in Chapter 10, I decided to call it "SS_{XY}" to make it easier to keep track of the components of this equation. So for this equation, we're back to using the original SS_{XY}, not the estimated $SS_{Y|X}$ we used for the standard error. Keeping that distinction clear, all three of these elements are found in Table 12.1. Substituting them into the equation, we can find R^2:

$$R^2 = (SS_{XY})^2/SS_X SS_Y$$

$$= (155.07)^2/(31.03)(803.97)$$

$$= 0.9639$$

R-square is a PRE measure of how well our independent variable explains our dependent variable. Just as we did with eta-square, we interpret the R^2 as the proportion of the variance in the dependent variable that is explained by the independent variable. So we say "The population of a state explains 96.39 percent of the variance in state expenditures among New England states." This is an incredibly strong association.

USE REGRESSIONS WITH TIME SERIES DATA

Frequently, we want to track how a variable changes over time. If we believe that the change is linear, then it is appropriate to use linear regression to measure that change. In this kind of time series analysis, the variable that is changing is our dependent variable, and time is our independent variable. It is a good idea to code our first point in time as zero, with each subsequent time period increasing by one. If you do this, then you would interpret the intercept as the value at the beginning of the time series and the slope as how that value changed on average during each of the subsequent time periods.

BOX 12.1 Numbers in the News

In 1996, ten prominent French female politicians called for a constitutional amendment that would require gender equity in the national legislature. These women had looked at how few women were being elected and projected that if the current pattern continued, it would be centuries before women could claim equal representation in France. The argument was sufficiently persuasive to lead to the passage of a constitutional amendment in 1999. As a result, political parties in France are now required to nominate equal numbers of men and women for office or face financial sanctions.[2] Although there are still more men elected than women, parity is projected in the nearer future, far sooner than it had been before the constitutional amendment.

A Political Example: Murder Rates

We began the chapter with the discussion about crime rates declining after the early 1990s. We can see that decline by regressing murder rates on the years since 1994. In this instance, the murder rate is our dependent variable, how many years it has been since 1994 is our independent variable. Figure 12.6 shows the scatter plot of these two variables. It is obvious that the murder rate did decline. But we can't just assume that there is a strong relationship between our dependent and independent variables. The questions we want to ask are, How much did it decline? Is it possible that the decline is just a random fluctuation? How strong is the relationship between the murder rate and time?

The results of the regression are found in Table 12.3. The model we describe yields the following equation:

$$\text{Number of Murders} = 19{,}859 - 333(\text{Number of Years since 1994})$$

$$(\text{s.e.} = 92)$$

In interpreting the results, we go through the same three-step process that we always do: describe the pattern, identify the statistical significance, and evaluate the strength of the relationship. To describe the pattern, we interpret the intercept and slope. The intercept indicates that in 1994, the number of murders was about 19,859. The slope indicates that

| Figure 12.6 | Number of Murders in the United States by Year |

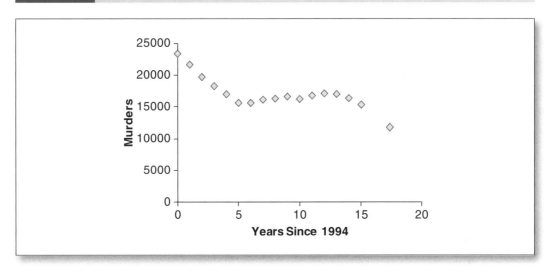

every year after 1994, the murder rate declined by an average of 333 murders. To identify the statistical significance, we compare the magnitude of the slope to its standard error. The standard error of 92 is less than half the slope, indicating that the relationship is statistically significant at the 0.05 level and we can reject the null hypothesis. Looking at Table 12.3, we can refine our results. The relationship is actually statistically significant at the 0.01 level— we are even more confident in rejecting the null hypothesis. To evaluate the strength of the relationship, we look at the R^2. In this case, the R^2 indicates that this decline over time explains 48.2 percent of the variation in murders in the United States.

| Table 12.3 | Number of Murders in the United States Regressed on Year |

Independent Variable	Unstandardized Coefficient (s.e.)	Standardized Coefficient
Constant	19,859	
Years Since 1994	333** (92)	−0.694**
		$R^2 = 0.482$

Source: Uniform Crime Reports, Federal Bureau of Investigation, 1994-2009. www.fbi.gov/about-us/cjis/ucr/ucr.

*prob. < 0.05

**prob. < 0.01

Table 12.3 is what a professional-looking regression table will look like. The column headed "Unstandardized Coefficient" gives the values we've been discussing in this chapter. The coefficient for the constant is B_0, or the intercept. The coefficient for Year is B_1, or the slope. Although political scientists will frequently place the standard error below the slope for you to use the 1.96 rule, more helpful is the stars designating the level of statistical significance. We have not discussed the contents of the column headed "Standardized Coefficient." This is also sometimes called the "beta." It is standardized in the same way that we standardized Pearson's r. Because this table is a bivariate relationship, the beta is identical to Pearson's r and so is constrained to be between −1.0 and 1.0. In this case, the standardized coefficient of −0.694 suggests that there is a strong relationship between the murder rate and the year. And, of course, the R^2 at the bottom indicates that year explains 48.2 percent of the variance in the murder rate.

INTERPRET REGRESSIONS WITH DICHOTOMOUS INDEPENDENT VARIABLES

Regressions are normally used for interval-level variables. But mathematically, a dichotomous variable can be used in measures of association for any level of measurement—you just need to be careful how you interpret it. For regression, you can use a dichotomous variable as an independent variable. The interpretation is easiest if you code the two categories with the numbers 0 and 1. (When a dichotomous variable is coded in this way, we call it a dummy variable.) Then when you interpret the slope, you say that the group you coded as 1 is different from the 0 group on the dependent variable by the amount of the slope.

A Political Example: Income and Gender

Table 12.4 shows income regressed on gender. The data come from the 2010 GSS with income coded in dollars and gender coded as 0 for men and 1 for women. The constant

Table 12.4 Income Regressed on Gender

Independent Variable	Unstandardized Coefficient (s.e.)	Standardized Coefficient
Constant	25,718.52	
Women	−7,694.12** (729.84)	−0.192**
n = 2899		$R^2 = 0.037$

Source: General Social Survey 2010, National Opinion Research Center, 2012. www3.norc.org/GSS+Website/Download/SPSS+Format/.

*prob. < 0.05

**prob. < 0.01

Variable: RealRInc is the respondent's income in 1972 dollars.

indicates that the men surveyed earned an average of $25,718.52. The slope indicates that women earned an average of $7,694.12 less than the men (or $18,024.40). This relationship is statistically significant at the 0.01 level—we can reject the null hypothesis. The R^2 indicates that gender explains 3.7 percent of the variance in income.

SUMMARIZING THE MATH: REGRESSION

When you are looking at the relationship between two interval-level variables, you can use regression to find the line that best describes that relationship. The intercept is interpreted as the value of the dependent variable that you would expect to see if the independent variable has a value of zero. The slope tells you how much the dependent variable changes, on average, for every unit increase in the independent variable. If the slope is positive, there is a positive relationship between the two variables; if it is negative, there is a negative relationship. The slope will have a significance value connected to it that is the statistical significance of the relationship. It will tell you whether or not you can reject the null hypothesis. Finally, regression tells you the R^2 for the relationship. This is a PRE measure of the proportion of variance in the dependent variable that is explained by the independent variable. To calculate these statistics, you will complete two work tables.

Describe the Pattern by Estimating the Line

The first work table will be identical to the one you used in calculating Pearson's r. It will have eight columns:

1. Identify the case.
2. Give the values of the independent variable. At the bottom, calculate the mean of X.
3. Give the difference between the observed value of X and the mean.
4. Square the difference for X. At the bottom, sum the values to get SS_X.
5. Give the values of the dependent variable. At the bottom, calculate the mean of Y.
6. Give the difference between the observed value of Y and its mean.
7. Square the difference for Y. At the bottom, sum the values to get SS_Y.
8. Multiply the difference for X (from column 3) and the difference for Y (from column 6). At the bottom, sum the values to get SS_{XY}.

From this table, you can calculate the slope and intercept of the line. The equation of the line is given as

$$Y = B_0 + B_1 X_1$$

To get the slope, use the equation:

$$B_1 = \frac{\sum(X-\bar{X})(Y-\bar{Y})}{\sum(X-\bar{X})^2}$$

$$= SS_{XY}/SS_X$$

To get the intercept, substitute your estimate of the slope along with the mean values of X and Y into the equation:

$$B_0 = \bar{Y} - B_1 \bar{X}_1$$

Finally, substitute the values of the slope and intercept into the general equation for a line to give your best estimate of the linear relationship between your dependent and independent variables.

$$Y = B_0 + B_1 X_1$$

You interpret the intercept as what you would expect your dependent variable to be if your independent variable has a value of zero. You interpret the slope as how much the dependent variable increases for every unit increase in the independent variable.

Identify the Statistical Significance with the Standard Error

To find the standard error, you need to complete a second work table. This will have six columns.

1. Identify the case.
2. Give the value of the independent variable (X).
3. Calculate the estimated value of the dependent variable ($\hat{Y}$) by substituting the value of X into the equation for the line you estimated for the relationship between your dependent and independent variables.
4. Give the actual value of the dependent variable (Y).
5. Get the difference between the actual and estimated values of Y ($Y - \hat{Y}$).
6. Square that difference ($Y - \hat{Y})^2$ and, at the bottom, sum all the squared differences to get the sum of squares of Y given X ($SS_{Y|X}$).

Once you've completed the table, you have the components you need to find the standard error of the slope. Use the equation

$$s^2_{B1} = SS_{Y|X}/(d.f.)(SS_X)$$

$$s.e. = \sqrt{s^2}$$

Take the square root of the result to get the standard error of the slope. Find the t value of the slope in comparison to its standard error to test the null hypothesis:

$$t = B_1/s.e.$$

Look up the value t (where d.f. = n − 2) to get the statistical significance. If your t-score is larger than the one given in the table, you can conclude that the slope is statistically significant and reject the null hypothesis.

Evaluate the Strength of the Association with R^2

The strength of the association is given by the statistic R^2. You have already calculated the elements that go into its measurement:

$$R^2 = (SS_{XY})^2/SS_X SS_Y$$

This is a PRE measure that you can interpret as the proportion of variance in the dependent variable that is explained by the independent variable.

A Political Example: Birth Rates and Abortion Rates

The argument in *Freakonomics* is that as the abortion rate increased after the Roe decision, the birth rate declined. The argument has particular relevance among teens, where a higher proportion of births would be unwanted. Table 12.5 shows how abortion rates and birth rates among teens have changed after *Roe v. Wade*. How does the abortion rate affect the birth rate among teens? Regress the birth rate on the abortion rate and complete the three-step analysis for the regression.

1. Describe the pattern by finding the equation of the line between the two variables.

To describe the pattern portrayed in these data, we want to find the line that best describes how the

Table 12.5 Teen Abortion and Birth Rates per 1,000

Year	Abortion Rate	Birth Rate
1976	34.3	52.8
1980	42.7	53.0
1981	42.9	52.2
1982	42.7	52.4
1983	43.2	51.4
1984	42.9	50.6
1985	43.5	51.0
1986	42.3	50.2
1987	41.8	50.6
1988	43.5	53.0
1989	42.0	57.3
1990	40.3	59.9
1991	37.4	61.8
1992	35.2	60.3

Source: The National Campaign to Prevent Teen and Unplanned Pregnancy, "National Pregnancy Rates for Teens, aged 15–19," 2012. www.thenationalcampaign.org/national-data/NPR-teens-15-19.aspx.

abortion rate influences the birth rate. If *Freakonomics* is correct, we expect the slope for the abortion rate to be negative, indicating that an increasing abortion rate will lead to a decreasing birth rate. To find the equation for the line, we use ordinary least squares regression. This requires us to complete the work table found in Table 12.6.

The slope is given by the equation:

$$B_1 = SS_{XY}/SS_X$$
$$= -95.115/125.615$$
$$= -0.76$$

Table 12.6 Teen Birth Rate Regressed on the Abortion Rate per 1,000

Year	Abortion Rate X	$X - \bar{X}$	$(X - \bar{X})^2$	Birth Rate Y	$Y - \bar{Y}$	$(Y - \bar{Y})^2$	$(X - \bar{X})(Y - \bar{Y})$
1976	34.3	−6.75	45.563	52.8	−1.24	1.538	8.370
1980	42.7	1.65	2.723	53.0	−1.04	1.082	−1.716
1981	42.9	1.85	3.423	52.2	−1.84	3.386	−3.404
1982	42.7	1.65	2.723	52.4	−1.64	2.690	−2.706
1983	43.2	2.15	4.623	51.4	−2.64	6.970	−5.676
1984	42.9	1.85	3.423	50.6	−3.44	11.834	−6.364
1985	43.5	2.45	6.003	51.0	−3.04	9.242	−7.448
1986	42.3	1.25	1.563	50.2	−3.84	14.746	−4.800
1987	41.8	0.75	0.563	50.6	−3.44	11.834	−2.580
1988	43.5	2.45	6.003	53.0	−1.04	1.082	−2.548
1989	42.0	0.95	0.903	57.3	3.26	10.628	3.097
1990	40.3	−0.75	0.563	59.9	5.86	34.340	−4.395
1991	37.4	−3.65	13.323	61.8	7.76	60.218	−28.324
1992	35.2	−5.85	34.222	60.3	6.26	39.188	−36.621
n = 14	Σ = 574.7	$SS_X =$	125.615	Σ = 756.5	$SS_Y =$	208.772	$SS_{XY} = -95.115$
d.f. = 14 − 2 = 12	$\bar{X} = 41.05$			$\bar{Y} = 54.04$			

Source: The National Campaign to Prevent Teen and Unplanned Pregnancy, "National Pregnancy Rates for Teens, aged 15-19," 2012. www.thenationalcampaign.org/national-data/NPR-teens-15-19.aspx.

Using that slope and the mean values of the two variables, we can find the intercept with the equation:

$$B_0 = \bar{Y} - B_1 \bar{X}_1$$
$$= 54.04 - (-0.76)(41.05)$$
$$= 85.12$$

This yields the equation

$$\text{Birth Rate} = 85.12 - 0.76(\text{Abortion Rate})$$

If the abortion rate were zero, we would expect that the teen birth rate would be 85.12 births per thousand teens. If the abortion rate increases by one, we expect the birth rate to decrease by 0.76.

2. Identify the statistical significance.

To find the statistical significance for a regression coefficient, we need to find the standard error of the slope. Doing so requires us to complete a second work table, shown in Table 12.7. In it, we find the predicted value of the dependent variable by substituting the value of the independent variable into the equation of the line. We then find the residual for each case by taking the difference between the observed and predicted values of the dependent variable. Next, we square the residuals and sum them to find the sum of squares of Y given X.

We now have the information we need to find the standard error of the slope:

$$s^2_{B1} = SS_{Y|X}/(d.f.)(SS_X)$$
$$= 136.93/(12)(125.615)$$
$$= 0.091$$

$$\text{s.e.} = \sqrt{s^2}$$
$$= \sqrt{0.091}$$
$$= 0.301$$

From there, we can find the t-score that we look up to determine the statistical significance.

$$t = B_1/\text{s.e.}$$
$$= -0.76/0.301$$
$$= -2.51$$
$$P(t = -2.51) < 0.01$$

Table 12.7 Finding the Residuals

Year	Abortion Rate X	Predicted Value $\hat{Y}$	Birth Rate Y	Residual $Y - \hat{Y}$	$(Y - \hat{Y})^2$
1976	34.3	59.05	52.8	−6.25	39.09
1980	42.7	52.67	53	0.33	0.11
1981	42.9	52.52	52.2	−0.32	0.10
1982	42.7	52.67	52.4	−0.27	0.07
1983	43.2	52.29	51.4	−0.89	0.79
1984	42.9	52.52	50.6	−1.92	3.67
1985	43.5	52.06	51	−1.06	1.12
1986	42.3	52.97	50.2	−2.77	7.68
1987	41.8	53.35	50.6	−2.75	7.57
1988	43.5	52.06	53	0.94	0.88
1989	42	53.20	57.3	4.10	16.81
1990	40.3	54.49	59.9	5.41	29.25
1991	37.4	56.70	61.8	5.10	26.05
1992	35.2	58.37	60.3	1.93	3.73
n = 14	d.f. = 12			$SS_{Y|X} =$	136.93

On the t table for 12 degrees of freedom for a two-tailed test, the t-score necessary for being statistically significant at the 0.05 level is 2.179. Because 2.51 is greater than 2.179, we know that the slope is statistically significant at the 0.01 level. We can reject the null hypothesis.

3. Evaluate the strength of the relationship.

The strength of the association is given by R^2:

$$R^2 = (SS_{XY})^2/SS_X SS_Y$$

$$= (-95.115)^2/(125.615)(208.772)$$

$$= 0.335$$

The R² indicates that the abortion rate explains 33.5 percent of the variance in the birth rate among teens.

USE SPSS TO ANSWER A QUESTION WITH BIVARIATE REGRESSION

Although regression has got to be one of the most useful statistical tools available to us, the mathematical computations involved are incredibly time consuming. In this chapter, I've tried to minimize the work by limiting the number of cases to a handful. But doing a regression with even as many as twenty cases would get exhausting, and usually, we want to analyze way more than twenty cases. Using SPSS makes regression analysis much more practical.

The command for regression is found under "Analyze." There are many kinds of regression. To get the OLS regression we've been discussing in this chapter, request the "Linear" option. At this point, all you need to do is correctly identify the independent variable and the dependent variable, and four different tables will appear in the output window. Before completing the command, one option that you have is to save the predicted value of the dependent variable so that you can compare this to the actual value. To do this, click on the "Save" button in the regression window. In the "Predicted Values" box, request the "Unstandardized" value. Then click on "Continue" to proceed with the request for the regression. This process is summarized in Box 12.2.

BOX 12.2 How to Run a Regression in SPSS

>Analyze
>>Regression
>>>Linear
>>>>Dependent = *dependent variable*
>>>>Independent = *independent variable*
>>>>Save
>>>>>Predicted Values:
>>>>>>Unstandardized
>>>>>>Continue
>>>>OK

Mathematically, you can have SPSS calculate the difference between the observed and expected values of your dependent variable using the "Compute Variables" command. This command allows you to calculate a new variable by performing mathematical functions on pre-existing variables. It is found as the first option under the

"Transform" button. When you open the window, it will have a "Target Variable" box in which you enter the name of the new variable. Then, in the "Numeric Expression" box, you can move variables over from the variable list and perform functions on them. To find the difference between your observed and predicted variables, you would move over the name of the dependent variable. Then, on the key pad below the "Numeric Expression" box, you would click on "-." Finally, you would move the variable for the expected value (PRE_1) over and click on "OK." If you move to the data window, you'll see that your new computed variable will be in the last column. Box 12.3 summarizes how to use the "Compute Variables" command to create a new variable. Keep in mind that you can use this command in many ways, using many different mathematical operations. If you click on a "Function Group," a list of functions in that group will open in the "Functions and Special Variables" box. Normally, though, you will use only the "Arithmetic" functions.

BOX 12.3 How to Compute a New Variable in SPSS Mathematically from an Old Variable

>Transform
    >Compute Variable
        Target Variable = *New Variable*
        Numeric Expression = *mathematical functions of old variable(s)*
    >OK

Once you have your regression output, you follow the three-stage process of analysis. To describe the pattern, look at the table labeled "Coefficients." This table will give you the values of the slope and the intercept. You'll want to look at the second column labeled "Unstandardized Coefficients." This column is divided into two: "B" and "Std. Error." In the first row, you'll see the value of the intercept. This is labeled "(Constant)" in the first column. Each of the subsequent rows will be labeled according to the variable name in the first column. (In Chapter 13, we'll be discussing how to do multivariate regression, where you include several independent variables in a single regression.) The corresponding slope for each will be in the second column. In describing the pattern that you see in a regression, you may or may not interpret the constant, depending on whether it is relevant to your analysis. Just remember that it is the same as the value you would predict for your dependent variable if your independent variable were zero. But you will certainly want to interpret the unstandardized coefficient for your independent variable. As the slope, this is the amount that you would expect your dependent variable to change for each unit increase of your independent variable.

Second, you need to identify the statistical significance for the pattern you described. To the right of the value of "B" is a column giving the standard error of the slope. Remember that we want the standard error to be less than half the size of the slope for the relationship to be statistically significant at the 0.05 level. The fifth column will actually translate that ratio into a t-score. But even more useful is the last column, labeled "Sig." This column makes unnecessary the tedious process of looking up the t-score on a t table. Instead, this column actually gives the statistical significance of the slope. You should keep in mind one important point here, though. SPSS will sometimes round the significance to 0.000. This does not mean that there is a probability of zero that the pattern we see is due to chance. So if you get output that looks like this, simply say that the relationship is statistically significant at the 0.01 level.

Third, you need to evaluate the strength of the relationship. The R^2 is found in the table called "Model Summary." The third column, labeled "R Square," gives this value. Remember that you interpret this as a PRE measure: The independent variable explains this proportion of the variance in the dependent variable.

From these tables, you will create a professional-looking regression table. As usual, it will have a title. For a regression, the title states the name of the dependent variable "regressed on" the independent variable(s). You will place headings over three columns: "Independent Variables," "Unstandardized Coefficients," and "Standardized Coefficients." The choice of whether to report the standardized or unstandardized coefficients is sort of a religious thing for sociologists and political scientists, so they will frequently report only their coefficient of preference without indicating which it is. This can get very confusing, so your best bet is to include both in your table with the clarifying heading of which each one is. You'll notice that SPSS gives a second name for each: "B" for the unstandardized coefficient and "Beta" (the Greek equivalent of the letter "b") for the standardized. These are both commonly used names for the two different kinds of coefficients.

After the headings, you'll fill in the rest of the table. Give the constant or intercept. Notice that the standardization process eliminates the need for an intercept in the "Standardized Coefficient" column. Then, list each of the independent variables (only one in this chapter, but more in the next) along with its two coefficients. Political scientists will normally give the standard error for the unstandardized coefficient. I suggest putting this in parentheses under the coefficient. (You do not give a standard error for the standardized coefficient.) Be sure to indicate the statistical significance for the coefficients—you'll use the same number of stars for both standardized and unstandardized coefficients. In the last row, include the sample size and the R^2. The R^2 is easy enough to find in "Model Summary," but the sample size is a little trickier. The "ANOVA" table gives the degrees of freedom in terms of "Regression," "Residual," and "Total." The "Residual" degrees of freedom is what we used earlier in the chapter when we looked up the statistical significance in a t table. If you'll recall, a bivariate regression has $n - 2$ degrees of freedom. So you could add two to the "Residual" degrees of freedom. Alternatively, I have an easier time just remembering to add one to the "Total" degrees of freedom. This is probably the better way to go, because once we get to Chapter 13, the degrees of freedom are going to change with the number of variables included in the regression. So just remember to add one to Total Degrees of Freedom listed in the ANOVA table to get the number of cases.

> **BOX 12.4 How to Create a Professional-Looking Regression Table**
>
> 1. As usual, the table is headed by a title. It takes the form of the name of the dependent variable "regressed on" the independent variable(s).
> 2. On the line after the title, insert a table with three columns and enough rows for the number of independent variables plus three.
> 3. The first column is headed "Independent Variables" and contains a list of the names of the independent variables in addition to "Constant" or "Intercept."
> 4. The second column is headed "Unstandardized Coefficients." For each variable and the constant, include the value of its "B" along with the appropriate number of stars to indicate its statistical significance. If you want, after you enter the "B," you can hit enter and give its standard error in parentheses directly under the slope.
> 5. The third column is headed "Standardized Coefficients." For each variable, include the value of its "Beta" along with the appropriate number of stars to indicate its statistical significance—these should be identical to the stars included for the standardized coefficient. Standardized coefficients never have an intercept, so leave that cell blank.
> 6. In the bottom line, include the number of cases (one more than the total degrees of freedom) and the R^2.
> 7. Using the border command in Word, place a single line at the top and bottom of the table, and a line after the column headings.
> 8. After the table, in a smaller font, include a key for the levels of statistical significance indicated by the number of stars. Also, if you use a single source for your data, indicate what it is.

After filling in the body of the regression table, polish it in the same way you have previously. Put lines at the beginning and the end and also under the column headings. Put any relevant information in a smaller font below the table. Be sure to include a key to the stars indicating statistical significance. But also indicate the source of the data if that is relevant. Box 12.4 summarizes the process of creating a professional-looking regression table.

A Political Example: State Expenditures and Population

It is February 2012 and Governor Chris Christie (R) of New Jersey and Governor Dannell Malloy (D) of Connecticut have been engaging in a war of words about the leadership skills (or lack thereof) of the other for the past year. You are interning in the office of Governor Malloy when Governor Christie criticizes him of being a typical tax-and-spend Democrat. You are assigned to evaluate where Connecticut and New Jersey lie in terms of spending in comparison with the rest of the country. Because total state expenditure is strongly influenced by the population of the state, you decide to regress spending on state population.

You will save the unstandardized value of what you would expect spending to be based entirely on population. If that is less than the actual amount spent, it might be reasonable to attack the state for overspending.

1. Open the data in SPSS and clean them.
   Once I've collected the data, I import them into SPSS, clean them, and begin my analysis.

2. Under "Analyze," "Regression," choose the "Linear" option. Identify your dependent and independent variables and request that the program "Save" the "Unstandardized" predicted value of the dependent variable.
   Figure 12.7 shows where I found the regression commands. Figure 12.8 shows the commands I used to regress state expenditures on state population and request that SPSS save the unstandardized estimate of expenditure.

3. Analyze the results.
   After requesting a regression, SPSS gives me several tables in the output window. Figure 12.9 shows the three tables I need to create a regression table for state expenditure. (My table is found in my memo.)

Figure 12.7 How to Find the Regression Commands

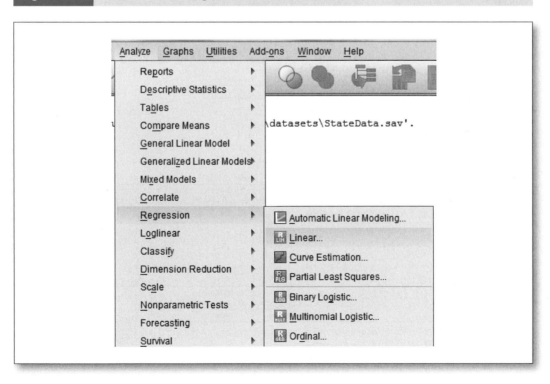

Figure 12.8 Regressing Expenditures on Population

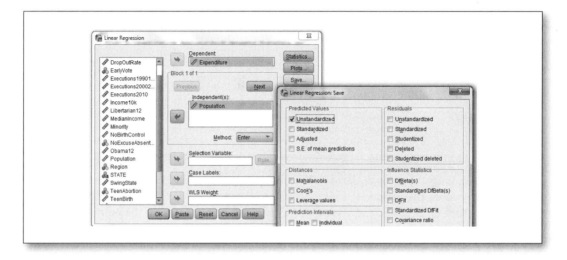

Figure 12.9 Output from Expenditure Regression

Model Summary[b]

Model	R	R Square	Adjusted R Square	Std. Error of the Estimate
1	.928[a]	.861	.858	6.009438

a. Predictors: (Constant), Population
b. Dependent Variable: Expenditure

ANOVA[a]

Model		Sum of Squares	df	Mean Square	F	Sig.
1	Regression	10744.899	1	10744.899	297.533	.000[b]
	Residual	1733.440	48	36.113		
	Total	12478.340	49			

a. Dependent Variable: Expenditure
b. Predictors: (Constant), Population

Coefficients[a]

Model		Unstandardized Coefficients		Standardized Coefficients	t	Sig.
		B	Std. Error	Beta		
1	(Constant)	-.389	1.149		-.339	.736
	Population	2.162	.125	.928	17.249	.000

a. Dependent Variable: Expenditure

4. If you want to look for outliers, go into the Data window to compare the observed and expected values of your dependent variable.
Once I have the data set up in a professional-looking table, I want to find out how Connecticut and New Jersey expenditures compare with what you would expect them to be based on population. I already had SPSS save the predicted value of expenditures from the regression. Switching to the Data window, I see this variable in the last column. SPSS has named it "PRE_1," which means that it is the first predicted value I have saved as a variable. I use the "Compute Variable" command under "Transform" to create a new variable that is the difference between "Expenditure" and "PRE_1." This process is shown in Figure 12.10.

Figure 12.10 Computing the Difference between Observed and Expected Spending

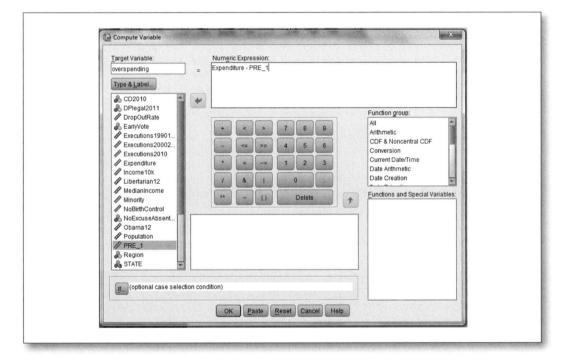

This new variable now shows up as the last column in the Data window. In the first column, I find the row for Connecticut and highlight it. Moving to the last column, I see that my new "Overspending" variable has a value of $10.59. This is shown in Figure 12.11. This means that it actually spends $10.59 billion more than you would expect based on its population. I then find New Jersey, highlight the line, and find that it spends $9.55 billion more than expected.

Figure 12.11 Finding Connecticut's Overspending

	TeenBirth	TeenAbortion	PRE_1	overspending
1	73	50	12	9.94669	-2.59
2	51	37	15	1.14584	4.30
3	39	58	17	13.42811	-5.06
4	30	59	9	5.92468	-1.45
5	75	39	26	80.15887	11.39
6	59	43	17	10.48729	-3.56
7	57	23	26	7.33023	(10.59)
8	33	44	27	1.55670	1.71

5. Write your memo.

Memo

To: Governor Dannell Malloy, Connecticut
From: T. Marchant-Shapiro, intern
Subject: State Expenditures
Date: February 24, 2012

You asked me to examine how much Connecticut spends in comparison to other states, especially New Jersey. In absolute terms, we would expect Connecticut to spend less than New Jersey simply because we have a smaller population. So I collected data on both state expenditures and state population to find how the spending of both states compares to what we would expect it to be based on population. I found that the accusation of overspending is probably an accurate description of both New Jersey and Connecticut.

Using regression analysis, I found, as expected, that population has a strong impact on state spending. As shown in the table below, on average, the spending of states increases by about $2.162 billion for every million extra residents. This averages to about $2,000 per resident. We can be 99 percent confident that there is a strong relationship between state spending and population. Indeed, the population of the states explains 86.1 percent of the variance in their spending.

If Connecticut and New Jersey exhibited average spending for their populations, you would expect them to spend $7.33 billion and $18.62 billion, respectively. In comparison to this expectation, Connecticut actually spends $17.921 billion and New Jersey actually spends $28.168 billion—both states spend about $10 billion more than expected. Although you would be justified in accusing Governor Christie of excessive spending for the population of his state, such an accusation would open you up for a similar attack.

State Expenditures Regressed on State Population

Independent Variable	Unstandardized Coefficient (s.e.)	Standardized Coefficient
Constant	−0.243	
Population	2.133** (0.110)	0.941**
n = 50		$R^2 = 0.886$

Source: National Association of State Budget Officers, *The Fiscal Survey of States* 2012. www.nasbo.org/sites/default/files/Spring%202012%20Fiscal%20Survey%20of%20States.pdf.

*prob. < 0.05

**prob. < 0.01

Your Turn: Bivariate Regression

YT 12.1 — Regress state expenditures on population for the industrial states found in Table 12.8. Remember that you will be using these data to set up two different work tables.

Table 12.8 Population and Expenditures of Certain Industrial States

	Population in millions	State Expenditures in $billions
IN	6.48	13.56
MI	9.88	8.25
NJ	8.79	28.03
NY	19.38	53.53
OH	11.54	27.19
PA	12.7	25.29

Source: National Association of State Budget Officers, *The Fiscal Survey of States* 2012. www.nasbo.org/sites/default/files/Spring%202012%20Fiscal%20Survey%20of%20States.pdf.

1. Find the line that best fits the data, using the following equations:

$$Y = B_0 + B_1 X_1$$

$$B_1 = \frac{\Sigma(X - \bar{X})(Y - \bar{Y})}{\Sigma(X - \bar{X})^2}$$

$$= SS_{XY}/SS_X$$

$$B_0 = \bar{Y} - B_1 \bar{X}_1$$

2. Identify the statistical significance by finding the standard error of the slope:

$$s^2_{B1} = SS_{Y|X}/(n-2)(SS_X)$$

$$s_{B1} = \sqrt{s^2}$$

$$t = B_1/s_{B1}$$

$$d.f. = n - 2$$

3. Evaluate the strength of the relationship with R^2, giving its PRE interpretation:

$$R^2 = (SS_{XY})^2/SS_X SS_Y$$

YT 12.2 After every census, the United States House of Representatives is reapportioned, with the goal of making the population in each congressional district equal. As a result, after the 2010 census, each state was allocated a certain number of seats for the next decade. We would expect that the number of seats would have a strong positive relationship with the population. Table 12.9 shows the number of seats allocated to each state regressed on the state's population. Analyze these data by identifying and interpreting the following statistics:

1. Intercept
2. Slope
3. Statistical Significance
4. R^2

YT 12.3 Since 1988, teen pregnancies in the United States have been declining. Table 12.10 shows the national pregnancy rates for 15- to 19-year-olds per

Table 12.9 Number of Congressional Districts Regressed on State Population, 2010 Census

Independent Variable	Unstandardized Coefficient (s.e.)	Standardized Coefficient
Constant	0.074	
Population 2010	1.400** (0.017)	−0.997**
n = 50		$R^2 = 0.993$

Source: U.S. Census Bureau, "Apportionment Data," *United States Census 2010*. www.census.gov/2010census/data/apportionment-data.php.

*prob. < 0.05
**prob. < 0.01

thousand teen girls regressed on time. Analyze these data by identifying and interpreting the following statistics:

1. Intercept
2. Slope
3. Statistical Significance
4. R^2

Table 12.10 Teen Pregnancy Rate Regressed on Number of Years since 1988

Independent Variable	Unstandardized Coefficient (s.e.)	Standardized Coefficient
Constant	118.050	
Years Since 1988	−2.791** (0.156)	−0.975**
n = 18		$R^2 = 0.950$

Source: The National Campaign to Prevent Teen and Unplanned Pregnancy, "National Pregnancy Rates for Teens, aged 15-19," 2012. www.thenationalcampaign.org/national-data/NPR-teens-15-19.aspx.

*prob. < 0.05
**prob. < 0.01

YT 12.4 Table 12.11 shows income regressed on race. How do blacks and whites compare in terms of income? Analyze these data by identifying and interpreting the following statistics:

1. Intercept
2. Slope
3. Statistical Significance
4. R^2

Table 12.11 Income Regressed on Race (Whites and Blacks)

Independent Variable	Unstandardized Coefficient (s.e.)	Standardized Coefficient
Constant	22,737.63	
Blacks	−5,998.67** (1042.12)	−0.111**
n = 2652		$R^2 = 0.012$

Source: General Social Survey 2010, National Opinion Research Center, 2012. www3.norc.org/GSS+Website/Download/SPSS+Format/.

*prob. < 0.05
**prob. < 0.01

Variable: RealRInc is the respondent's income in 1972 dollars.

Note: These data do not include those of "Other" race.

Apply It Yourself: Analyze Influences on Corruption

The year is 2009 and you are interning at the Organization of Economic Cooperation and Development. Your office is preparing recommendations to update the 1997 Anti-Bribery Convention. As part of it, you are asked to prepare an analysis of how corruption is influenced by the freedom of countries. You decide to regress Transparency International's measure of corruption[3] on Freedom House's Index of freedom.[4] In addition to finding the relationship between the two variables, you decide to identify those countries that are significantly more corrupt than their level of freedom would predict.

1. Open "WorldData" and clean "TransparencyIndex," "PoliticalRights," and "CivilLiberties." Be sure to set the −9 values to missing.

2. Use the "Compute" command to create a single variable for Freedom House by adding the values of PoliticalRights and CivilLiberties together.

   >Transform
   >   >Compute Variable
   >   Target Variable="FreedomHouse"
   >   Numeric Expression = "PoliticalRights + CivilLiberties"
   >   >OK

3. Regress the TransparencyIndex on the FreedomHouse index. Be sure to save the predicted value so you can see which countries have higher levels of corruption than you would expect.

   >Analyze
   >   >Regression
   >   >Linear
   >   Dependent=TransparencyIndex
   >   Independent=FreedomHouse
   >   >Save
   >   Predicted Values:
   >   >Unstandardized
   >   >Continue
   >   >OK

4. Find the countries that have higher than expected levels of corruption by computing a new variable that gets the difference between the observed and expected levels of corruption. Find the name of the predicted variable in the heading to the last column in the Data window. The first time you do this, the new variable will be named "PRE_1".

   >Transform
   >   >Compute Variable
   >   Target Variable="MoreCorrupt"
   >   Numeric Expression = "TransparencyIndex – PRE_1"
   >   >OK

5. The outliers will be the cases that have corruption values more than four points above the expected values. Use the "Select If" command to find them, and then the "Analyze," "Reports," and "Case Summaries" commands to identify them. Be sure to request the country name here.

>Data
>   >Select Cases
>   >If condition is satisfied
>   >If
>   "MoreCorrupt" → "MoreCorrupt >= 4"
>   >Continue
>   >OK
>Analyze
>   >Reports
>   >Case Summaries
>   "Country" "FreedomHouse" "TransparencyIndex" "PRE_1" → Variables
>   >OK

6. Write your memo. Be sure to include a professional-looking regression table and also the SPSS output for grading purposes.

Key Terms

Coefficient (p. 346)
Dummy variable (p. 356)
Residual (p. 347)

R-square (p. 353)
Variable (p. 346)

CHAPTER 13

Multiple Regression

The Final Frontier

If I walk out my front door, I have two directions I can go: north or south. If I were a crow, though, I'd have more options. I could still head north, this time flying past the end of my street to Sleeping Giant, my favorite place. Or I could fly in any of the 360° on a compass. If, however, I were the captain of the starship *Enterprise*, I would add a third dimension to my options because I could fly through space. On occasion, I would even travel through time, adding a fourth dimension to my repertoire.

Similarly, as we've progressed through this book, we've been adding dimensions to our statistical analyses of variables. We began with the unidimensional analysis of a single variable with measures of central tendency and dispersion. We then added a second dimension by looking at the relationship between a dependent and an independent variable. We moved into three-dimensional space when we added a control variable with a three-way contingency table. In the previous chapter, we took a quick step back into a two-dimensional plane when we examined bivariate regression. But regression is capable of moving into a three-dimensional realm as well.

What's interesting about regression is that when it does so, it doesn't matter which is the independent variable and which is the control variable. When we did a three-way contingency table, you'll recall that although its setup made it easy to see the impact of the independent variable for each of the categories of the control variable, it was not very easy to see how the control variable influenced the dependent variable. In contrast, with multiple regression, we get a slope for our independent variable controlling for our control variable, but we also get a slope for our control variable controlling for our independent variable. Mathematically, it doesn't matter which is which, so we end up calling them both independent variables.

A second interesting feature of multiple regression is that we can control for more than one variable. These controls expand the way we can depict relationships to multiple dimensions. With one control variable, we get an equation for a line in three dimensions; with two control variables, the line travels through four dimensions (like the starship *Enterprise* when it's traveling through time). But although we are limited conceptually to being able to picture four dimensions because we live in a four-dimensional world (forward/back, left/right, up/down, and past/future), regression can describe even more dimensions than that. Regression can give us an equation of a line that finds the slope between many possible independent variables and the dependent variable, simultaneously controlling for all the others. As with bivariate regression, we can describe the pattern we see by looking at the slopes connected to each of the independent variables in the multiple regression. Those slopes will each have a different standard error, allowing us to identify which variables have statistically significant relationships with the dependent variable. In evaluating the strength of the relationship, we will once again look at the R^2 to see how much of the variance in the dependent variable is explained by the model that contains all of the independent variables. In addition, we can look at the standardized regression coefficients (or betas) to see which of them explains the most variance in the dependent variable.

USING REGRESSION TO CONTROL FOR OTHER VARIABLES

Frequently in the social sciences, there is a group of highly interrelated variables that are difficult to distinguish in a causal way. At the individual level, education and income are highly correlated and are also both important explanatory variables in political attitudes. At an international level, individual education and income aggregate to national literacy and wealth as predictors of national political attributes like political stability. Regression is a good tool for disaggregating the impact of each variable, independent of the other, at both the individual and the national level.

Let's begin by looking at the national level. The stability of a country, operationalized here with the State Fragility Index (SFI), is a function of a set of several interrelated factors: education, wealth, corruption, and income inequality. Table 13.1 shows the correlation matrix for measures of these variables. The cells of the matrix show the Pearson's r for each pair of variables. Remember that in Chapter 10, we saw that the cells going down the diagonal have a correlation of one. This perfect positive correlation shows the not surprising fact that each variable is perfectly correlated with itself. The triangles on each side of the diagonal are mirror images of each other because Pearson's r is a symmetrical measure of association—the correlation of GDP with literacy is the same as the correlation between literacy and GDP. Pearson's r doesn't assume that one variable is the independent and the other is the dependent variable so the math is identical regardless of the order.

In this particular correlation matrix, you'll notice that all of the variables are strongly correlated with each other, some positively and some negatively. Because we will be using the State Fragility Index as our measure for state instability (our dependent variable), let's look at the first row. All of the variables have a statistically significant relationship with our dependent variable. Just in terms of bivariate relationships, corruption and literacy have the highest correlations with instability. But you can see that these variables are highly correlated. Is it possible that the bivariate relationships we see here are inaccurate because we haven't controlled for the others? For example, because GDP has a negative relationship with SFI but a positive relationship with governmental transparency, the underlying relationship between SFI and corruption might be masked.

Table 13.1 Correlation Matrix for State Stability Factors

Pearson's r (n)	SFI	Female Literacy	Per Capita GDP (in $1000s)	Transparency Index	Income Share of Top 10%
SFI	1 (163)	−0.743** (158)	−0.664** (163)	−0.714** (162)	0.357** (121)
Female Literacy	−0.743** (158)	1 (182)	0.459** (182)	0.464** (172)	−0.251** (121)
GDP	−0.664** (163)	0.459** (182)	1 (194)	0.781** (179)	−0.431** (123)
Transparency Index	−0.714** (162)	0.434** (172)	0.781** (172)C	1 (179)	−0.355** (123)
Income Share of Top 10%	0.357** (121)	−0.251** (121)	−0.431** (123)	−0.355** (123)	1 (123)

Sources: Marshall, Monty G. and Benjamin R. Cole, "State Fragility Index and Matrix 2011." *Center for Systemic Peace*, 2012. www.systemicpeace.org/SFImatrix2011c.pdf; CIA World Factbook, 2012. www.cia.gov/library/publications/the-world-factbook/fields/2103.html; "GDP per capita, PPP (constant 2005 international $)" *GapMinder*. 2010. Accessed 17 November 2012. http://www.gapminder.org/data/; Transparency International, *Corruption Perceptions Index 2011*. http://www.transparency.org/cpi2011/in_detail; "Income Share of Richest 10%." *GapMinder*. 2010. Accessed 17 November 2012. http://www.gapminder.org/data/.

*prob. < 0.05

**prob. < 0.01

To tease out the independent effects of each variable on the instability of the country, we regress our dependent variable (the SFI) on our independent variables. I am not going to make you do the math. In general, I strongly believe that doing the math gives you intuitions about how to interpret the different statistics. But in the case of multiple regression, I'll leave the math for graduate students; for undergraduates, I think the process of doing the math is more tedious than enlightening.

Let me just give you a feel for how the math goes. The basic approach is to regress all of the variables on each other. Then, for each individual variable, you track both the direct impact on the dependent variable and the indirect impacts through the other variables. If the indirect effects inflate the relationship, they get subtracted out. But if the indirect effects mask the relationship, they get added back in. The resulting regression gives you what are sometimes called partial slopes because they include only the impact of each individual variable, controlling for the influences of all the other variables.

Regressing SFI on all the other variables, we get the regression shown in Table 13.2. In general, we can interpret this in much the same way we did with the bivariate relationship. As with all analyses, we describe the pattern, identify the statistical significance, and evaluate the strength of the relationship.

Table 13.2   State Instability Regressed on Corruption, Wealth, Literacy, and Income Inequality

Independent Variable	Unstandardized Coefficient (s.e.)	Standardized Coefficient
Constant	22.783	
Transparency Index	−0.811** (0.270)	−0.270**
GDP per Capita (in $1000s)	−0.138* (0.058)	−0.250*
Female Literacy Rate	−0.124** (0.015)	−0.481**
Income Share of the Top 10%	0.001 (0.048)	0.001
n = 120		$R^2 = 0.727$

Sources: Marshall, Monty G. and Benjamin R. Cole, "State Fragility Index and Matrix 2011." *Center for Systemic Peace,* 2012. www.systemicpeace.org/SFImatrix2011c.pdf; CIA World Factbook, 2012. www.cia.gov/library/publications/the-world-factbook/fields/2103.html; "GDP per capita, PPP (constant 2005 international $)" *GapMinder.* 2010. Accessed 17 November 2012. http://www.gapminder.org/data/; Transparency International, *Corruption Perceptions Index 2011.* http://www.transparency.org/cpi2011/in_detail; "Income Share of Richest 10%." *GapMinder.* 2010. Accessed 17 November 2012. http://www.gapminder.org/data/.

Describing the pattern in a multiple regression is not quite the same as in a bivariate regression. In the context of a multiple regression, we lose much of our ability to interpret the constant. It is still the value of the dependent variable we would expect to see if all the independent variables were zero, but what does that mean? Not much unless it is possible for a case to actually have a value of zero on all of the variables. So usually, in **multiple regression**, we focus on the interpretation of the slopes. Remember that the slope is the change in the dependent variable for every unit increase in the independent variable. This means that the magnitude of the slope is dependent on the size of the scales for the dependent and independent variables. As a result, we cannot really compare the slopes; we can only interpret them individually. For example, the Transparency Index is a 10-point scale for which zero is the score of the most corrupt state. The State Fragility Index is a 25-point scale, where 25 is the most fragile state. The regression tells us that, controlling for the other variables, for every point increase in the Transparency Index, we expect the Fragility Index to decrease by 0.811 points. We can do the same description for each of the variables.

Although the standard interpretation of the slopes may be interesting, it is probably more important to describe the pattern that you see in the signs of the coefficients. In Table 13.2, the negative slopes indicate that increasing female literacy, GDP per capita, and governmental transparency all decrease the levels of instability within states. The positive coefficient for Income Share of the Top 10% indicates that income inequality increases instability. All of these coefficients are consistent with our understanding of how these variables influence state instability. Most importantly, because we controlled for all of the other variables in the model, we know that these relationships are not spurious. (Unless, of course, we've excluded another variable for which we need to control.)

In the second stage of analysis, we identify the statistical significance of the slope for each of the variables. Transparency and Female Literacy are both statistically significant at the 0.01 level. GDP per capita is statistically significant at the 0.05 level. For all three of these variables, we can reject the null hypothesis. However, once we've controlled for these three variables, income inequality is no longer statistically significant—we cannot reject the null hypothesis. Apparently, the correlation between fragility and income inequality is spurious, due only to the correlation of income inequality with the other three explanatory variables. Now that we've controlled for those other variables, the original correlation has evaporated. It is education, wealth, and corruption that are the primary factors driving instability.

We can begin the third stage of analysis in much the same way we did with bivariate regression. For multiple regression, R^2 gives the proportion of variance in the dependent variable explained by all of the variables in our model. In this case, the model explains 72.7 percent of the variance in state fragility. In addition to the R^2, we can compare the standardized coefficients. Because they have been standardized, we can compare the magnitude of these coefficients to see which variables explain the most variance in the dependent variable. In this case, because Female Literacy has the largest standardized coefficient, we know that it explains the most variance in state fragility. Corruption and GDP both have betas about the same size, so we can conclude that they explain about the same amount of variance.

Two words of caution here. First, although in a bivariate regression the standardized coefficient is the same as Pearson's r, this is not the case in multiple regression. On occasion (if one variable has a strong positive relationship with the dependent variable and another a strong negative relationship), the standardized coefficient can exceed a magnitude of one. Second, although the magnitude of the standardized coefficient indicates which variable has the most explanatory power for the dependent variable, this is not to say that the value of the standardized coefficient is the amount of variance that a particular variable explains. The PRE explanation applies only to the R^2 and describes only the explanatory power of the full model, not any individual variable.

THE ASSUMPTIONS OF REGRESSION

Regression is an amazingly powerful statistical tool. But its power is dependent on the data meeting certain assumptions. In this section, we'll be looking at how we know whether these assumptions are being met, as well as how to cope with violations of them. If data meet these assumptions, we say that the data meet the **Gauss-Markov assumptions**. If they are met, we say that "**OLS** is BLUE," which means that Ordinary Least Square regression (the kind we have been doing) gives us the Best Linear Unbiased Estimate of the slope. By "estimate," we mean that we are making a guess as to the relationship between our dependent and independent variables. By "linear," we mean that we are fitting a straight line to the data. By "unbiased," we mean that the estimate is not consistently too small or too big. By "best," we mean that the line fits the data with the smallest amount of error around it.

The Gauss-Markov theorem can be presented in different ways, but for clarity's sake, I'm going to summarize it with six assumptions:

1. The variables have an interval level of measurement.
2. The relationship between the dependent variable and each independent variable is linear.
3. The model is correctly specified.
4. Two or more independent variables are not collinear.
5. The errors in predicting the dependent variable have a mean of zero.
6. The errors in predicting the dependent variable are homoscedastic (as opposed to heteroscedastic), meaning that the distribution of the errors doesn't consistently increase or decrease across values of the dependent variable.

Gauss-Markov Assumption 1: Interval-Level Variables

The variables need to have an interval level of measurement. This assumption should make sense intuitively, because when you did the math to calculate the slope and intercept, you had to calculate the mean for both the dependent and independent variables. If either of those variables was nominal or ordinal, you couldn't have done that because the numbers assigned to the categories don't mean anything. For the variables to be interval (or ratio) level, the values need to be ordered and the scale also needs to have units of the same size. The easy way to check this is to ask the question: Do the numbers actually mean something? Or are they just assigned to the categories for ease in communicating with the computer? Remember that we interpret the slope as "a one-unit increase in the independent variable leads to this much unit increase in the dependent variable." In order for this interpretation to make sense, a unit needs to mean the same thing regardless of where you are on the scale.

Possible Solutions. There are ways to use regression analysis with nominal or ordinal variables. Which solutions you use depend on whether the categorical variable is your dependent variable or your independent variable. If your dependent variable is nominal or ordinal, you have two options. First, if the variable is dichotomous, you can perform a logit or probit regression (or nested logit if it has more than two categories). This kind of regression assumes that your dependent variable can't be less than zero or greater than one. That means that it assumes a curved relationship between your dependent and independent variables. There are two problems with this option: (1) I'm not going to teach you how to do them, and (2) they are tricky to interpret. Your second option is to collapse your interval-level independent variable into categories and analyze the relationship using a contingency table.

If your independent variable is nominal or ordinal, you can use analysis of variance. But that doesn't give you the kind of freedom regression did in terms of controlling for other variables. If you really want to use regression, you can remember that regression allows you to treat dichotomous independent variables as if they were interval level. In order to include a variable that has three or more categories, you simply transform it into a series of dichotomous (or dummy) variables. To do this, you choose one of the categories as your baseline against which you want to compare all the other categories. You then make a **dummy variable** for each of the other categories. Code all the cases within each category as one in its corresponding dummy variable but zero in all of the others. In the end, you will have one fewer dummy variables than you had categories in the original variable. You would run a multiple regression, including all of the dummies as independent variables. You would interpret the slope of each as the difference observed in the dependent variable between members of that category and your baseline (the group that was coded as zero on all the dummies).

An Example: Race. The National Election Studies ask respondents to place individuals and groups on a feeling thermometer from 0 (very cold) to 100 (very warm). One of the groups placed on this thermometer is "Blacks." Suppose you wonder whether an individual's race affects how warmly he or she feels toward blacks in general. We could hypothesize that blacks probably feel warmest toward their own group; whites, coldest; and other minorities, somewhere in between. Because our independent variable, race, is nominal, we cannot use it directly in a regression. But we can change its categories into dummy variables and include them in a multiple regression.

The race variable has categories for "White," "Black," "Hispanic," "Asian," and "Other." In order to change a categorical variable into dummies, you first need to choose one baseline category. I choose "White" as my baseline. I then create dummy variables for "Black," "Hispanic," "Asian," and "Other." In my regression, I regress the feeling thermometer for blacks on all four of those dummy variables. For any given slope, the regression will hold members of the other categories constant, which means that the only cases left to analyze are whites and the members of that particular group. Because I chose whites as my baseline, I will interpret each slope as the difference between members of that category and whites. The results are shown in Table 13.3.

Table 13.3 Feeling Thermometer for Blacks Regressed on Race

Independent Variables	Unstandardized Coefficients	Standardized Coefficients
Constant	66.079	
Black	19.783**	0.415**
Hispanic	4.637**	0.088**
Asian	−1.387	−0.009
Other	10.695**	0.063**
n = 2055		$R^2 = 0.160$

Source: American National Election Study, *ANES 2008 Time Series Study*. www.electionstudies.org/studypages/2008prepost/2008prepost.htm.

*prob. < 0.05

**prob. < 0.01

Normally, with multiple regression, the intercept doesn't mean much. But if you have only dummy variables in the regression, it actually does. Remember that the way to interpret the intercept is as the value of the dependent variable you expect if all the independent variables are zero. In this case, you know that when all the dummy variables are zero, the respondent is white. So the constant is the average feeling thermometer for blacks among the white respondents. On average, whites place blacks at 66° on the feeling thermometer. The slopes for each of the other categories indicate how much warmer (or colder) that group feels toward blacks than the white average of 66°. On average, blacks place themselves 20° warmer than whites, for a total of 86°. Hispanics place blacks 5° warmer, for a total of 71°, Asians place blacks 1° colder (65°), and other races place them 11° warmer (76°). Although the difference between Asians and whites is not statistically significant, the differences between whites and the other groups are each statistically significant at the 0.01 level. Except for Asians, we can reject the null hypothesis. The **betas** indicate that being black explains the most variance in the feeling thermometer. The R^2 of 0.160 indicates that race explains 16.0 percent of the variance in feelings toward blacks.

Gauss-Markov Assumption 2: Linear Relationship

It probably seems obvious that because the purpose of regression is to find an equation describing a linear relationship between a dependent variable and an independent variable, the underlying relationship needs to be linear. If the two variables have a curvilinear relationship, a line is not going to do a good job of describing the relationship. The purpose of this section is, first, to describe how to verify that the relationship is linear, and, second, to describe some techniques to modify the data so that there is a **linear relationship**.

To check the linearity of a relationship, do a scatter plot of the two variables. If the data points fall in line, you are fine using the variables to do a regression. If it looks curvilinear, then you will do a mathematical operation on the independent variable corresponding to the shape of the curve to correct for it. There are three main curves you might see: squared, exponential, and log linear.

An Example of a Linear Relationship. Figure 13.1 shows the scatter plot for a linear relationship. There are two commonly used measures of government fragility: The Failed States Index created by The Fund for Peace and the State Fragility Index created by The Center for Systemic Peace. If the two indexes are reliable, we would expect them to be both highly correlated and linearly related. Figure 13.1 shows the scatter plot for the values of the two indexes. Not only do they have a correlation of 0.892, the scatter plot shows a fairly clear linear relationship of the kind we want for variables included in regression analysis.

What to Do for a Squared Relationship. Figure 13.2 shows a squared relationship. If your scatter plot looks either like the valley shown in Figure 13.2 or a mountain (imagine flipping it over), you would use the Compute command to get the square of your independent variable. In your regression, include both the original variable and its square. If your scatter plot looks like a valley, the squared value of X will have a positive coefficient; if it looks like a mountain, it will have a negative coefficient. Interpreting the meaning of

Figure 13.1 State Fragility Index 2011 by Failed States Index 2012

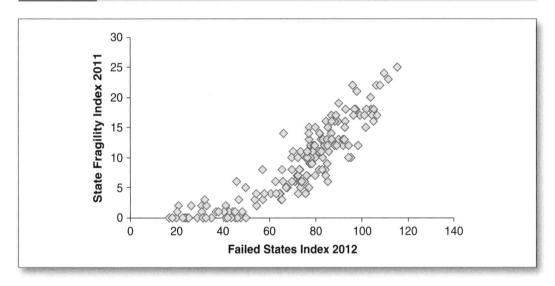

the coefficients isn't as direct as usual (a one-unit increase in the independent variable leads to B increase in the dependent variable), but at least you are able to control for this variable and even to interpret the impact it has on the dependent variable through the beta.

Figure 13.2 A Squared Relationship

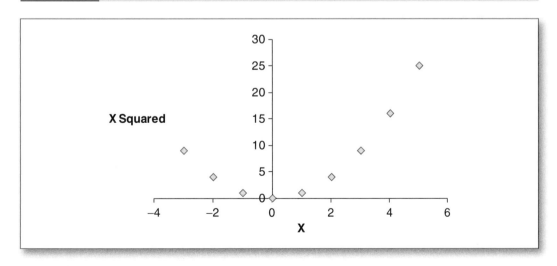

An Example of a Squared Relationship. For example, when you look at the instability of countries, you find that very free countries are very stable because they are responsive to the needs of their citizens. But very repressive countries can also be fairly stable because they keep such tight control on their countries. In contrast, transitional countries are in a position of not yet being satisfactory in terms of policy but being free enough to allow revolts. As a result, those countries in the middle of a democracy scale are more instable than those on either end of the scale. Because the relationship between freedom and stability looks like a mountain, it is curvilinear in a **squared relationship**. I measure freedom by adding Freedom House's measures of Civil Liberty and Political Rights.[1] I use the FSI measure of instability for 2012.[2]

Figure 13.3 Failed States Index 2012 by Freedom House Score 2012

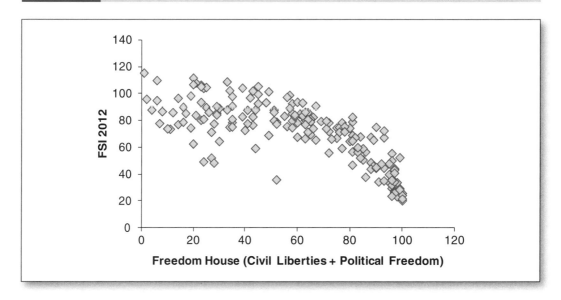

Figure 13.3 shows the scatter plot of this relationship. It is clear that as countries become more democratic, their fragility decreases. But you can also see that the least democratic countries are slightly more stable than those in the middle of the scale.

From the regression found in Table 13.4, you wouldn't notice any problems with the analysis. The unstandardized coefficient indicates that for every point on the scale a country becomes more democratic, its fragility decreases by half a point. This relationship is statistically significant at the 0.01—we can reject the null hypothesis. The R^2 indicates that democracy explains 54.9 percent of the variance in stability. But the scatter plot in Figure 13.3 gives some concern. Let's redo the regression, including the democracy variable squared as well as in its original version. This yields the regression in Table 13.5. With the squared variable, the coefficients become difficult to interpret. But the negative value on the Squared Democracy

Table 13.4 Fragility Regressed on Democracy

Independent Variables	Unstandardized Coefficients	Standardized Coefficients
Constant	105.005	
Democracy	−0.581**	0.741**
n = 178		$R^2 = 0.549$

Sources: Freedom House, *Freedom in the World 2012*. http://freedomhouse.org/report/freedom-world/freedom-world-2012; Fund for Peace, *The Failed States Index 2012: The Indicators* 2012. http://ffp.statesindex.org/indicators.

*prob. < 0.05

**prob. < 0.01

variable indicates that the curve is a mountain rather than a valley. Both the original variable and the squared variable are statistically significant, which indicates that it is important to include both variables—the relationship is curved. And the R^2 has increased to 0.676. By including the squared variable, we have increased the explanatory power of the model by 12.7 percentage points.

Table 13.5 Fragility Regressed on Squared Democracy

Independent Variables	Unstandardized Coefficients	Standardized Coefficients
Constant	80.209	
Democracy	0.648**	0.827**
Squared Democracy	−0.011**	−1.608**
n = 178		$R^2 = 0.676$

Sources: Freedom House, *Freedom in the World 2012*. http://freedomhouse.org/report/freedom-world/freedom-world-2012; Fund for Peace, *The Failed States Index 2012: The Indicators* 2012. http://ffp.statesindex.org/indicators.

*prob. < 0.05

**prob. < 0.01

What to Do for an Exponential Relationship. Figure 13.4 shows an **exponential relationship**. As X increases, Y increases slowly at first and then more rapidly—like the letter "J." If your scatter plot looks like this, the solution is to take the log of the dependent variable to pull its extreme values down in line with the other data. There are two log functions in the compute command of SPSS: log and ln. The first is the log base 10 of the variable; the second is called the natural log or log base e. (The letter e symbolizes a mathematical constant for a number that is approximately 2.72.) The latter is called the natural log

Figure 13.4 Exponential Relationship

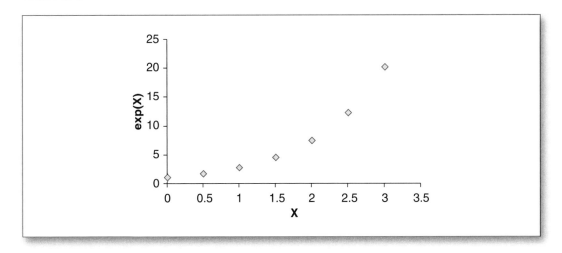

because it gives a pattern that repeats itself frequently in nature. And we've found that using it tends to give better results in straightening out curvilinear political events as well. So take the natural log of your dependent variable and use it in your regression.

An Example of an Exponential Relationship. The economic development of a country has a strong impact on its infant mortality. The more rural a country is, the higher its infant mortality rate. Figure 13.5 show the scatter plot of these two variables. To a certain degree,

Figure 13.5 Infant Mortality Rate by Percentage Rural

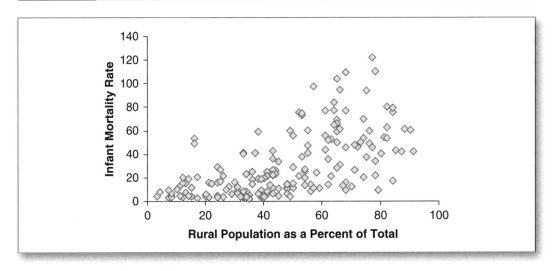

having a higher rural population doesn't make too much of a difference. But once half of the population lives in rural areas, the infant mortality really increases, following the J curve of Figure 13.4.

Table 13.6 Infant Mortality Regressed on Rural Population

Independent Variables	Unstandardized Coefficients	Standardized Coefficients
Constant	−3.914	
Percent Rural	0.726**	0.619**
n = 179		$R^2 = 0.383$

Sources: CIA, "Infant Mortality Rate," *The World Factbook* 2012. www.cia.gov/library/publications/the-world-factbook/fields/print_2091.html; "Percentage Living in Rural Areas," *Nation Master* 2012. www.nationmaster.com/graph/peo_per_liv_in_rur_are-people-percentage-living-rural-areas#source.

*prob. < 0.05

**prob. < 0.01

If we regress infant mortality on rural population, we get the regression shown in Table 13.6. The intercept doesn't make much sense: If a country has no rural population, you would expect the infant mortality rate to be −4. That, of course, isn't possible, but because you are fitting a line to a curvilinear relationship, the result is that you can get a negative value. For every percent increase in rural population, the infant mortality increases by 0.726. This relationship is statistically significant at the 0.01 level—we can reject the null hypothesis. The proportion of the population that lives in rural areas explains 38.3 percent of the variance in infant mortality.

Because the underlying relationship is curvilinear, you get a better estimate of the relationship if you can model it as a curve instead of a line. Once again, you can do this by performing a mathematical function on one of the variables. With an exponential relationship, the best way to do this is to take the natural log of the dependent variable. This will have the effect of decreasing the magnitude of the high values of the dependent variable. As a result, it mushes the curve down into a line. Figure 13.6 shows the natural log of infant mortality plotted against the percent of urban dwellers. You can see that the relationship looks much more linear than in Figure 13.5.

Table 13.7 gives this logged dependent variable regressed on rural population. The constant now makes more sense: If a country has no urban population, the natural log of its infant mortality is 1.554. If we raise this to e^x, we find that the infant mortality is 4.73, which is very close to the minimum infant mortality rate. The unstandardized coefficient is hard to interpret because the amount that infant mortality increases would depend on where on the scale you are. But you can see that there is a statistically significant positive relationship. You can reject the null hypothesis. What is most important to see here is that the R^2 has increased by 0.06—this model explains 6 percentage points more variance than the linear model. Modeling the curve really does make for a better estimate of the relationship.

Figure 13.6 Natural Log of Infant Mortality Rate by Percentage Rural

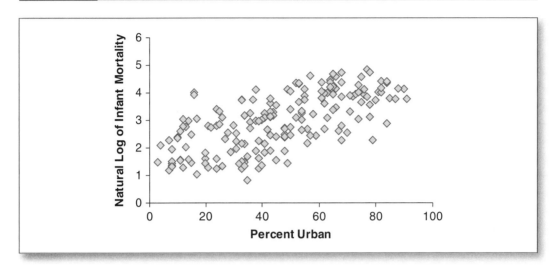

Table 13.7 Natural Log of Infant Mortality Regressed on Rural Population

Independent Variables	Unstandardized Coefficients	Standardized Coefficients
Constant	1.554	
Percent Rural	0.030**	0.666**
n = 179		$R^2 = 0.443$

Sources: CIA, "Infant Mortality Rate," *The World Factbook* 2012. www.cia.gov/library/publications/the-world-factbook/fields/print_2091.html; "Percentage Living in Rural Areas," *Nation Master* 2012. www.nationmaster.com/graph/peo_per_liv_in_rur_are-people-percentage-living-rural-areas#source.

*prob. < 0.05

**prob. < 0.01

What to Do for a Log Linear Relationship. Figure 13.7 shows a **log linear relationship**. As X increases, Y increases quickly at first and then more slowly—curving over like a lower case "r." If your scatter plot looks like this, the solution is to use the "Compute" command to get the natural log of your independent variable and include only it in your regression. This has the effect of pulling the latter data in so that those values you see that are high on the independent variable get mushed in close enough to the left that they are pulled in line with the other data.

An Example of a Log Linear Relationship. Life expectancy is a variable that various hardships can lower, but if you remove those hardships, you can only increase life expectancy so

Figure 13.7 Log Linear Relationship

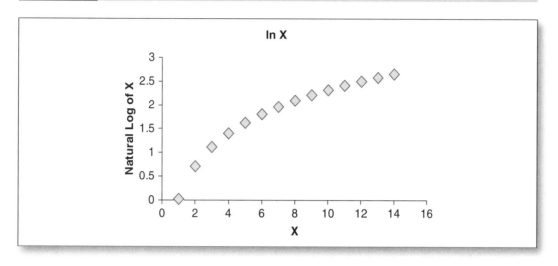

much. It tends to peak at around eighty years. For example, poor countries tend to have lower life expectancies, and increasing the GDP can have a dramatic impact on the average life expectancy. In contrast, among countries with adequate wealth, increasing the GDP does not make much difference in the life expectancy. Figure 13.8 shows a scatter plot of the average life expectancy by the GDP per capita in 2011. It looks like the lower case r shape that we know describes a logged relationship.

Figure 13.8 Life Expectancy by GDP Per Capita

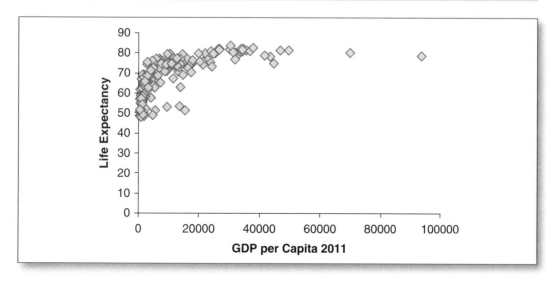

We could try to fit these data with a straight line. Table 13.8 shows life expectancy regressed on GDP per capita. The intercept indicates that if GDP is zero, we would expect life expectancy

Table 13.8 Life Expectancy Regressed on GDP per Capita (in $thousands)

Independent Variables	Unstandardized Coefficients	Standardized Coefficients
Constant	64.043	
GDP per capita	0.432**	0.612**
n = 178		$R^2 = 0.375$

Sources: "Life Expectancy at Birth (Years)," *GapMinder* 2011. www.gapminder.org/data/; "GDP per Capita PPP," *GapMinder* 2011. www.gapminder.org/data/.

*prob. < 0.05

**prob. < 0.01

to be sixty-four years. This is much higher than the actual minimum life expectancy—mostly because we are trying to fit a line to curved data. The slope indicates that for every increase of $1,000 in GDP per capita, we expect life expectancy to increase about half a year. This relationship is statistically significant at the 0.01 level—we can reject the null hypothesis. The model explains a respectable 37.5 percent of the variance in life expectancy.

But to get a better fit, we need to straighten out the curve. To do this, we take the natural log of GDP. Figure 13.9 shows this relationship. Although this isn't perfectly straight, it is an improvement over Figure 13.8.

Figure 13.9 Life Expectancy by Natural Log of GDP

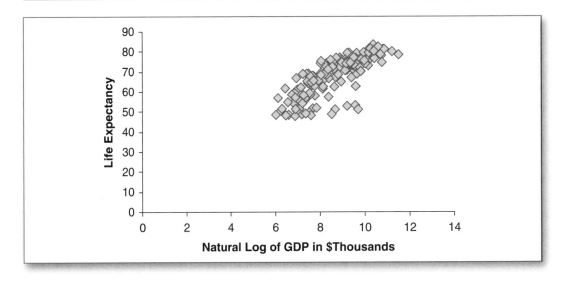

We should get a better fit with this model. Table 13.9 shows the regression. By fitting the curve, the constant has come down to a more reasonable minimum life expectancy of 57 years. The slope is positive, as it should be, and statistically significant; we can reject the null hypothesis. More important, the R^2 has increased to 0.644. This means that fitting a curve to the data increased the explanatory power of the model; it explains 27.3 percentage points more variance in life expectancy. As usual, if the relationship is not linear, a linear model did not give the best estimate of the relationship between the two variables. We can get a better estimate by taking the shape of the curve into account.

Table 13.9 Life Expectancy Regressed on Logged GDP per Capita (in $thousands)

Independent Variables	Unstandardized Coefficients	Standardized Coefficients
Constant	57.749	
Natural Log GDP	6.312**	0.804**
n = 178		$R^2 = 0.644$

Sources: "Life Expectancy at Birth (Years)," *GapMinder* 2011. www.gapminder.org/data/; "GDP per Capita PPP," *GapMinder* 2011. www.gapminder.org/data/.

*prob. < 0.05

**prob. < 0.01

BOX 13.1 Numbers in the News

After the shooting of elementary school children in Newtown, Connecticut, in December 2012, attention throughout the country turned to gun violence. First Lady Michelle Obama drew attention to the murder rates in Chicago when she spoke at the funeral of Hadiya Pendleton, a fifteen-year-old girl who had performed at the inauguration the month before. In spite of having some of the strictest gun control legislation in the country, Chicago had one of the highest gun violence rates in 2012. But in 2013, the rate declined again. Mayor Rahm Emanuel took credit for the policies that led to the 2013 decline, but not everyone agrees with his description of the pattern. Those who oppose gun control suggest that the 2013 data have been taken out of context and so should not be included in the description of a trend. Others point to the national decline in murder rates since the early 1990s and suggest that the high rate of violence in 2012 was an anomaly.[3] In the end, the way in which you specify the model of gun crime in Chicago influences your interpretation of the 2012 and 2013 data points and thus your forecast of future trends.

Gauss-Markov Assumption 3: Model Correctly Specified

Specification is how you model the relationship in terms of what independent variables you choose to include as causal factors influencing your dependent variable. In Chapter 11, we discussed the importance of controlling for variables that might be correlated with both our dependent and independent variables. Sometimes, excluding them will cause a spurious relationship to appear statistically significant; other times, excluding them will mean that a real relationship is hidden. In each of these situations, regression will give a poor estimate of the relationship. The beauty of regression is that you can include lots of control variables—there is no reason not to make sure the model is properly specified (meaning all relevant variables are included) so that OLS regression can give you the best possible estimate of the relationship between your dependent and independent variables.

There are only two caveats here. First, the number of variables you can include in the model is constrained by the number of cases you have. Remember in algebra when your teacher told you that you need two equations to solve for two variables? Mathematically, you need at least as many cases as you have independent variables. Realistically, if you want to get statistically significant relationships, you want a lot more cases than you have independent variables.

Second, if you include many independent variables, you will want to include the "Adjusted R^2" instead of the R^2. Every variable you include will necessarily increase the proportion of variance explained by the model. (Unless you include a variable with extra missing cases; in that case, having fewer cases might decrease the R^2.) Because you could artificially inflate the R^2 by including random variables, the adjusted R^2 puts a fudge factor into its calculus to compensate for controlling for too many variables. My personal bias is to include only those independent variables you have a theoretical justification for including. If your model is theoretically sound, I don't see any reason you shouldn't report the original (unadjusted) R^2.

Gauss-Markov Assumption 4: Non-Collinear

Variables are collinear when they have a very high correlation. If you have two (or more) collinear independent variables, this can cause problems mathematically in calculating the regression coefficients. Remember that the point of multivariate regression is to be able to control for one variable when estimating the relationship between the dependent variable and a different independent variable. For example, education and income are very highly correlated and both tend to have a lot of impact on various political views. We use regression so that we can hold income constant and see the impact of just education. And we want to hold education constant to see the impact of income. But if two independent variables are too highly correlated, then when you hold the first one constant, the second one is going to end up a constant, too. If it doesn't vary, you cannot estimate a slope.

Identifying Collinearity. This problem will show up most often if you forget the proper procedures while you are creating dummy variables. Remember that you are supposed to choose one category to hold as the baseline and create a dummy variable for each of the other categories. If you forget and mistakenly make a dummy for all the categories, you will end up with a **collinearity** problem. If you include all the categories in the regression, you will put the computer in the position of needing to calculate a slope when it is holding everything constant. Some programs will spit out an error message; SPSS will randomly choose one of the dummy variables to exclude and you will waste hours wondering why one is missing.

Less often, you will have a set of variables that makes it difficult, but not impossible, for the computer to calculate the regression coefficients. Some people call this "multicollinearity" because multiple variables are very highly correlated. Because you can't have collinearity without more than one variable, I find the "multi" part redundant and choose not to use it. The violation of this assumption, however, does not yield biased regression coefficients. OLS is still BLUE with collinearity problems. The impact of collinearity is to increase the standard errors around the coefficients, making it difficult to gain statistical significance. So if you include education and income in a regression and someone criticizes you for violation of the collinearity assumption, don't worry about it. If the relationships are statistically significant, then you can be confident in your results. If, however, you run a regression with independent variables that you know are highly correlated and get results that are not statistically significant, then look for collinearity. Maybe you have two (or more) independent variables that are so highly correlated that the computer cannot distinguish them statistically. It is possible that you need to exclude one of them.

A Political Example: Political Instability. Earlier in this chapter, we found that many factors contribute to the instability of a country. Corrupt governments tend to face instability. Table 13.10 replicates Table 13.2 in its model of the factors leading to instability. Remember

Table 13.10 State Instability Regressed on Corruption, Wealth, Literacy, and Income Inequality

Independent Variable	Unstandardized Coefficient (s.e.)	Standardized Coefficient
Constant	22.783	
Transparency Index	−0.811** (0.270)	−0.270**
GDP per Capita (in $1000s)	−0.138* (0.058)	−0.250*
Female Literacy Rate	−0.124** (0.015)	−0.481**
Income Share of the Top 10 Percent	0.001 (0.048)	0.001
n = 120		$R^2 = 0.727$

Sources: Marshall, Monty G. and Benjamin R. Cole, "State Fragility Index and Matrix 2011." *Center for Systemic Peace,* 2012. www.systemicpeace.org/SFImatrix2011c.pdf; CIA World Factbook, 2012. www.cia.gov/library/publications/the-world-factbook/fields/2103.html; "GDP per capita, PPP (constant 2005 international $)" *GapMinder.* 2010. Accessed 17 November 2012. http://www.gapminder.org/data/; Transparency International, *Corruption Perceptions Index 2011.* http://www.transparency.org/cpi2011/in_detail; "Income Share of Richest 10%." *GapMinder.* 2010. Accessed 17 November 2012. http://www.gapminder.org/data/.

*prob. < 0.05

**prob. < 0.01

that corruption, wealth, and literacy are all statistically significant at the 0.01 level. Suppose I am wondering about the measure I am using for GDP and decide to use a different one that the World Bank produced in 2010. Table 13.11 expands the model to include the alternative measure of GDP. Instead of replacing the original GDP variable, I mistakenly included both it and the alternative GDP variable in my model because I am under the misapprehension that two is better than one.

Table 13.11 State Instability Regressed on Corruption, Wealth in 2010 and 2011, Literacy, and Income Inequality

Independent Variable	Unstandardized Coefficient (s.e.)	Standardized Coefficient
Constant	22.645	
Transparency Index	−0.783** (0.289)	−0.261**
GDP per Capita (in $1000s), 2011	−0.345 (0.319)	−0.628
GDP per Capita (in $1000s), 2010	0.202 (0.311)	0.371
Female Literacy Rate	−0.123** (0.015)	−0.480**
Income Share of the Top 10 percent	0.003 (0.049)	0.003
n = 118		$R^2 = 0.726$

Sources: Marshall, Monty G. and Benjamin R. Cole, "State Fragility Index and Matrix 2011." *Center for Systemic Peace*, 2012. www.systemicpeace.org/SFImatrix2011c.pdf; CIA World Factbook, 2012. www.cia.gov/library/publications/the-world-factbook/fields/2103.html; "GDP per capita, PPP (constant 2005 international $)" *GapMinder*. 2010. Accessed 17 November 2012. http://www.gapminder.org/data/; "Alternative GDP per Capita, PPP WB." *GapMinder*. 2010. Accessed 17 November 2012. http://www.gapminder.org/data/; Transparency International, *Corruption Perceptions Index 2011*. http://www.transparency.org/cpi2011/in_detail; "Income Share of Richest 10%." *GapMinder*. 2010. Accessed 17 November 2012. http://www.gapminder.org/data/.

*prob. < 0.05
**prob. < 0.01

By comparing Table 13.10 with Table 13.11, you can see what happens when you add a collinear variable to the model. The other variables keep the same levels of statistical significance with approximately the same coefficients. But for the two GDP variables, the standard errors explode. Because the two variables are collinear, it is hard to talk about one of them varying when the other is held constant. As a result, the pattern is not statistically significant and we cannot reject the null hypothesis. When you are doing statistical analysis, one clue that you might have two or more collinear variables is that the beta is large, but the variable is not statistically significant. In this case, it should strike you as odd that, according to the betas, the per capita GDP for 2011 explains more variance than any other variable but it is not statistically significant. You should feel comfortable excluding the GDP for 2010 from the model and going with the regression found in Table 13.10 instead.

Gauss-Markov Assumption 5: Errors Have a Mean of Zero

If you examine the residuals (otherwise known as the error terms—how far the estimate of the dependent variable is from its actual value) in an OLS regression, their mean will always be zero. It is an assumption of OLS. If, in reality, the errors do not average zero, then OLS has made a faulty assumption and its estimate will be biased. One time this will happen is if the underlying relationship is curvilinear rather than linear. If you estimate a line for curvilinear data, at the tails the estimate will systematically have error terms that are either usually positive or usually negative. Trying to impose a line with errors that average zero will lead the estimate to be biased. So it is important to deal with curvilinear relationships.

Gauss-Markov Assumption 6: Errors are Homoscedastic

Homoscedastic means that when you look at a scatter plot of the relationship between two variables, the spread of the dots on either side of the line stays the same width as you move from right to left. If one end of the line is much wider than the other (sort of funnel shaped) we call that **heteroscedastic**. If you look back to Figure 13.6, you'll see the data form a ribbon: They are homoscedastic. In contrast, if you look at Figure 13.5, you'll see the data look like a slanted tornado: This relationship is heteroscedastic.

In calculating the standard error, OLS regression assumes that the errors are distributed in the same way across all values of your independent variable. If the data are heteroscedastic, the estimate of the slope may well be unbiased—it will still find the center of the data. But you are not nearly as confident in the estimate as OLS indicates. Figure 13.10 shows the relationship between the unemployment rate of countries and their income inequality (measured as the percent of total income earned by the richest 10 percent). In the lower left corner, you are very sure that countries where the richest 10 percent earn only about 20 percent of the total income will have low unemployment rates. But as you move up and to the right, you'll see that as income inequality increases, its unemployment rate does not increase at uniform rates. Among highly unequal countries, there are both low and high unemployment rates. The standard error in the right-hand side of the graph is much higher than on the left. The single standard error measured with regression fails to account for this variability. If you were to draw lines in these data, they would all begin in the lower right corner, but there would be a whole range of slopes that you could draw through that point that would still fit the data. In contrast, Figure 13.6 is so tight, there are a limited number of slopes you could draw through those data. In OLS, you want your data to look like a ribbon—to be homoscedastic. There isn't really a way to fix heteroscedastic

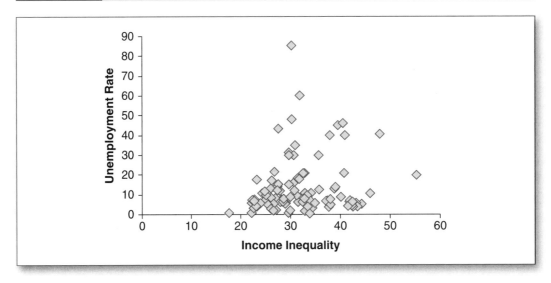

Figure 13.10 Unemployment Rate by Income Inequality

data unless the cause is an underlying curvilinear relationship. But if it looks heteroscedastic, remember that your coefficients are not biased. You do, however, need to question the statistical significance given by the regression.

SUMMARIZING THE PROCESS: MULTIPLE REGRESSION

Multiple regression is the appropriate statistical measure of association between interval-level variables controlling for one or more other interval-level variables. It gives us a multiple dimensional equation for a line that parses out the impact of each independent variable on the dependent variable controlling for all of the others. The entire model has a single constant, and each independent variable has its own unstandardized coefficient, standard error, standardized coefficient, and statistical significance. In addition, the entire model has a single R^2.

Analyzing Multiple Regression

As usual, we follow the three-step pattern of analyzing the results.

1. We describe the pattern by looking at the slopes connected with each independent variable. Each can be interpreted as the change in the dependent variable for each unit increase in the independent variable, controlling for the other variables. It is important to look at the signs of the slope to see whether they show the same pattern we expected to see.

2. We identify the statistical significance of each variable separately. Each slope has its own standard error, which translates into its own t-score, which finally determines its statistical significance.

3. As with bivariate regression, the entire model produces an R^2 that can be interpreted as the proportion of variance in the dependent variable explained by all the variables included in the model. In addition, the strength of the relationship can be evaluated for the different variables by comparing their standardized coefficients or betas. Those variables with the largest betas explain the most variance in the dependent variable.

The Assumptions of Regression

Regression's ability to control for multiple variables simultaneously makes it a very useful measure of association. But, as with all statistics, regression makes assumptions about the data. Although it may be tempting to use regression as an all-purpose tool, if we violate those assumptions, the results end up being less than optimal. Ordinary least squares regression is the best linear unbiased estimate of a slope if the Gauss-Markov assumptions have been met. We say "OLS is BLUE" if the following assumptions are true:

1. The variables have an interval level of measurement.
2. The relationship between the dependent variable and each independent variable is linear.
3. The model is correctly specified.
4. Two or more independent variables are not collinear.
5. The errors in predicting the dependent variable have a mean of zero.
6. The errors in predicting the dependent variable are homoscedastic (as opposed to heteroscedastic), meaning that the distribution of the errors doesn't consistently increase or decrease across values of the dependent variable.

These assumptions need to be met for OLS to produce the best estimate of the underlying relationship. But one of the nice things about regression is that it is **robust**. Although these assumptions need to be met to get the best estimate, OLS produces good estimates even with minor violations of its assumptions. For example, even with collinearity problems, the estimates are unbiased; they just have larger standard errors than they would otherwise. Similarly, violating the assumptions about the residuals will tend to influence the standard errors rather than the slopes. The first three assumptions are the ones that are most important. It doesn't make sense to fit a line to data that are not linear. And leaving out important variables can have a huge impact on the quality of your results.

A Political Example: Teen Pregnancy

In April 2012, the Centers for Disease Control and Prevention issued a press release announcing that the teen birth rate has declined since its previous high in 1991. It attributed the decline to two factors: a decline in sexual activity by teens and an increase in the use of contraceptives among sexually active youth. Which of these two factors is the more important predictor of teen pregnancy rates? Table 13.12 shows the regression of teen pregnancy rates for the states against the percent of high school girls who report ever having sexual intercourse and the percent who report using some kind of contraceptive during their previous sexual encounter.

Table 13.12 State Teen Pregnancy Rates Regressed on the Percent of Teens Who are Sexually Active and the Percent Who Do Not Use Contraceptives

Independent Variable	Unstandardized Coefficient (s.e.)	Standardized Coefficient
Constant	7.948	
Sexual Activity	0.670* (0.333)	0.272*
No Contraceptives	1.982*** (0.450)	0.596***
n = 32		$R^2 = 0.484$

Source: Centers for Disease Control and Prevention, "Youth Online: High School YRBS," 2011. http://apps.nccd.cdc.gov/youthonline/App/Default.aspx.

*prob. < 0.10
**prob. < 0.05
***prob. < 0.01

1. The slopes indicate that for every percentage point increase in teens who have engaged in sexual intercourse, the state's teen pregnancy rate increases by 0.67. Of those teens who are sexually active, as the number of them who do not use contraceptives increases by one percentage point, the teen pregnancy rate increases by almost two percentage points. As expected, both behaviors are positively related with the pregnancy rate.

2. Although sexual activity is statistically significant at the 0.10 level, the non-use of contraceptives is statistically significant at the 0.01 level. For it, we can reject the null hypothesis.

3. This model explains 48.4 percent of the variance in state-level teen pregnancy rates. The betas indicate that the non-use of contraceptives explains more variance in pregnancy rates than sexual activity.

USE SPSS TO ANSWER A QUESTION WITH MULTIPLE REGRESSION

The difference between conducting bivariate and multiple regression in SPSS is a simple matter of adding more independent variables. You should be able to do that without further repetition from me. When you do so, you'll notice that there is still the model summary table with the R^2. If you decide to use a lot of independent variables in your model, you should report the adjusted R^2 so that the value is not artificially inflated by including so many variables. Once again, you'll need to use the ANOVA table to get the number of cases. The simple way to find n is to add one to the total degrees of freedom. Then, in the coefficients table, you'll see that there is only one constant that has an unstandardized coefficient (B) but no standardized coefficient (beta). But there is a separate row for each of the independent variables you entered along with Bs and betas and statistical significance for each. Remember that the statistical significance is identical, whichever kind of coefficient you choose to report.

One tool available in SPSS that can be useful in understanding the relationships between the variables is the command to get a correlation matrix. Although the goal of multiple regression is to find a relationship between two variables controlling for other things, it is good to put that in the perspective of what the relationships look like without controls. The correlation matrix gives you a series of bivariate correlations between not only the dependent and each of the independent variables, but also all of the independent variables. We discussed how to get a correlation matrix in Chapter 10. As a reminder, to get a correlation matrix, go to "Analyze," "Correlate," and "Bivariate." You can bring as many variables as you want over to the "Variable" box (although it will get clunky to read if you have more than about five variables). Just remember that Pearson's r (the statistic given in the correlation matrix) needs to have interval-level variables. It is worth going across the row for the dependent variable to see which of the independent variables has the highest (and lowest) correlations with it.

What is more important at this point is learning how to address the assumptions of OLS regression. In this section, we'll address three main skills: creating dummy variables so that you can include variables that do not have an interval level of measurement, producing scatter plots to see whether relationships are linear, and producing scatter plots of the residuals to see if they have means of zero and are homoscedastic.

Dummy Variables

If you have an ordinal- or nominal-level variable that you want to include in your regression, you cannot include it as is because it violates the assumption of regression that each variable is on a scale. Without a scale, you cannot calculate the means that are necessary for the math of regression. And without a scale, the notion of a linear relationship does not make sense. But regression makes allowances for dichotomous independent variables. This means that you can take a categorical variable and translate it into a series of dummy variables, where each one is connected with one of the categories of the variable. As you do this, all of the cases are coded "0" except the cases in the appropriate category that are coded "1." The one trick is that you need choose one of the categories as the baseline against which you will compare each of the other categories. You will not create a dummy variable for this baseline category—it will be 0 on all of the dummy variables.

To create the dummy variables, go into "Define Variable Properties" for the categorical variable. Identify the category you want to be the baseline, and for each of the other categories, take notes on its value. Make sure that any cases you do not want included in the analysis are set to missing at this point. Then use "Transform," "Recode into Different Variable" to create new variables, each of which is connected to one of the categories of the original variable (except, of course, your baseline category). You will repeat this process for each of the dummy variables. For each, you will set the value of the appropriate category equal to "1." You will then set the "System- or User-Missing" cases to "System-Missing." Finally, at the bottom of the left-hand column, you will select "All Other Values" and set them equal to "0." In the end, you will have one fewer dummy variables than you had categories in your variable. Go into "Define Variable Properties" and make sure that each variable has the correct number of cases in each category.

When you do your regression analysis, be sure to include each of these dummy variables. Keep in mind that if you mistakenly make dummy variables for all of the categories, SPSS will realize that it can't perform its calculations because the set of variables is perfectly collinear, and it will randomly choose one of the dummy variables to exclude. That excluded variable becomes the baseline by default.

Bivariate Scatter Plots

Before conducting your regression, it is a good idea to get scatter plots of each of your interval-level independent variables with your dependent variable to make sure that the relationship is linear. The command for a scatter plot is found under "Graphs," "Legacy Dialogs," and "Scatter/Dot." Choose the "Simple Scatter" option and click on "Define." Pull your dependent variable over into the "Y-Axis" box and the independent variable into "X-Axis." This process is summarized in Box 13.2.

BOX 13.2   How to Get a Scatter Plot in SPSS

>Graphs
>>Legacy Dialogs
>>>Scatter/Dot
>>>>Simple Scatter
>>>>Define
"*Dependent Variable*"→Y-Axis
"*Independent Variable*"→X-Axis

In the resulting scatter plot, you should see a ribbon of data points that extends in a straight line. If that is the case, you can include this variable in the regression as it is. If you see a curve, you will need to perform the mathematical function described in the chapter using the "Transform," "Compute" commands. One note here: If the scale on one of your variables has a limited number of possible values, the scatter plot is going to look funky. For each of those values, you will see a ribbon of dots along its range. If that is the case, the scatter plot is not likely to be very helpful in determining whether or not the relationship is linear.

Residual Scatter Plots

When you perform the regression, one of the options available is "Plots." You can do a series of scatter plots here, but for basic purposes, you will want to place your predicted value of the dependent variable onto the x-axis and the residual on the y-axis. I suggest using the standardized values of each of these: "ZPred" and "ZResid." This process is summarized in Box 13.3. This will transform the two variables into z-scores so that the resulting scatter plot will range from −3 to +3 on each of the axes. What you want to see is a ribbon of data centered on the horizontal axis where X = 0.

BOX 13.3   How to Run a Multiple Regression in SPSS with Scatter Plot for Residuals

>Analyze
>>Regression
>>>Linear
>>>>Dependent=*Dependent Variable*
>>>>Independent=*Independent Variable(s)*
>>>>\>Save
>>>>>Predicted Values:
>>>>>\>Unstandardized
>>>>>\>Continue
>>>>\>Plots
>>>>>*ZRESID→Y
>>>>>*ZPRED→X
>>>>>\>Continue
>>>>\>OK

An SPSS Application: The Relationship of Multiple Factors on State Instability

It is fall 2011, and you are interning at the Center for Global Development. After the instability of the Arab Spring, many at the institute are trying to address the factors that lead to state fragility. In a conversation with your boss, Todd Moss, vice president for programs at the Center, you mention your understanding of the importance of both education and governmental transparency. He suggests you do a statistical analysis of the relationship, but also include a couple of other factors, including controls for the strength of the economy, income inequality, and the level of freedom of a country. Because freedom influences instability in a non-linear way (the least stable governments are those that are partially free), he suggests including controls for Freedom House's designation of states as free, part free, and not free.

1. Open the data and clean them, making sure that missing data are so designated.

2. Get a correlation matrix for all your interval-level variables.
   Under "Analyze" and "Correlate," I choose the "Bivariate" option as shown in Figure 13.11. I then choose my interval-level variables (FemaleLiteracy, GDPPPP2011, IncomeShareTop10%, sfi2011, and TransparencyIndex) to include in the matrix as shown in Figure 13.12. The output correlation matrix is in Figure 13.13.

Figure 13.11 Requesting a Correlation Matrix

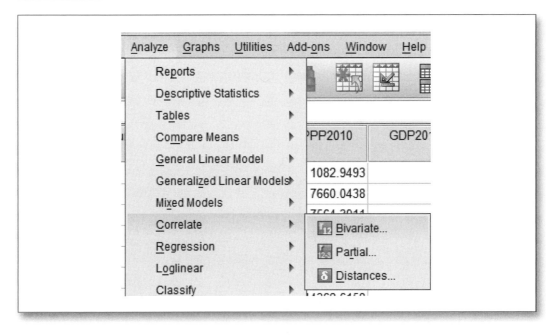

CHAPTER 13 Multiple Regression: The Final Frontier   **405**

Figure 13.12 Selecting Variables for a Correlation Matrix

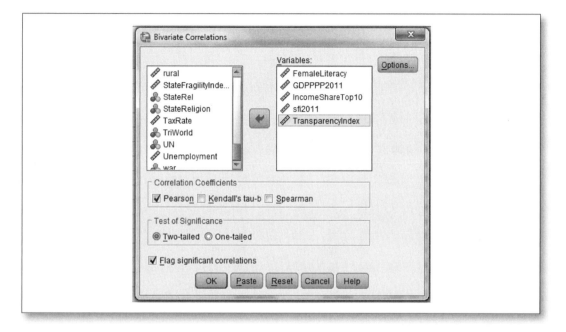

Figure 13.13 Output Correlation Matrix

Correlations

		FemaleLiteracy	GDPPPP2011	IncomeShareTop10%	sfi2011	TransparencyIndex
FemaleLiteracy	Pearson Correlation	1	.335**	.068	-.232**	.321**
	Sig. (2-tailed)		.000	.345	.001	.000
	N	194	194	194	194	194
GDPPPP2011	Pearson Correlation	.335**	1	-.156*	-.376**	.536**
	Sig. (2-tailed)	.000		.030	.000	.000
	N	194	194	194	194	194
IncomeShareTop10%	Pearson Correlation	.068	-.156*	1	.438**	.240**
	Sig. (2-tailed)	.345	.030		.000	.001
	N	194	194	194	194	194
sfi2011	Pearson Correlation	-.232**	-.376**	.438**	1	.120
	Sig. (2-tailed)	.001	.000	.000		.095
	N	194	194	194	194	194
TransparencyIndex	Pearson Correlation	.321**	.536**	.240**	.120	1
	Sig. (2-tailed)	.000	.000	.001	.095	
	N	194	194	194	194	194

**. Correlation is significant at the 0.01 level (2-tailed).
*. Correlation is significant at the 0.05 level (2-tailed).

3. Check for the linearity of the relationships by getting scatter plots.
   Figure 13.14 shows the location of the commands for a scatter plot. Because I want only a two-way scatter plot, I click on the "Simple Scatter" option and "Define" button. Then I designate the dependent variable for the y-axis and the independent variable for the x-axis as shown in Figure 13.15.

Figure 13.14 Location of Commands for a Scatter Plot

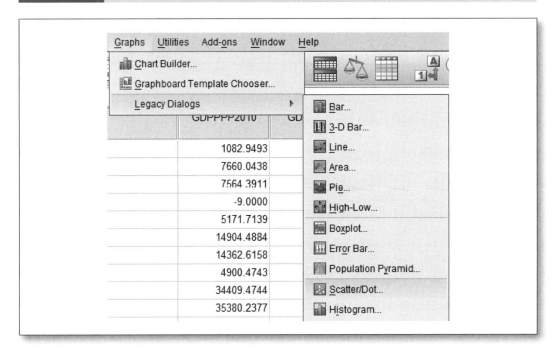

Figure 13.16 shows the scatter plot for the relationship between instability and freedom. Dr. Moss is clearly correct that it is curvilinear.

4. Create any dummy variables as necessary.
   I decide to overcome the non-linearity problem by using Freedom House's categorization of countries as free, part free, and not free to create dummy variables. I choose to make "Free" the baseline and create dummy variables for the other two categories. Figure 13.17 shows the process of creating the "Not Free" dummy.

5. Run the regression with all of the independent variables, including the dummies. Request a scatter plot of the predicted value and the residuals.
   This process is shown in Figure 13.18.

CHAPTER 13 Multiple Regression: The Final Frontier **407**

Figure 13.15 Designating the Dependent and Independent Variables

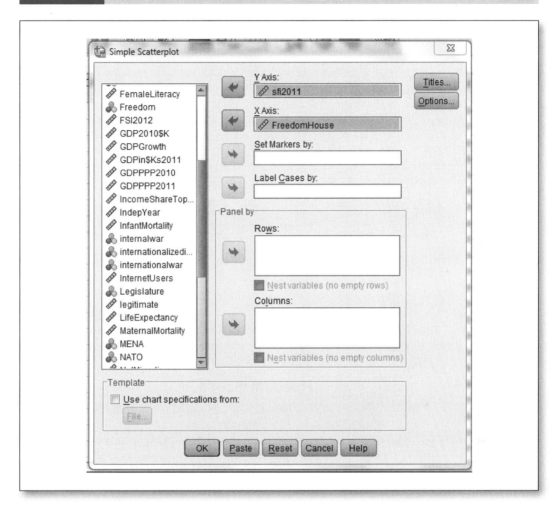

6. Examine the scatter plot to make sure that the residuals are distributed in a ribbon across the middle of the graph where X = 0.
   The scatter plot shown in Figure 13.19 satisfies those requirements.

7. Use the output tables to create a professional-looking regression table.
   The output tables are found in Figure 13.20. Notice that each independent variable has its own unstandardized and standardized coefficients along with the corresponding level of statistical significance. My regression table is in the resulting memo.

Figure 13.16 Scatter Plot for Instability by Freedom Index

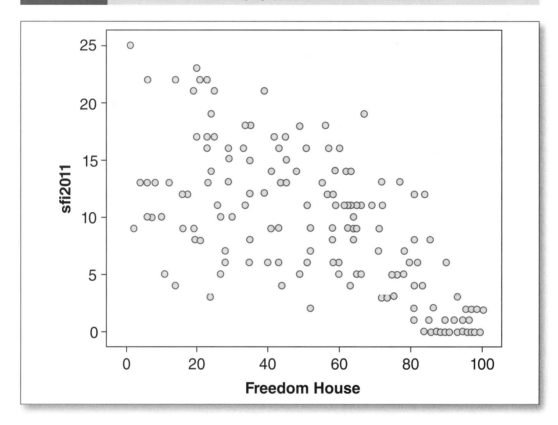

8. Write your memo using the three-step process of analyzing your results.

Memo

To: Todd Moss, Vice President for Programs, CGD
From: T. Marchant-Shapiro, intern
Subject: State Fragility
Date: October 2, 2011

You asked me to investigate the factors that predispose a state to instability. Using the State Fragility Index as my measure for state instability, as well as other measures from such reputable sources as the World Bank, Transparency International, and Freedom House, I analyzed which factors have the most impact on state fragility, controlling for the others. I found that the female literacy rate of a country explains the most variance in state fragility.

The table below shows state fragility regressed on various factors that affect the stability of states. This regression indicates that increasing the per capita GDP, female literacy rate, and governmental transparency all decrease the overall fragility of a country. In contrast, income inequality (measured as the income share of the top 10 percent), as well as being designated by Freedom House as either Not Free or Part Free, increase the level of fragility. Female literacy and level of freedom are statistically significant at the 0.01 level, and GDP is statistically significant at the 0.05 level. For each of these variables, we can be confident that the relationship we are observing is not due to chance. The fact that the transparency variable is not statistically significant when the two freedom variables are included suggests that much of the previously observed correlation between fragility and corruption is due to the correlation between freedom and corruption. Apparently, freedom is an antecedent variable that affects both corruption and fragility, but is the driving force behind the fragility of corrupt states. The model as presented explains an impressive 78.5 percent of the variance in state fragility. Of these variables, female literacy explains the most variance.

It appears that if our goal is to decrease the fragility of states, we would be well advised to pursue means to increase female literacy as our primary goal. In addition, it appears that attempts to increase freedom would be more efficacious than dealing with corruption if our primary concern is stability. Although researchers here at the Center for Global Development are investigating all of these issues, we might be well advised to reallocate resources to focus on those factors that have the most impact on state fragility.

Figure 13.17 Creating a Dummy Variable for Not Free States

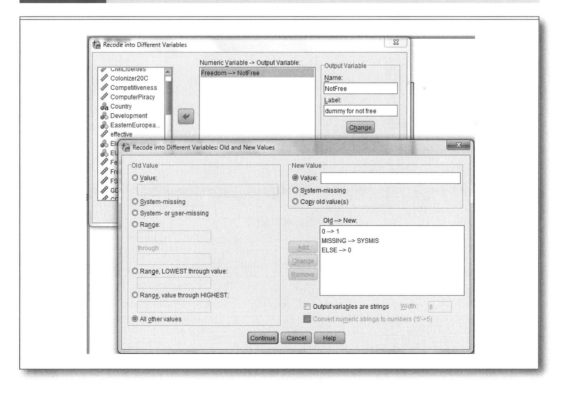

Figure 13.18 Requesting a Multiple Regression

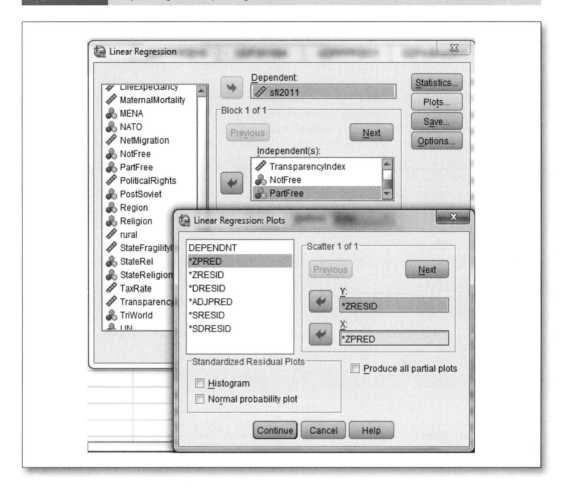

State Instability[4] Regressed on Corruption,[5] Wealth,[6] Literacy,[7] Income Inequality,[8] and Freedom[9]

Independent Variable	Unstandardized Coefficient (s.e.)	Standardized Coefficient
Constant	17.442	
Transparency Index	−0.395 (0.264)	0.131
GDP per Capita (in $1000s)	−0.129* (0.052)	−0.234*

Independent Variable	Unstandardized Coefficient (s.e.)	Standardized Coefficient
Female Literacy Rate	−0.113** (0.014)	−0.438**
Income Share of the Top 10%	0.028 (0.044)	0.030
Not Free	4.577** (0.826)	−0.322**
Part Free	2.344** (0.728)	0.189**
n = 120		$R^2 = 0.785$

*prob. < 0.05
**prob. < 0.01

Figure 13.19 Scatter Plot of the Standardized Residuals by Predicted Value

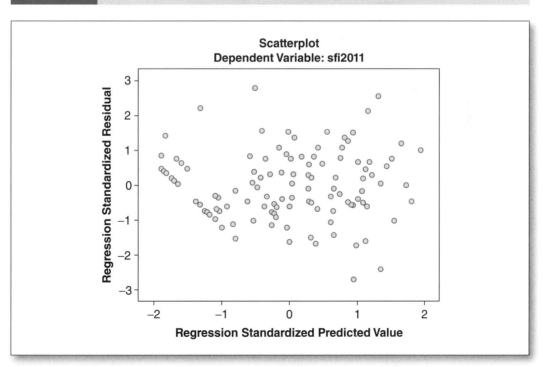

Figure 13.20 Regression Output

Model Summary[b]

Model	R	R Square	Adjusted R Square	Std. Error of the Estimate
1	.886[a]	.785	.774	2.876

a. Predictors: (Constant), dummy for Part Free, FemaleLiteracy, IncomeShareTop10%, TransparencyIndex, dummy for not free, GDPPPP2011
b. Dependent Variable: sfi2011

ANOVA[a]

Model		Sum of Squares	df	Mean Square	F	Sig.
1	Regression	3412.028	6	568.671	68.744	.000[b]
	Residual	934.764	113	8.272		
	Total	4346.792	119			

a. Dependent Variable: sfi2011
b. Predictors: (Constant), dummy for Part Free, FemaleLiteracy, IncomeShareTop10%, TransparencyIndex, dummy for not free, GDPPPP2011

Coefficients[a]

Model		Unstandardized Coefficients		Standardized Coefficients	t	Sig.
		B	Std. Error	Beta		
1	(Constant)	17.442	2.133		8.179	.000
	FemaleLiteracy	-.113	.014	-.438	-8.134	.000
	GDPPPP2011	.000	.000	-.234	-2.463	.015
	IncomeShareTop10%	.028	.044	.030	.622	.535
	TransparencyIndex	-.395	.264	-.131	-1.496	.137
	dummy for not free	4.577	.826	.322	5.540	.000
	dummy for Part Free	2.344	.728	.189	3.220	.002

Your Turn: Multiple Regression

YT 13.1 The 2008 National Election Study asks the respondent to rank several individuals and groups on a 100-point Feeling Thermometer (where 0 = cold and 100 = warm). Among these groups are conservatives, Christian fundamentalists, the military, big business, rich people, environmentalists, illegal immigrants, and liberals. If you regress how warmly respondents feel toward conservatives on their warmth toward the other groups, you get the regression shown in Table 13.13. Analyze these findings using the three-step process.

1. Describe the pattern.
2. Evaluate the statistical significance.
3. Assess the strength of the relationship.

CHAPTER 13 Multiple Regression: The Final Frontier

Table 13.13 Feeling Thermometer for Conservatives Regressed on Feeling Thermometers for Christian Fundamentalists, the Military, Big Business, Rich People, Environmentalists, Illegal Immigrants, and Liberals

Independent Variable	Unstandardized Coefficient (s.e.)	Standardized Coefficient
Constant	23.246	
Christian Fundamentalists	0.255** (0.018)	0.301**
The Military	0.156** (0.020)	0.163**
Big Business	0.128** (0.021)	0.141**
Rich People	0.219** (0.022)	0.228**
Environmentalists	−0.022 (0.022)	−0.024
Illegal Immigrants	−0.040* (0.016)	−0.052*
Liberals	−0.124** (0.021)	−0.132**
n = 1749		$R^2 = 0.353$

Source: American National Election Study, *ANES 2008 Time Series Study*. www.electionstudies.org/studypages/2008prepost/2008prepost.htm.

*prob. < 0.05
**prob. < 0.01

YT 13.2 The 2008 National Election Study asks the respondent to rank several individuals and groups on a 100-point Feeling Thermometer (where 0 = cold and 100 = warm), including the group "Illegal Immigrants." It also has a variable for the race of the respondent with categories for several races. Table 13.14 shows the feeling thermometer for Illegal Immigrant regressed on dummy variables for the different race and ethnic groups the respondent could be—white is the baseline category for which there is not a dummy variable. Interpret this regression.

1. Describe the pattern (including the interpretation of the intercept).
2. Evaluate the statistical significance.
3. Assess the strength of the relationship.

Table 13.14 Feeling Thermometer for Illegal Immigrants Regressed on Dummy Variables for Race or Ethnicity of Respondent

Independent Variable	Unstandardized Coefficient (s.e.)	Standardized Coefficient
Constant	35.095	
Black	12.964** (1.351)	0.205**
Asian	14.437** (4.444)	0.066**
Hispanic	29.299** (1.490)	0.418**
Other Non-White	13.516** (3.016)	0.092**
n = 2035		$R^2 = 0.167$

Source: American National Election Study, *ANES 2008 Time Series Study*. www.electionstudies.org/studypages/2008prepost/2008prepost.htm.

*prob. < 0.05

**prob. < 0.01

YT 13.3 — What do I mean when I say "OLS is BLUE?" (hint: Gauss-Markov)

Apply It Yourself: Evaluate the Impact of Multiple Factors on the 2012 Presidential Election

It is November 2012, and you are interning at the National Urban League, an organization dedicated to empowering African Americans. After losing his bid for the vice presidency, Paul Ryan has explained the outcome of the election as being rooted in high urban turnout. Your boss, Chanelle Hardy, senior vice president for Policy, asks you to do a statistical analysis evaluating whether race or urban vote had more impact on the Democratic win in 2012. You collect data about how each state voted[10] along with demographic data about the percent of the population that lives in urban areas and the percent that can be considered minorities because they are not non-Hispanic whites.[11] In addition, you decide to control for the region of the state.

1. Open "StateData" in SPSS and use "Define Variable Properties" to clean the data, making sure that any missing data are so designated. You will be using the variables Obama12, Urban, Minority, and Region. You are going to need to recode Region into dummy variables, so take notes on which categories are associated with which numerical values.

2. Get a correlation matrix between the interval-level variables to see the bivariate relationships between each. Look for how strongly the vote for Obama is correlated with the other two variables, but also how strongly they are correlated with each other.

   >Analyze
   >>Correlate
   >>>Bivariate
   >>>>Obama12→Variables
   >>>>Urban→Variables
   >>>>Minority→Variables
   >>>>>OK

3. Get a separate scatter plot for the dependent variable with each of the independent variables. For percent urban, it will look like the following. For percent minority, replace the urban variable with the minority variable in the x-axis.

   >Graphs
   >>Legacy Dialogs
   >>>Scatter/Dot
   >>>>Simple Scatter
   >>>>Define
   >>>>>Obama12→Y-Axis
   >>>>>Urban→X-Axis

   Both scatter plots appear to have a slightly positive slope that confirms the positive correlations in the matrix. On the Urban scatter plot, you can see that at low, medium, and high levels of urban population, there are states that supported Obama at various levels. There may not be much going on here, but the relationship does not look curvilinear.

4. Create dummy variables for Region. To do this, choose one category to be the baseline against which all of the other regions will be compared. I suggest choosing the Northeast as your baseline, so you will not create a dummy variable for it. The following command will give you a dummy variable for states in the West. You will need to modify it for the South (Region=3) and the Midwest (Region=4).

   >Transform
   >>Recode into Different Variables
   >>>Region→New Variable
   >>>Output Variable: Name=West
   >>>>Change
   >>>>Old and New Values
   >>>>>1→1
   >>>>>Add
   >>>>>System- or user-missing→System-missing
   >>>>>Add
   >>>>>All other values→0

>Add

>Continue

>OK

Repeat this process to create a dummy variable for the South and the Midwest, clicking on "Reset" to clear the previous coding. Then go into "Define Variable Properties" to make sure the dummy variables look correct with the right number of cases in the values of 0 and 1.

5. Once all the variables look correct, you can run your regression. Remember that vote for Obama is the dependent variable and percent urban and minority along with the dummy variables for region are all independent variables. Be sure to request a scatter plot of the residuals and the predicted value.

>Analyze

>Regression

>Linear

Dependent=Obama12

Independent=Urban

Minority

West

South

Midwest

>Plots

*ZRESID→Y

*ZPRED→X

>Continue

>OK

6. Check the scatter plot to make sure that it looks like a ribbon running through the middle of the graph. Save the graphs and tables for grading purposes, then type the results into a professional-looking table.

7. Write your memo.

Key Terms

Betas (p. 384)
Collinearity (p. 394)
Dummy variable (p. 383)
Exponential relationship (p. 387)
Gauss-Markov assumptions (p. 381)
Heteroscedastic (p. 397)
Homoscedastic (p. 397)
Linear relationship (p. 384)

Log linear relationship (p. 390)
Model (p. 378)
Multiple regression (p. 380)
OLS (p. 381)
Robust (p. 399)
Specification (p. 393)
Squared relationship (p. 386)

CHAPTER 14

Understanding the Numbers

Knowing What Counts

The final round of the 2004 World Series of Poker Tournament of Champions pitted Annie Duke against her brother Howard Lederer. Annie had a pair of sixes and Howard had a pair of sevens. Both of them knew that Howard had an 82 percent chance of winning the hand, whereas Annie had only an 18 percent chance of winning. Both stayed in, expecting Howard to win. But when the dealer turned over a six, Annie ended up winning the $2,000,000 prize. Good poker players are able to weigh the pot odds (depending on how much money is on the line) against the hand odds (how good a hand they have). Doing so, they know that although there is no guarantee they might win a particular hand, in the long run, they will earn more than they lose. Poker players learn to handle losing hands because, as Annie puts it, that "embracing of uncertainty does some really wonderful things for you."[1]

Similarly, as we gather data to analyze political phenomena, we understand that there is uncertainty involved. We learn to gather data in the best way we can, understanding that although we might draw mistaken conclusions in the short run, in the long run, the picture we paint of political life will, in general, be accurate. In this book, I've focused on three key elements that go into that painting: measurement, univariate statistics, and multivariate statistics. At this point, you should be familiar with each of them separately; the purpose of this chapter is to put them together so that you can see the big picture of how we use statistics to understand how politics works.

BOX 14.1 Numbers in the News

A political campaign always needs to decide how to use its contributions most effectively. Traditionally, media spending for presidential campaigns has been directed toward viewers of the evening news because, demographically, this older, more politically interested population is more

(Continued)

(Continued)

likely to vote than any other group of viewers. A more refined analysis of more extensive data, however, allows for a more focused media strategy. If a poll indicates that the candidate needs to appeal to more women or more young voters, the campaign can use Nielsen ratings to identify which TV shows the subgroup is more likely to watch. During the 2012 presidential campaign, Obama staffers were able to refine this approach even more. Linking up to the friends of Facebook supporters, the campaign was able to identify 15 million Americans who were not currently Obama supporters, but who were persuadable. For tens of thousands of those individuals, the campaign was able to find actual viewing histories from a new competitor to Nielsen, Rentrak. Because they were able to change the unit of analysis from groups of likely voters to voters who were persuadable in this specific election, the campaign was able to spend much less on television advertising than the Romney campaign, while simultaneously getting more coverage with the voters who ended up determining the outcome of the election.[2]

MEASUREMENT

Politics is not an appropriate topic for polite conversation because it is simultaneously incredibly important and incredibly normative. As political scientists, we avoid discussing what policies ought to be adopted because the value assumptions inherent in such normative debates are not subject to evidence. Instead, as scientists, we gather evidence in order to describe political realities. By taking a step away from a discussion of "good" and "bad" decisions, we can focus instead on the possible causes of an event or the possible effects of a particular decision. Such empirical analyses are very valuable to policymakers, making their job less emotional and more rational. Political science relies heavily on statistics as a systematic way to analyze that evidence. But statistical analysis requires that we measure political concepts numerically.

Each of the chapters in this book has featured an example of "Numbers in the News"—events in which numbers made a difference politically. As you look at these (and as you notice numbers in the news yourself), you'll see that before the numbers could even be collected, key political actors needed to identify the characteristics that are relevant to the political phenomena in which they were interested. For example, in this chapter, the "Numbers in the News" looks at how exposure to media ads affected support for President Obama in the 2012 election. The two characteristics of exposure to ads and support for Obama are interesting precisely because they varied across the population. Obama staffers assumed that variation in exposure led to variations in support. Similarly, when we begin political analysis, we need to conceptually define the important variables in the political process that interest us.

We observe the variation in those characteristics by comparing different cases. Traditionally, the unit of analysis for campaigns was likely voters. Obama's staffers narrowed down their population to persuadable voters. As political scientists, when we measure a characteristic, we may do so for individual people or organizations or countries. The important thing is to identify (and measure) the individual cases that interact in producing the political phenomena that we are interested in understanding. Our first task of measurement is to identify those cases (or units of analysis). Second, looking at the interaction, we define the salient factors conceptually. What do we mean when we say a case has a particular characteristic? Third, we define the variable operationally. How will we measure that characteristic?

In the end, we are going to need to defend our measurement on two fronts. First, is it a valid measure? We want there to be a logical connection between the concept we found politically interesting and our measurement of that characteristic for the cases that we are studying. If the two are not closely connected, any conclusion that we draw from our data will be thrown into question. Second, is it a reliable measure? Our instructions on how to measure the concept need to be clear enough that repeated measures yield consistent results.

The decisions that you make in measuring a variable affect the conclusions that you can draw from your data. For example, although the congressional approval rating is at an all-time low (only one in ten Americans approve of Congress), voters tend to approve of their own representative.[3] You need to make sure you draw conclusions in accordance with the unit of analysis you have chosen. If you are looking at national-level data, you cannot draw conclusions about citizens. (Doing so is called an ecological fallacy and can result in an aggregation bias to your results.) Similarly, if you focus on a particular aspect of the concept in your measurement, keep that in mind as you draw your conclusions. For example, frequently we operationalize freedom in terms of the presence of fair elections in a country. Using that measurement, supporters of Egyptian president Mohamed Morsi defended his regime as free. But his opponents were able to overthrow him in the summer of 2013 because that electoral freedom did not translate into civil liberties, another aspect which might be considered part of the attribute of freedom.

UNIVARIATE STATISTICS

The purpose of measuring an attribute for a set of cases is to aggregate those measures into a description of the population. As a result, once we've measured a concept, we want to know how it is distributed across our cases. We look at two aspects of that distribution. First, we look at measures of central tendency, as discussed in Chapter 3. Second, we look at measures of dispersion, as discussed in Chapter 4. Both of these types of statistics allow us to describe the distribution of our variable for our cases.

How to choose a measure of central tendency or a measure of dispersion depends on the level of measurement of our variable. For mathematical purposes, we always assign numbers to the different possible values of an attribute. But those numbers can have more or less meaning. Sometimes, the numbers are simply assigned to different categories that have no underlying order. For example, the Obama campaign might have wanted to have a measure for the gender or ethnicity of persuadable voters. We call this a nominal level of measurement. Sometimes, the numbers imply both categorization and order, but have no scale. For instance, in order to find the congressional approval rating, Gallup asked respondents how much confidence they had in Congress—a great deal, quite a lot, some, or very little? Although this measure was ordered from more to less confidence, the numbers assigned to each category did not measure a uniform amount of confidence. We call this an ordinal level of measurement. Sometimes, the numbers classify, order, and place on a scale, but a value of zero is not connected to the total absence of the characteristic. For example, the change in approval for Congress can be negative, as it was in 2013 when it was − 3, having declined from 13 to 10 percent. We call this an interval level of measurement. If the numbers classify, order, and place on a scale, and zero is connected to the total absence of the characteristic (so there cannot be a negative value), we call this a ratio level of measurement. Because approval ratings are the percent of respondents who indicate that they have a great deal or quite a lot of confidence, they range from zero to 100 percent. As a result, it is fair to conclude that twice as many Americans approve of labor unions (with a 20 percent approval rating) as approve of Congress (with a 10 percent approval rating). For statistical purposes, I do not distinguish between interval and ratio levels of measurement. As long as a measure uses a uniform scale, you can perform the same statistical functions.

Measures of central tendency identify the average value of the variable. How we find the average will depend on the level of measurement of the variable because those variables with higher levels of measurement allow us to do more mathematically. To describe the average of a variable measured on a scale (either interval or ratio), we use the statistical measure called the mean. This adds up all the values of the variable for all of the cases and then divides by the number of cases. When we use the word "average," we normally are talking about the mean. But if the data are not on a scale, we cannot add them up. So if we have ordinal data, the word "average" means something different. In that case, we mean the value for the middle-most case. Statistically, this is called the median. To find it, you need to put all the cases in order and, coming in from both sides, find the middle case. Without order, the numbers have even less meaning. Thus, for nominal-level variables, the only way you can describe an average is by reporting the value that occurs most often. This is called the mode.

Sometimes, the average does a better job of describing the distribution than others. To see how well the average describes the data, we look at measures of dispersion. With interval-level variables, we frequently report the statistic called the standard deviation. With this measure, we compare each case with the average to see how far away it is. We then square all those distances to make them positive. If we total all those values, we get the sum of the squared distances, which we can then divide by the number of cases to get the average sum of squares—a number that we call the variance. If we then take the square root of the variance, we get the standard deviation. This math assumes that we have all the data points for an entire population, though. If we did this math for only a sample, it would underestimate the dispersion of the population because a sample is necessarily less spread out than the population. So to calculate the standard deviation for a sample, we would divide the sum of squares by one less than the number of cases in our sample to boost the variance up just a bit. And then we would again take the square root of the variance to get the standard deviation. The standard deviation is a single number that contains a lot of information about how widely the cases are distributed around the mean. If the distribution is normally distributed, about two-thirds of the cases are contained in the sandwich of one standard deviation below to one standard deviation above the mean. Fully 95 percent of cases are sandwiched between two standard deviations on each side of the mean.

But if the data are not measured on a scale, or if they are not normally distributed, neither the mean nor the standard deviation does a good job of describing the data. In that case, we sometimes report the data in terms of where certain percentiles fall on the distribution. Sometimes, we report the interquartile range as the values between which the middle half of the cases fall. Sometimes, we will report the "five-point summary," which gives the minimum, second quartile, median, third quartile, and the maximum. Both of those measures give us a good idea of what the distribution looks like; they just are not as succinct as the standard deviation.

MULTIVARIATE STATISTICS

Usually, as political scientists, we want to do more than describe a political characteristic. Usually, we are interested in explaining why things happen the way they do. This means that we need to look at the relationship between two (or more) variables. We begin the process by hypothesizing a causal relationship that we think explains the event in question. To analyze the relationship between multiple variables, we follow a three-step process. First, we describe the pattern that we see in the relationship between the variables. Second, we identify the statistical significance of the relationship. Third, we evaluate the substantive significance of the relationship. The way in which we complete each step will be determined by the level of measurement of the variables.

Hypotheses

A hypothesis is a statement of the causal relationship between two variables. It needs to be theoretically driven. It needs to be formulated in terms of how an independent variable (the cause) affects a dependent variable (the effect). And it needs to be empirically testable. How to state a hypothesis correctly depends on the level of measurement of the two variables. If they both have order (with ordinal, interval, or ratio levels of measurement), then you state the hypothesis in terms of the direction of the relationship. If increases in one are associated with increases in the other, you would say, "There is a positive (or direct) relationship between the dependent variable and the independent variable." If increases in one are associated with decreases in the other, you would say, "There is a negative (or inverse) relationship between the dependent variable and the independent variable." If we think that the current trend in congressional approval is a part of a long-term pattern, we might hypothesize that there is a negative relationship between congressional approval ratings and time. But if one or both of the variables have a nominal level of measurement, they cannot have increases, which means it doesn't make sense to refer to a positive or negative relationship. In that case, you split your independent variable into two groups (even if it is an interval-level variable) and state that the first group is more likely to have a particular value of the dependent variable than the second group. Obama staffers hypothesized that persuadable voters who saw more Obama campaign ads were more likely to vote for Obama than those who saw fewer ads.

Closely connected to the hypothesis is the null hypothesis. This is the statement that there is no relationship between the dependent variable and the independent variable: "There is no relationship between congressional approval ratings and time" or "There is no relationship between voter preference and exposure to campaign ads." When we do statistical analysis, we are actually testing the null hypothesis. That means that we begin by assuming that there is no relationship between the two variables and that any pattern that we see is simply due to chance. We need overwhelming evidence to convince us to reject the null hypothesis. But even then, we know that random things do happen, so we never say that we have "proven" our hypothesis. No matter how confident we are, we may be wrong. If we incorrectly reject the null hypothesis (a false positive finding), we say that we have committed a Type I error. If we incorrectly fail to reject the null hypothesis, we have committed a Type II error.

Describing the Pattern

After stating our hypothesis (and the corresponding null hypothesis), we collect the data for the variables and describe the pattern that we see between them. If the variables have categorical values (either nominal or ordinal), we will usually present their relationship in a contingency table. In a contingency table, the independent variable should be placed at the top of the table, with its values forming the columns. The dependent variable should be placed to the left of the table, with its values forming the rows. The cells of the contingency table contain the column percent for that cell (or 100 times the number of cases in the cell divided by the number of cases in the column). This means that the percentages in each column total 100 percent. The far right-hand column contains what we call the marginals, which are the percentage of all of the cases that have each value of the dependent variable. Giving columns percentages standardizes the numbers so that you can compare across the rows to see the pattern. If there is no relationship, you would expect that the overall percentage in a row would be the same as the percentages found across that row for each of the categories of the independent variable. If there is a relationship, then these percentages will vary. To describe the pattern in a contingency table, you would choose one of the rows of the table, and compare the percentages across it.

If one or both of your variables has an interval level of measurement, you can still present the relationship in a contingency table by collapsing its full range into a limited number of categories—in essence, by turning it into an ordinal-level variable. You could then describe the pattern in the same way you would if you had categorical variables. Alternatively, if the independent variable is categorical and the dependent variable is interval, you can do an analysis of variance and report the mean value of the dependent variable for each of the values of the independent variable.

But if you have two interval-level variables, you also have another option available to you. Because both variables are on a scale, you can graph the data and depict the relationship as a line. When you took algebra, you got used to thinking about a line as the relationship between two variables. At that time, you gave the equation of the line as

$$y = mx + b$$

In statistics, we change that equation a little and say

$$Y = B_0 + B_1 X_1$$

But the elements of the equation of the line you get with regression have the same interpretation as they did in algebra. To describe the pattern you find with a regression, you interpret the intercept (B_0) as the value of Y (the dependent variable) we would expect to see if X (the independent variable) has a value of zero. Then you would describe the slope (B_1) as how Y (the dependent variable) changes for every unit increase in X (the independent variable). If the slope is positive, the data suggest a positive relationship; if negative, a negative relationship.

Identifying the Statistical Significance

Statistical significance answers the question: Is there a relationship? We know that not every pattern we see is a reflection of an underlying real relationship. Sometimes, we just get wacky data that don't actually mean anything. For example, in 1980, the band Styx released a song called "Eddie," in which they begged Edward "Ted" Kennedy not to run for president that year. Their fear was that if elected, he would fall victim to the twenty-year curse that had taken the life of his brother and every other president elected in a year ending in zero since William Henry Harrison died after his 1840 election. Fortunately, after his 1980 election, Ronald Reagan seems to have broken the "curse" by surviving his tenure in office. Because random things happen, the next step after describing the pattern is to ask, "What is the probability that the pattern we observed is due to chance?" A small probability makes us more confident in our hypothesis; a large probability makes us less confident. As social scientists, we require a high level of confidence. Normally, we want to be 95 percent confident. That means that we will reject the null hypothesis only if there is a 5 percent or less chance of observing this pattern randomly. If the probability is less than 5 percent, we say that the relationship is statistically significant at the 0.05 level, and we conclude that we can reject the null hypothesis.

The way we find that probability depends on the levels of measurement of our variables. If we have an interval-level variable for which we want to compare a sample mean to a population mean, we do a means test by determining the t-score of the sample and finding its associated probability. If we want to compare an interval-level variable across multiple groups, we use analysis of variance that uses an F test to find the probability of finding the difference if the groups come from the same underlying distribution. If we have two interval-level variables, and so are able to use regression, we calculate the probability by comparing the magnitude of the slope to its standard error with an F test. If the slope is twice as big as the standard error, we know that it is unlikely to be due to chance and so we reject the null hypothesis. All of these determinations of statistical significance require us to be able to calculate the mean and standard deviation, which means we need interval-level variables.

If we cannot calculate the mean, we need to take a different approach to find the probability of observing a pattern due to chance. One such approach is found in the calculation of the statistic chi-square. Chi-square can be calculated for data presented in a contingency table. It compares the observed frequency for each cell with the frequency we would expect to see if there were no relationship between the two variables (which we call the expected frequency). If the two variables are independent, we indicated above that the column percent should be constant across each row. The expected value for each cell is the proportion of cases found in that row multiplied by the number of cases in the column. The chi-square for the table is the difference between the observed frequency for the cell and its expected frequency, squared, and summed for all the cells in the table. You can look up your value of chi-square in order to find the statistical significance for your data. If it is statistically significant, you can be confident that a relationship exists.

Evaluating the Substantive Significance

If the statistical significance indicates that a relationship is likely to exist, we can then find the substantive significance. Substantive significance answers the question: How strong is the relationship? Statisticians use the word "significance" to refer to statistical significance, but when most people refer to a significant relationship, they usually mean that the relationship is strong. This kind of substantive significance can be measured using various measures of association. Most measures of association are constrained so that a value of zero implies no association and a value of one implies a perfect association. Usually, they are somewhere between these two extremes. As usual, the appropriate measure of association depends on the level of measurement of your variables.

If one or both of your variables are nominal, you have two options. The statistic Cramer's V modifies chi-square to compensate for the size of the table and the number of cases and to constrain it to being less than one. Because most measures of association are constrained to be between zero and one, they can be interpreted in similar ways. There is not much going on if the association is less than 0.10, so we say that the relationship is very weak. Because we are describing messy social relationships, it is hard to get an association approaching one. So if we have an association that is greater than 0.30, we will usually describe it as a strong relationship.

Alternatively, you can also use lambda (given by the Greek letter λ) to measure an association if at least one of the variables has a nominal level of measurement. Lambda asks the question, "How well does knowing the value of the independent variable for a particular case allow you to predict the value of the dependent variable?" Just looking at the dependent variable, if you wanted to predict what value any given case had, you would be best off always predicting the mode. You'd be right for all the cases that have the modal value but wrong for all the others. But if for a case you know its value for the independent variable, you could refine your prediction so that you would predict the modal value for that subset of cases. Again, you would make some errors in your predictions. Lambda compares the number of errors you would make in predicting without knowing the independent variable

Table 14.1 Overview of Statistical Measures and Their Analytical Purpose

	Statistical Measure	Analytical Purpose
Univariate Statistics		Describes a single variable
	Measures of central tendency	Identifies the average value of the variable
	Measures of dispersion	Identifies how well the average describes the distribution of a variable
Multivariate Statistics		Analyzes the relationship between more than one variable
Statistical Significance		Identifies the probability that the null hypothesis is correct
	t-test	Compares a sample mean to a population mean for an interval-level variable
	ANOVA (F test)	Compares the distribution of an interval-level variable across multiple groups
	Chi-square	In a contingency table, compares observed and expected frequencies for each cell
Substantive Significance		Measures the strength of the relationship
	Cramer's V	Measure of association when at least one variable has a nominal level of measurement
	Lambda	Measure of association when at least one variable has a nominal level of measurement; has a PRE interpretation: the proportion of errors in the dependent variable explained by the independent variable
	Gamma	Measure of association for two ordinal-level variables
	ANOVA (eta and eta-squared)	Measure of association when the independent variable is categorical; eta-squared has a PRE interpretation: the proportion of variance in the dependent variable explained by the independent variable
	Pearson's r	Measure of association for two interval-level variables
	Regression	Identifies a linear relationship controlling for multiple independent variables; R-square has a PRE interpretation: the proportion of variance in the dependent variable explained by the model

with the number of errors you would make if you do know it. If you eliminate all the errors, you end up with a lambda of one. If you don't eliminate any, you end up with a lambda of zero. Lambda is in a special class of measures of association called proportional reduction in error (or PRE) measures. For PRE measures, instead of translating the number into a word describing the strength of the relationship, you use a more specific interpretation. For example, you interpret the value of lambda as the proportion of errors in the dependent variable explained by the independent variable.

If you have two ordinal-level variables, you can measure their association with a statistic called gamma (given by the Greek letter γ). With ordered variables, you can talk about a relationship being direct or inverse. Gamma looks at each case and finds how many other cases it can be paired with in support of a direct relationship. It calls these "concordant pairs." Then it looks at each case and finds how many other cases it can be paired with in support of an inverse relationship. It calls these "discordant pairs." Gamma takes the difference between the number of concordant and discordant pairs and then divides it by the total of concordant plus discordant pairs. If you only have concordant pairs, you end up with a gamma of 1.0, indicating a perfect direct relationship between the two variables. If you have equal numbers of concordant and discordant pairs, you end up with a gamma of zero, indicating that the two variables are unrelated. If you only have discordant pairs, you end up with a gamma of -1.0, indicating a perfect inverse relationship.

Given an occasion in which your independent variable is categorical (either nominal or ordinal) and your dependent variable has an interval level of measurement, you would use ANOVA to analyze the relationship. It has two measures of association connected to it: eta (the Greek letter η) and eta-squared (η^2). Eta-squared is the ratio of the between group sum of squares (the sum of squares explained by being in the groups designated by the categories of your independent variable) and the total sum of squares (ignoring the independent variable). It is a PRE measure of the proportion of variation in the dependent variable explained by the independent variable. If you take the square root of eta-squared, you get eta, which you interpret in the same way that you usually interpret measures of association.

If you have an interval-level independent variable and a categorical dependent variable, describing the strength of the relationship is trickier. Technically, you should use some variant of logit or probit (two statistical tests that we haven't covered in this book and that are difficult to interpret). An easier solution is to collapse your independent variable into an ordinal-level variable, get a contingency table, and use the appropriate measure of association for it as an ordinal-level variable—either gamma or Cramer's V.

Where you have two interval-level variables, Pearson's r is the correlation between them. It takes the covariance of the two variables and standardizes it by dividing by the square root of the product of the sum of squares of the first variable and the sum of squares of the second variable. Like gamma, Pearson's r can be either positive or negative, reflecting either a positive or negative linear relationship. And like most measures of association Pearson's r is constrained to have a magnitude less than or equal to one.

Alternatively, you can use regression, which will allow you to control for multiple independent variables. There are two things you can do with regression to evaluate the strength of the relationship. First, you can compare the standardized coefficients (or betas) for your variables to identify which of them has the most influence on the dependent variable. Second, you should report the R^2, which is a PRE measure you can interpret as the proportion of variance in the dependent variable explained by the model (the group of independent variables you included in the regression).

KEEPING THE NUMBERS MEANINGFUL

As I summarized the statistics you learned to calculate in this book, I focused on how the choice of the proper statistic depends on the level of measurement of your variables. As the statistics assume a specific level of measurement, they make other assumptions as well. For example, gamma assumes not only that the two variables are ordinal, but also that the relationship between them shows a pattern that is either direct or inverse. In describing the pattern of the relationship, if you find that the values of your dependent variable do not consistently either increase or decrease, gamma would not do a very good job of summarizing the relationship—you would be better off choosing Cramer's V as your measure of association. Pearson's r and regression assume not only a relationship that either increases or decreases, but also that the increase (or decrease) is linear in nature. Violating the linearity assumption can, in turn, lead to violations of other assumptions we make when we do a regression analysis. The consequence of violating the Gauss-Markov assumptions given in Chapter 13 is that your description of the relationship is not going to be very accurate. In Chapter 13, I gave various suggestions for how to compensate for violations of the assumptions. If you get into a situation where you've got a problem, fix it so that your numbers remain meaningful.

One of the Gauss-Markov assumptions actually applies to all measures of association: Your model needs to be correctly specified. Keep in mind that political factors tend to be interrelated. If you model a political process as being a simple interaction between your independent and dependent variables, you might not get an accurate measure of their relationship. If there is another variable that is highly connected to both variables, its effect on the dependent variable will get tied into the measure of the association between your two variables. One possible consequence is that the relationship you report may actually be spurious—a mirage of not controlling for the actual cause of your dependent variable. Alternatively, if the two variables affect the dependent variable in opposite ways, excluding the control variable may end up hiding a real underlying relationship between your independent and dependent variables. Either way, if there is another factor that is highly correlated with both of your variables, you need to control for it. You can do this either by creating a three-way contingency table or by including multiple independent variables in a multiple regression analysis.

A correctly specified model needs to be thorough and to correctly identify the dependent variable. None of the statistics you've learned in this book can determine which variable is the cause and which variable is the effect. They only measure the association between them. Some of the statistics do not require you to designate the dependent variable. For example, chi-square and Pearson's r are symmetrical; they give you the same number regardless of the identification of the dependent variable. But other statistics, such as regression, assume that you modeled the process correctly. This is one reason why it is so important to base your hypothesis on a theory. The theoretical framework for your hypothesis requires you to explain why you believe that your independent variable affects your dependent variable. If you want your results to be meaningful, you need to take the time to consider the possibility that you might have reversed the causal relationship in your model.

EMBRACING THE UNCERTAINTY

My husband's favorite comic strip shows Dilbert and Dogbert reading a news report, "Our town hasn't had a murder since 1957." Dilbert, ever the optimist, responds, "We're safe forever." But the pessimistic Dogbert concludes, "We're due."[4] Neither character has fully embraced the probabilistic nature of life. Dilbert is correct that there are underlying causal relationships controlling both past and future events. This is the principle upon which all science is built: Our systematic observation of the world allows us to better understand why it works the way it does. But Dogbert is correct that unexpected things do happen. Even in nature, a 500-year flood will occur. The trick is that it doesn't always occur in the 500th year. If unexpected events are a problem in the hard sciences, they are even more of a problem in the social sciences, where the human ability to choose makes prediction even harder.

But our inability to forecast with certainty whether a murder will occur this year should not deter us from using data to understand the world—we just need to embrace the fact that our understanding is probabilistic in nature. It is too facile to equate statistics with lies. And it is too facile to reject a number because "63 percent (or any other percent) of statistics are made up on the spot." You now have what it takes to look a statistic in the eye and face it down. You know how to evaluate the quality of a measurement in terms of its validity and reliability. You understand that correlation does not prove causation—and with any luck you've stopped using the dreaded "P" word outside the confines of the deductive reasoning you use in your math or philosophy classes. You understand the importance of asking the question of whether there is a relationship between your dependent and independent variables. And you've learned that the answer to that question is expressed in probabilistic terms as the level of statistical significance. You've learned that answering the question "How strong is the association?" requires you to understand how the variables are measured as well as the nature of their relationship.

As you've analyzed real-world data, I hope you have also learned how fun such analysis can be. There are a host of statistical tests we have not covered—maybe you'll decide to go to graduate school so you can learn them. For example, I've covered only one type of regression, ordinary least squares (OLS) regression. But it is not the only kind. I've mentioned logit and probit as being useful kinds of regression when your dependent variable is not interval level. If you have a dichotomous dependent variable, you would want to model the relationship as an s-curve that approaches zero to the left and one to the right—logit and probit do that. Alternatively, a different approach to estimating a linear relationship is called maximum likelihood estimate (or MLE). Where OLS regression defines the best estimate of a line as the one that minimizes the squared error, MLE defines best as the estimate that is most likely. Instead of assuming (as OLS does) that the mean and standard deviation are constant, MLE assumes that different samples reflect different means and standard deviations. If the Gauss-Markov assumptions are correct, OLS and MLE produce the same results, but if not, MLE's results are better.

Even without going on to more advanced statistical training, though, at this point, you know enough to be an intelligent consumer of data, and that gives you political power. If you've worked hard this semester, you should also have learned some skills that will allow you to analyze your own data, skills that can give you professional power. Keep what you've learned in mind as you take your remaining political science classes. Your statistical training should open windows of understanding as you read about the research of others and doors of opportunity as you conduct your own original research.

Appendix 1

Tips for Professional Writing

1. Write a Memorandum
2. Create a Frequency Table
3. Create a Table for ANOVA
4. Create a Contingency Table
5. Create a Three-Way Contingency Table
6. Create a Professional-Looking Regression Table

One of the most important skills you learn in college is how to communicate through writing. This is a skill you'll use for the rest of both your personal and professional lives. But the professional writing you do is unlikely to be in the format of the term papers you write most often in college; rather, it is more likely to take the form of a memo. The goal of the memo is to communicate information clearly and briefly. This appendix begins by summarizing how to write a memo. And because the most efficient way to communicate numerical information is typically in a table, the rest of this appendix focuses on how to augment your writing by creating professional-looking tables for the different statistical analyses included in this book.

WRITE A MEMORANDUM

A. The Heading
  1. To: Name and Title
  2. From: Your Name
  3. Date: Date Sent
  4. Subject: Keyword the nature of the assignment
B. The Introduction
  1. Briefly describe the assignment
  2. Describe the data you will use
  3. Summarize your results in one sentence

C. The Body
   1. Analyze the data
D. The Conclusion
   1. Summarize the data
   2. Explain how it answers the original question
   3. Describe any broader implications

CREATE A FREQUENCY TABLE IN WORD

1. Give it a title describing its content.
2. After the title, insert a table of the appropriate size: columns = 3; rows = 2 + the number of categories in your variable.
   >Insert
      >Table
         >Insert Table
            Number of columns = 3
            Number of rows = 2 + number of categories of variable
3. In the first row, label the three columns: *Variable Label*, "Frequency," "Percent"
4. In the first column, list the categories of the variable.
5. In the second column, give the frequencies of those categories.
6. In the last column, give the "Valid" percentages—do not include the missing cases unless this is actually relevant to your analysis.
7. In the last row, give the totals for the columns: "Total," "*number of valid cases*," "100.0%."
8. Draw the appropriate lines.

   Highlight the entire table:
   >Borders and Shading *(the arrow, not the icon)*
      >Borders and Shadings
         Settings
            >None
         Style
            The single line should be highlighted.
         Preview
            Click above and below the table to add lines there.
         >OK

Highlight the first line (which has the column labels):

\>Borders and Shading *(the arrow, not the icon)*

>Bottom Border

9. Below the last line use a smaller font to add the data source and any other clarifying information.

CREATE A TABLE FOR ANOVA

1. Give it a title describing its contents.
2. Insert a table of the appropriate size: columns = 2, rows = 4 + number of categories of independent variable.

    \>Insert

    >Table

    >Insert Table

    Number of columns = 2

    Number of rows = 4 + number of categories of independent variable

3. In the first row, label the two columns: *Independent Variable Label*, "Average" *Dependent Variable Label*
4. In the first column, list the categories of the independent variable, followed by the word "Total."
5. In the second column, give the means of those categories.
6. In the row under "Total," indicate your sample size "n = ."
7. Under the second column, identify eta and eta-squared. Place stars next to these values to designate the level of significance of the relationship.
8. Draw the appropriate lines.

Highlight the entire table:

\>Borders and Shading *(the arrow, not the icon)*

>Borders and Shadings

Settings

>None

Style

The single line should be highlighted.

Preview

Click above and below the table to add lines there.

>OK

Highlight the first line (with column labels):

>Borders and Shading *(the arrow, not the icon)*
>>Bottom Border

9. Below the last line, use a smaller font to first give a key for the stars indicating the level of significance "*prob. < 0.05." Second, add the data source and any other clarifying information.

CREATE A CONTINGENCY TABLE

1. Give the table a title in the form of the name of the "*dependent variable*" "by" the "*independent variable.*"
2. Insert a table with as many rows as it has categories in the dependent variable plus four. It has as many columns as it has categories in the independent variable plus two.
3. In the first row, type the name of the dependent variable in the first column, each of the categories of the independent variables in the next columns, and the word "Total" in the last column.
4. Below the name of the dependent variable in the first column, place each of the categories of the dependent variable. In the row third from the bottom type "Total"; second from the bottom, "n"; and leave the first column of the bottom row empty.
5. Calculate the column percents for each cell by dividing the cell count by the column count and multiplying by one hundred.
6. Include marginals for both the rows and the columns—in both percent and number of cases.
7. In the last column of the last line of the table, place the measure of association along with the appropriate number of stars for the statistical significance.
8. Place lines after the title, after the column headings, and at the end using the "Borders and Shadings" command in Word.
9. Give any relevant information in a smaller font below the bottom line, including a key to the number of stars for statistical significance.

CREATE A THREE-WAY CONTINGENCY TABLE

1. Give the table a title in the form of the name of the "*dependent variable*" "by" the "*independent variable*" "controlling for" the "*control variable.*"
2. As with a regular contingency table, the independent variable is found in the columns; the dependent variable, in the rows. Insert a table with three more columns (c) than the number of categories in the independent variable (IV). That is, $c = IV + 3$.

3. In addition to two rows for column headings, each section will need enough rows for the number of categories of the dependent variable (DV) plus one for the number of cases and one for the measure of association. Think of the number of categories of the control variable as "CV." To get the number of rows (r), use the following equation: $r = (DV + 2)CV + 2$.

4. The first two rows contain headings. The first column heading should identify the control variable; the second, the dependent variable. In the middle columns, place the categories of the independent variable in the second row and center the label of the independent variable in the first row above the categories. The last column of the second row should be headed "Total" for the row marginals.

5. Below the heading in the second column, give the categories of the dependent variable followed by two blank lines. Repeat this as many times as there are categories in the control variable.

6. Go to each time you repeated the first category of the dependent variable. Consecutively place each of the categories of the control variable to the left in the first column.

7. Within the cells, you will include the column percent within each of the sections of the table.

8. Include the row marginals for each section as you would for a regular contingency table, with both the row percent and the number of cases in parentheses.

9. Instead of the complete column marginal, in the second to the last row of each section just include the column n in parentheses.

10. In the lower right-hand cell for each section, give the value of the measure of association for that section with the appropriate number of stars to indicate statistical significance.

11. Divide the table by lines: at the top and bottom, below the column headings, and between each subsection.

12. Below the table, in a smaller font, give the information necessary to understand it: the data source, the key for statistical significance, and the question wording.

CREATE A PROFESSIONAL-LOOKING REGRESSION TABLE

1. As usual, the table is headed by a title. It takes the form of the name of the dependent variable "regressed on" the independent variable(s).

2. On the line after the title, insert a table with three columns and enough rows for the number of independent variables plus three.

3. The first column is headed "Independent Variables" and contains a list of the names of the independent variables in addition to "Constant" or "Intercept."

4. The second column is headed "Unstandardized Coefficients." For each variable and the constant, include the value of its "B" along with the appropriate number of stars to indicate its statistical significance. If you want, after you enter the "B," you can hit enter and give its standard error in parentheses directly under the slope.

5. The third column is headed "Standardized Coefficients." For each variable, include the value of its "beta" along with the appropriate number of stars to indicate its statistical significance—these should be identical to the stars included for the standardized coefficient. Standardized coefficients never have an intercept, so leave that cell blank.

6. In the bottom line, include the number of cases (one more than the total degrees of freedom) and the R^2.

7. Using the border command in Word, place a single line at the top and bottom of the table, and a line after the column headings.

8. After the table, in a smaller font, include a key for the levels of statistical significance indicated by the number of stars. Also, if you use a single source for your data, indicate what it is.

Appendix 2

How to Use SPSS

1. Import an Excel Data File
2. Collapse Variables
3. Select Cases
4. Compute a New Variable Mathematically from an Old Variable
5. Get a Frequency of a Variable
6. Calculate z-Scores
7. Do a Means Test
8. Conduct an Analysis of Variance
9. Get a Correlation Matrix
10. Get a Cross Tabulation
11. Get a Scatter Plot
12. Run a Regression

Much of this book focused on teaching you the math behind the statistical tests that political scientists are most likely to use. Understanding the math is indispensable for building the intuitions you need to use statistics properly. But in your professional life, you are unlikely to need to calculate statistics by hand; you are much more likely to use a statistical package to do the math for you. SPSS is the most commonly used statistics package used by social scientists in business and in government. This appendix includes the commands that you are most likely to use.

IMPORT AN EXCEL DATA FILE

Find IBM SPSS Statistics in your computer's programs and double click to open it.

At this point, your first window opens called "Output [Document]." This is the window where any statistical results that you request will be reported. Superimposed on this window is a box asking if you want to open an existing data source. Since you want to import a new Excel file, you want to cancel out of this box.

What would you like to do?

>Cancel

>File

>Open

>Data

Find the appropriate folder

Files of type= "Excel" (*.xls, *.xlsx, *.xlsm)

File Name = *younamedit.xlsx*

>Open

✓Read variable name from first row of data

>OK

At this point, your second window opens called "Untitled [Dataset]." This window will show your data. But to be able to use it in the future, you will need to save the data into SPSS format. Make sure you are in the Data Window and do the following:

>File

>Save As

File Name=*make sure you are in the right folder and give it a name*

>OK

RECODE VARIABLES

After opening your data in SPSS:

>Transform

>Recode into Different Variables

"*Variable*"→Input Window

Variable Name="*collapsed variable name*" *(only 8 characters)*

Variable Label="*collapsed variable label*" *(longer and you can use punctuation)*

>Change

>Old and New Values

>Range, Lowest through Value

"*upper limit of lowest range*"

Value="1"

>Add

>Range

"*lower limit of second range*"

"*upper limit of second range*"

Value="2"
>Add
>Range, Value through Highest
"lower limit of highest range"
Value="3"
>Add
>Continue
>OK

At this point, it is a good idea to make sure that the new variable is properly collapsed. Using the "Define Variable Properties" command, pull up the new variable: Make sure that the recoded categories are all there and include the correct number of cases. You will also want to enter the ranges into the value label box for each category.

SELECT CASES

After opening your data in SPSS and cleaning them:

>Data
>Select Cases
>If condition is satisfied
>If
"Variable"→*"mathematical operators for condition"*
For example: "ZMurder > 2"
>Continue
>OK

At this point, if you enter the Data window, you will see slashes through the case numbers for all cases not meeting your criteria.

COMPUTE A NEW VARIABLE MATHEMATICALLY FROM AN OLD VARIABLE

>Transform
>Compute Variable
Target Variable = *New Variable*
Numeric Expression = *mathematical functions of old variable(s)*
>OK

GET A FREQUENCY OF A VARIABLE

After opening your data in SPSS and cleaning them:

>Analyze
>>Descriptive Statistics
>>>Frequencies
>>>>"*Variable*"→Variable
>>>>>Statistics
>>>>>>Mean
>>>>>>Median
>>>>>>Mode
>>>>>>Range
>>>>>>Standard Deviation
>>>>>>Percentiles "25" "75"
>>>>>>Continue
>>>>OK

At this point, the frequency of the variable will show up in the "Output" window.

CALCULATE Z-SCORES

After opening your data in SPSS and cleaning them:

>Analyze
>>Descriptive Statistics
>>>Descriptives
>>>>"*Variable*"→Variable
>>>>> Save Standardized Values as Variables
>>>>OK

At this point, the variable will show up in the last column of the data window with a "Z" in front of the name.

DO A MEANS TEST

After opening your data in SPSS and cleaning your variables, get the population mean for the variable:

>Analyze

>Descriptive Statistics
>>Descriptives
>>>"*Variable*"→Variable
>>>Options
>>>>Mean
>>>>Continue
>>>OK

Select the cases that belong to the group you want to compare to the population.
>Data
>>Select Cases
>>>If condition is satisfied
>>>If
>>>>"*GroupIdentifier*"→"*GroupIdentifier=appropriate value*"
>>>Continue
>>>OK

Compare the group to the population mean for the variable.
>Analyze
>>Compare Means
>>>One-Sample T-Test
>>>>"*Variable*"→Test Variable
>>>>Test Value="*population mean*"
>>>OK

CONDUCT AN ANALYSIS OF VARIANCE

After opening your data in SPSS, cleaning them, and selecting the appropriate cases:
>Analyze
>>Compare Means
>>>Means
>>>>"*Dependent Variable*"→Dependent Variable
>>>>"*Group Variable*"→Independent Variable
>>>>Options
>>>>>ANOVA Table and eta
>>>>>Continue
>>>>OK

GET A CORRELATION MATRIX

>Analyze
>>Correlate
>>>Bivariate
>>>>*First Variable*→Variables
>>>>*Second Variable*→Variables
>>>>*Third Variable*→Variables
>>>>. . .
>>>OK

GET A CROSS TABULATION

After opening and cleaning your data in SPSS:
>Analyze
>>Descriptive Statistics
>>>Crosstabs
>>>>"*Dependent Variable*"→Row Variable
>>>>"*Independent Variable*"→Column Variable
>>>>"*Control Variable (if desired)*"→Layer 1 of 1
>>>>Cells
>>>>>Observed
>>>>>Column
>>>>>Continue
>>>>Statistics
>>>>>Chi-square
>>>>>Phi and Cramer's V
>>>>>*any other statistics that you might want*
>>>>OK

GET A SCATTER PLOT

>Graphs
>>Legacy Dialogs
>>>Scatter/Dot

>Simple Scatter
>Define
"*Dependent Variable*"→Y-Axis
"*Independent Variable*"→X-Axis

RUN A REGRESSION

>Analyze
>Regression
>Linear
Dependent=*Dependent Variable*
Independent=*Independent Variable(s)*
>Save
Predicted Values:
>Unstandardized
>Continue
>Plots
*ZRESID→Y
*ZPRED→X
>Continue
>OK

Appendix 3

z Table

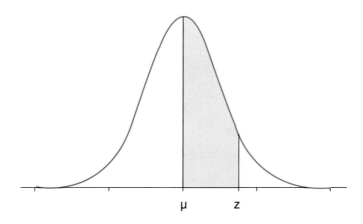

The cells contain the proportion between the mean and the z-score.

z	0.00	0.01	0.02	0.03	0.04	0.05	0.06	0.07	0.08	0.09
0.0	0.0000	0.0040	0.0080	0.0120	0.0160	0.0199	0.0239	0.0279	0.0319	0.0359
0.1	0.0398	0.0438	0.0478	0.0517	0.0557	0.0596	0.0636	0.0675	0.0714	0.0753
0.2	0.0793	0.0832	0.0871	0.0910	0.0948	0.0987	0.1026	0.1064	0.1103	0.1141
0.3	0.1179	0.1217	0.1255	0.1293	0.1331	0.1368	0.1406	0.1443	0.1480	0.1517
0.4	0.1554	0.1591	0.1628	0.1664	0.1700	0.1736	0.1772	0.1808	0.1844	0.1879
0.5	0.1915	0.1950	0.1985	0.2019	0.2054	0.2088	0.2123	0.2157	0.2190	0.2224
0.6	0.2257	0.2291	0.2324	0.2357	0.2389	0.2422	0.2454	0.2486	0.2517	0.2549
0.7	0.2580	0.2611	0.2642	0.2673	0.2704	0.2734	0.2764	0.2794	0.2823	0.2852

(Continued)

(Continued)

z	0.00	0.01	0.02	0.03	0.04	0.05	0.06	0.07	0.08	0.09
0.8	0.2881	0.2910	0.2939	0.2967	0.2995	0.3023	0.3051	0.3078	0.3106	0.3133
0.9	0.3159	0.3186	0.3212	0.3238	0.3264	0.3289	0.3315	0.3340	0.3365	0.3389
1.0	0.3413	0.3438	0.3461	0.3485	0.3508	0.3531	0.3554	0.3577	0.3599	0.3621
1.1	0.3643	0.3665	0.3686	0.3708	0.3729	0.3749	0.3770	0.3790	0.3810	0.3830
1.2	0.3849	0.3869	0.3888	0.3907	0.3925	0.3944	0.3962	0.3980	0.3997	0.4015
1.3	0.4032	0.4049	0.4066	0.4082	0.4099	0.4115	0.4131	0.4147	0.4162	0.4177
1.4	0.4192	0.4207	0.4222	0.4236	0.4251	0.4265	0.4279	0.4292	0.4306	0.4319
1.5	0.4332	0.4345	0.4357	0.4370	0.4382	0.4394	0.4406	0.4418	0.4429	0.4441
1.6	0.4452	0.4463	0.4474	0.4484	0.4495	0.4505	0.4515	0.4525	0.4535	0.4545
1.7	0.4554	0.4564	0.4573	0.4582	0.4591	0.4599	0.4608	0.4616	0.4625	0.4633
1.8	0.4641	0.4649	0.4656	0.4664	0.4671	0.4678	0.4686	0.4693	0.4699	0.4706
1.9	0.4713	0.4719	0.4726	0.4732	0.4738	0.4744	0.4750	0.4756	0.4761	0.4767
2.0	0.4772	0.4778	0.4783	0.4788	0.4793	0.4798	0.4803	0.4808	0.4812	0.4817
2.1	0.4821	0.4826	0.4830	0.4834	0.4838	0.4842	0.4846	0.4850	0.4854	0.4857
2.2	0.4861	0.4864	0.4868	0.4871	0.4875	0.4878	0.4881	0.4884	0.4887	0.4890
2.3	0.4893	0.4896	0.4898	0.4901	0.4904	0.4906	0.4909	0.4911	0.4913	0.4916
2.4	0.4918	0.4920	0.4922	0.4925	0.4927	0.4929	0.4931	0.4932	0.4934	0.4936
2.5	0.4938	0.4940	0.4941	0.4943	0.4945	0.4946	0.4948	0.4949	0.4951	0.4952
2.6	0.4953	0.4955	0.4956	0.4957	0.4959	0.4960	0.4961	0.4962	0.4963	0.4964
2.7	0.4965	0.4966	0.4967	0.4968	0.4969	0.4970	0.4971	0.4972	0.4973	0.4974
2.8	0.4974	0.4975	0.4976	0.4977	0.4977	0.4978	0.4979	0.4979	0.4980	0.4981
2.9	0.4981	0.4982	0.4982	0.4983	0.4984	0.4984	0.4985	0.4985	0.4986	0.4986
3.0	0.4987	0.4987	0.4987	0.4988	0.4988	0.4989	0.4989	0.4989	0.4990	0.4990

Appendix 4

t Table

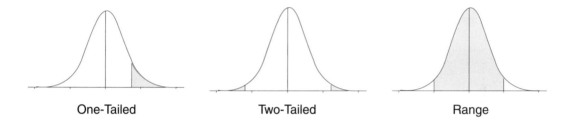

One-Tailed   Two-Tailed   Range

The cells contain the minimum t-score necessary to achieve the level of significance in the heading.

	Proportion in One Tail *(positive or negative)*					
	0.25	**0.1**	**0.05**	**0.025**	**0.01**	**0.005**
d.f.	**Proportion in Two Tails Combined**					
(n − 1)	**0.50**	**0.2**	**0.10**	**0.05**	**0.02**	**0.01**
1	1.000	3.078	6.314	12.706	31.821	63.657
2	0.816	1.886	2.920	4.303	6.965	9.925
3	0.765	1.638	2.353	3.182	4.541	5.841
4	0.741	1.533	2.132	2.776	3.747	4.604
5	0.727	1.476	2.015	2.571	3.365	4.032
6	0.718	1.440	1.943	2.447	3.143	3.707
7	0.711	1.415	1.895	2.365	2.998	3.499
8	0.706	1.397	1.860	2.306	2.896	3.355
9	0.703	1.383	1.833	2.262	2.821	3.250
10	0.700	1.372	1.812	2.228	2.764	3.169
11	0.697	1.363	1.796	2.201	2.718	3.106

(Continued)

(Continued)

d.f. (n − 1)	Proportion in One Tail *(positive or negative)*					
	0.25	0.1	0.05	0.025	0.01	0.005
	Proportion in Two Tails Combined					
	0.50	0.2	0.10	0.05	0.02	0.01
12	0.695	1.356	1.782	2.179	2.681	3.055
13	0.694	1.350	1.771	2.160	2.650	3.012
14	0.692	1.345	1.761	2.145	2.624	2.977
15	0.691	1.341	1.753	2.131	2.602	2.947
16	0.690	1.337	1.746	2.120	2.583	2.921
17	0.689	1.333	1.740	2.110	2.567	2.898
18	0.688	1.330	1.734	2.101	2.552	2.878
19	0.688	1.328	1.729	2.093	2.539	2.861
20	0.687	1.325	1.725	2.086	2.528	2.845
21	0.686	1.323	1.721	2.080	2.518	2.831
22	0.686	1.321	1.717	2.074	2.508	2.819
23	0.685	1.319	1.714	2.069	2.500	2.807
24	0.685	1.318	1.711	2.064	2.492	2.797
25	0.684	1.316	1.708	2.060	2.485	2.787
26	0.684	1.315	1.706	2.056	2.479	2.779
27	0.684	1.314	1.703	2.052	2.473	2.771
28	0.683	1.313	1.701	2.048	2.467	2.763
29	0.683	1.311	1.669	2.045	2.462	2.756
30	0.683	1.310	1.697	2.042	2.457	2.750
60	0.679	1.296	1.671	2.000	2.390	2.660
100	0.677	1.290	1.660	1.984	2.364	2.626
1000	0.675	1.282	1.646	1.962	2.330	2.581
∞	0.674	1.282	1.645	1.960	2.326	2.576
	50%	80%	90%	95%	98%	99%
	Confidence Level for Margin of Error for Range					

Appendix 5

Chi-Square Table

Cells contain the chi-square necessary to get the given probability in the tail.

$$\text{d.f.} = (c - 1)(r - 1)$$

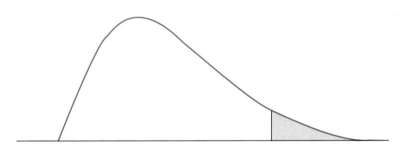

Probability of Tail

d.f.	0.10	0.05	0.01	0.001
1	2.71	3.84	6.63	10.83
2	4.61	5.99	9.21	13.82
3	6.25	7.81	11.34	16.27
4	7.78	9.49	13.28	18.47
5	9.24	11.07	15.09	20.51
6	10.64	12.59	16.81	22.46
7	12.02	14.07	18.48	24.32
8	13.36	15.51	20.09	26.12
9	14.68	16.92	21.67	27.88

(Continued)

(Continued)

d.f.	0.10	0.05	0.01	0.001
10	15.99	18.31	23.21	29.59
11	17.28	19.68	24.72	31.26
12	18.55	21.03	26.22	32.91
13	19.81	22.36	27.69	34.53
14	21.06	23.68	29.14	36.12
15	22.31	25.00	30.58	37.70
16	23.54	26.30	32.00	39.25
17	24.77	27.59	33.41	40.79
18	25.99	28.87	34.81	42.31
19	27.20	30.14	36.19	43.82
20	28.41	31.41	37.57	45.31
21	29.62	32.67	38.93	46.80
22	30.81	33.92	40.29	48.27
23	32.01	35.17	41.64	49.73
24	33.20	36.42	42.98	51.18
25	34.38	37.65	44.31	52.62
26	35.56	38.89	45.64	54.05
27	36.74	40.11	46.96	55.48
28	37.92	41.34	48.28	56.89
29	39.09	42.56	49.59	58.30
30	40.26	43.77	50.89	59.70
40	51.81	55.76	63.69	73.40
50	63.17	67.50	76.15	86.66
60	74.40	79.80	88.38	99.61
80	96.58	101.90	112.30	124.80
100	118.50	124.30	135.80	149.40

Notes

Chapter 1

1. Mark Twain, *Chapters from my Autobiography XX*, April 1904 (Salt Lake City, Utah: Project Gutenberg, 2006), accessed May 14, 2013, http://www.gutenberg.org/files/19987/19987-h/19987-h.htm#CHAPTERS_FROM_MY_AUTOBIOGRAPHY_XVI.
2. Darrell Huff, *How to Lie with Statistics* (New York: Norton, [1954] 1982).
3. Huff, *How to Lie,* 87.
4. Lao-Tzu, *The Sayings of Lao-Tzu,* trans. Lionel Giles (1905), accessed September 6, 2013, http://www.sacred-texts.com/tao/salt/salt08.htm.
5. Max Weber, Part III, Chapter 6. In *Economy and Society,* trans. and ed. G. Roth and C. Wittich (New York: Bedminster Press, [1921] 1968), 650-78, accessed June 7, 2011, http://www.faculty.rsu.edu/~felwell/TheoryWeb/readings/WeberBurform.html.
6. Daniel M. Butler and David E. Broockman, "Do Politicians Racially Discriminate Against Constituents? A Field Experiment on State Legislators," *American Journal of Political Science* 55 (2001): 463-77.
7. Richard F. Fenno, *Home Style: House Members in Their Districts* (Boston: Little, Brown, 1978).
8. Gary C. Jacobson, "The Effects of Campaign Spending in House Elections: New Evidence for Old Arguments," *American Journal of Political Science* 34 (1990): 334-62.
9. Daniel J. Coffey, "More Than a Dime's Worth: Using State Party Platforms to Assess the Degree of American Party Polarization," *PS: Political Science and Politics* 44 (2001): 331-37.
10. Robert D. Putnam, "Bowling Alone: America's Declining Social Capital," *Journal of Democracy* 6 (1995): 65-78.
11. Joshua Green, "The Other War Room," *Washington Monthly* (April 2002), accessed May 15, 2013, http://www.washingtonmonthly.com/features/2001/0204.green.html.
12. See, for example, Tom Ripley, "Where Crashes Occur," Studio One Networks, accessed December 14, 2010, http://www.drivingtoday.com/wfor/features/archive/crashes/index.html.
13. Huff, *How to Lie,* 123-37.
14. Neil Gross, "The Indoctrination Myth," *New York Times,* March 3, 2012, accessed March 5, 2012, http://www.nytimes.com/2012/03/04/opinion/sunday/college-doesnt-make-you-liberal.html.
15. Gross.

Chapter 2

1. Alan Schwarz, *The Numbers Game: Baseball's Lifelong Fascination with Statistics* (New York: Thomas Dunne Books, 2004).
2. Theodor Geisel, *Dr. Seuss's Sleep Book* (New York: Random House, 1962).
3. Census History Staff, "History," U.S. Census Bureau, June 8, 2011, accessed June 10, 2011, http://www.census.gov/history/.
4. Bureau of the Census, *Heads of Families at the First Census of the United States Taken in the Year 1790: Vermont* (Washington, D.C.: Government Printing Office, 1907), accessed June 10, 2011, http://www.census.gov/prod/www/abs/decennial/1790.html.
5. Lewis Carroll, *Through the Looking Glass, and What Alice Found There* (Ann Arbor, Michigan: Macmillan, [1880] 1966), 124.
6. Jacobellis v. Ohio, 378 U.S. 184 (1964).
7. Miller v. California, 415 U.S. 15 (1974).
8. Scott Hensley, "Study Linking Childhood Vaccine and Autism Was Fraudulent," All Things Considered, National Public Radio, January 6,

2011, accessed June 10, 2011, http://www.npr.org/blogs/health/2011/01/06/132703314/study-linking-childhood-vaccine-and-autism-was-fraudulent.
9. C.f. Michael Coppedge, John Gerring, David Altman, Michael Bernhard, Steven Fish, Allen Hicken, Matthew Kroenig, Staffan I. Lindberg, Kelly McMann, Pamela Paxton, Holli A. Semetko, Svend-Erik Skaaning, Jeffrey Staton, and Jan Teorell, "Conceptualizing and Measuring Democracy: A New Approach," *Perspectives on Politics* 9 (2011): 247–67.
10. Bureau of the Census, *Heads of Families,* 8.
11. Barbara Kiviat, "Should the Census Be Asking People if They Are Negro?" *Time,* January 23, 2010, accessed June 13, 2011, http://www.time.com/time/nation/article/0,8599,1955923,00.html.
12. U.S. Bureau of Labor Statistics, "How the Government Measures Unemployment," February 2009, accessed June 10, 2011, http://www.bls.gov/cps/cps_htgm.pdf.
13. Ann Marie Ryan, Mark J. Schmit, Diane L. Daum, Stephane Brutus, Sheila A. McCormick, and Michell Haff Brodke, "Workplace Integrity: Differences in Perceptions of Behaviors and Situational Factors," *Journal of Business and Psychology* 12 (1997): 67–83.
14. Shankar Vedantam, "Walking Santa, Talking Christ: Why Do Americans Claim to Be More Religious Than They Are?" *Slate,* December 22, 2010, accessed June 10, 2011, http://www.slate.com/id/2278923/.
15. J. Quin Monson, Kelly D. Patterson, and Jeremy C. Pope, "The Campaign Context for Partisanship Stability," in *State of the Parties,* ed. Daniel J. Coffey and John C. Green (Lanham, Maryland: Rowman and Littlefield, 2011), 271–88.
16. Mark Memmott, "Job Growth Beats Forecasts; Unemployment Rate Is 7.9 Percent," National Public Radio, November 2, 2012, accessed May 24, 2013, http://www.npr.org/blogs/thetwo-way/2012/11/02/164160789/october-unemployment-rate#0930.

Chapter 3

1. W. P. Kinsella, *Shoeless Joe* (Boston: Houghton Mifflin, 1982).
2. *Field of Dreams,* directed by Phil Robinson (Universal City, California: Universal Pictures, 1989).
3. "Career Leaders for Slugging Average," *Baseball Almanac,* accessed January 31, 2011, http://www.baseball-almanac.com/hitting/hislug1.shtml.
4. "Joe Jackson," *The Baseball Page,* accessed January 27, 2012, http://www.thebaseballpage.com/players/jacksjo01.php.
5. "George Carlin Quotes," *Goodreads,* accessed November 24, 2012, http://www.goodreads.com/quotes/43852-think-of-how-stupid-the-average-person-is-and-realize.
6. Jonathon Fahey, "US Boom Transforming Global Oil Trade," *Associated Press,* May 14, 2013, accessed May 24, 2013, http://news.yahoo.com/us-boom-transforming-global-oil-211450125.html.
7. "Data in Gapminder World," *Gapminder for a Fact-Based Worldview,* accessed June 18, 2012, http://www.gapminder.org/data/.
8. World Bank, *Millennium Development Goals,* 2012, accessed January 27, 2012, http://data.worldbank.org/data-catalog/millennium-development-indicators.

Chapter 4

1. *Moneyball,* directed by Bennett Miller (Los Angeles: Columbia Pictures, 2011).
2. David Segal, "Is Law School a Losing Game?" *The New York Times,* January 8, 2011, accessed February 7, 2011, http://www.nytimes.com/2011/01/09/business/09law.html?_r=1&scp=2&sq=law%20school&st=cse.
3. Chris Moody, "Summer Blitz: GOP Outside Groups Spending More than $19 Million on Anti-Obama Ads This Week," *Yahoo News,* June 20, 2012, accessed June 21, 2012, http://news.yahoo.com/blogs/ticket/summer-blitz-gop-outside-groups-spending-more-19-164330644.html.
4. Dylan Stableford, "U.S. Census: Minority Babies Now Majority, Surpassing Whites for First Time," *Yahoo News,* May 17, 2012, accessed June 26, 2012, http://news.yahoo.com/blogs/lookout/u-census-minority-babies-now-majority-surpassing-whites-144319476.html.
5. Valerie Strauss, "U.S. Scores on International Test Lowered by Sampling Error: Report," *The Washington Post,* January 15, 2013, accessed July

3, 2013, http://www.washingtonpost.com/blogs/answer-sheet/wp/2013/01/15/u-s-scores-on-international-test-lowered-by-sampling-error-report/.

Chapter 5

1. *21*, directed by Robert Luketic (Culver City, California: Sony Pictures, 2008).
2. "Sports Betting: Billy Walters," *60 Minutes*, January 16, 2011, accessed July 26, 2012, http://www.cbsnews.com/video/watch/?id=7253011n.
3. Kevin Poulson, "Interpol Issues 'Red Notice' for Arrest of WikiLeaks' Julian Assange," *Wired*, November 30, 2010, accessed June 3, 2013, http://www.wired.com/threatlevel/2010/11/assange-interpol/.
4. Federal Bureau of Investigation, "Table 8. Offenses Known to Law Enforcement by State by City, 2010," *Uniform Crime Report*, 2012, accessed July 27, 2012, http://www.fbi.gov/about-us/cjis/ucr/crime-in-the-u.s/2010/crime-in-the-u.s.-2010/tables/10tbl08.xls/view.
5. "Law School Rankings by Median Salary," *Internet Legal Research Group*, 2009, accessed June 3, 2013, http://www.ilrg.com/rankings/law/median.php/1/asc/LawSchool.

Chapter 6

1. Andy Bull, "Ye Shiwen's World Record Olympic Swim 'Disturbing,' Says Top US Coach," *The Guardian*, July 30, 2012, accessed July 31, 2012, http://www.guardian.co.uk/sport/2012/jul/30/ye-shiwen-world-record-olympics-2012.
2. *A Civil Action*, directed by Steven Zaillian (Burbank, California: Touchstone Pictures, 1998).
3. Jonathan Harr, *A Civil Action* (New York: Vintage, 1996).
4. "Chinese Olympic Swimmer Ye Shiwen Denies Doping," *BBC News*, July 31, 2012, accessed July 31, 2012, http://www.bbc.co.uk/news/world-asia-19058712.
5. Christie Aschwanden, "The Change in Mammogram Guidelines," *Los Angeles Times*, March 7, 2011, accessed August 3, 2012, http://articles.latimes.com/print/2011/mar/07/health/la-he-breast-cancer-mammography-20110307.
6. "American Cancer Society Responds to Changes to USPSTF Mammography Guidelines," *American Cancer Society*, November 16, 2009, accessed August 3, 2012, http://pressroom.cancer.org/index.php?s=43&item=201.
7. Nick Stango, "Top 25 Nate Silver Facts," *Gizmodo*, November 7, 2012, accessed November 24, 2012, http://gizmodo.com/5958549/top-25-nate-silver-facts.
8. David Salsburg, *The Lady Tasting Tea: How Statistics Revolutionized Science in the Twentieth Century* (New York: W. H. Freeman, 2001).
9. Marjorie Connelly and Megan Thee-Brenan, "Gallup Identifies Factors That Led to Its G.O.P. Skew in November," *The New York Times*, June 4, 2013, accessed June 5, 2013, http://thecaucus.blogs.nytimes.com/2013/06/04/gallup-identifies-factors-that-led-to-its-g-o-p-skew-in-november/?emc=tnt&tntemail0=y.
10. "2009 Job Patterns for Minorities and Women in Private Industry," *2009 EEO-1 National Aggregate Report*, 2009, accessed February 28, 2011, http://www1.eeoc.gov/eeoc/statistics/employment/jobpat-eeo1/2009/index.cfm.
11. Nathan Vardi, "America's Most Affluent Neighborhoods," *Forbes*, January 18, 2011, accessed January 20, 2011, http://www.forbes.com/2011/01/18/americas-most-affluent-communities-business-beltway.html.
12. National Center for Education Statistics, "NAEP Data Explorer," *Institute of Education Sciences*, accessed June 6, 2013, http://nces.ed.gov/nationsreportcard/naepdata/.
13. Timothy Bates, "Driving While Black in Suburban Detroit," *Du Bois Review* 7 (2010): 133–50.
14. Frank Newport, Jeffrey M. Jones, and Lydia Saad, "Final Presidential Estimate: Obama 55%, McCain 44%," *Gallup*, November 2, 2008, accessed August 10, 2012, http://www.gallup.com/poll/111703/final-presidential-estimate-obama-55-mccain-44.aspx.
15. Vince Blaser, "African First Ladies Convene in L.A. to Tackle Health Issues," *The Global Health Blog*, April 21, 2009, accessed August 4, 2012, http://theglobalhealthblog.blogspot.com/2009/04/african-first-ladies-convene-in-la-to.html.
16. Jeffrey M. Jones, "Catholics' Approval of Obama Little Changed," *Gallup Politics*, February 14, 2012, accessed February 25, 2012, http://www.gallup.com/poll/152636/Catholics-Approval-Obama-Little-Changed.aspx.

Chapter 7

1. Baker Library Historical Collections, "The Human Relations Movement: Harvard Business School and the Hawthorne Experiments," *Harvard Business School*, 2012, accessed October 6, 2012, http://www.library.hbs.edu/hc/hawthorne/.
2. Donald McNeil, Jr., "4 Germs Cause Most of Infants' Severe Diarrhea," *New York Times*, May 20, 2013, accessed June 7, 2013, http://www.nytimes.com/2013/05/21/health/4-germs-cause-most-of-infants-severe-diarrhea.html.
3. Richard Rhodes, *Dark Sun: The Making of the Hydrogen Bomb* (New York: Simon and Schuster, 1995).
4. American National Election Study, ANES 2008 Time Series Study, accessed October 6, 2012, http://www.electionstudies.org/studypages/2008prepost/2008prepost.

Chapter 8

1. Michael Nunez, "Obama Hosts Google Hangout, 'Enhanced' State of the Union," *International Business Times*, January 23, 2012, accessed August 20, 2012, http://www.ibtimes.com/articles/286199/20120123/obama-hosts-google-hangout-enhanced-state-union.htm.
2. Barack Obama, "Remarks by the President in State of the Union Address," *The White House*, January 24, 2012, accessed August 20, 2012, http://www.whitehouse.gov/the-press-office/2012/01/24/remarks-president-state-union-address.
3. The White House, "Blueprint for an America Built to Last," *Slideshare*, 2012, slide 42, accessed August 20, 2012, http://www.slideshare.net/whitehouse/state-of-the-union-enhanced-graphics.
4. The White House, "Blueprint for an America Built to Last," slide 13.
5. Tamara Keith, "Select Senators Stall Budget Process," *National Public Radio*, May 30, 2013, accessed June 10, 2013, http://www.npr.org/2013/05/30/187227970/select-senators-stall-budget-process.
6. "List of Legislatures by Country," *Wikipedia, The Free Encyclopedia*, October 9, 2012, accessed October 9, 2012, http://en.wikipedia.org/wiki/List_of_national_legislatures.
7. Monty G. Marshall and Benjamin R. Cole, *Polity IV Project*, 2012, accessed August 11, 2012, http://www.systemicpeace.org/polity/polity4.htm.

Chapter 9

1. "Math Model Helps Predict Olympic Medal Winners," *NPR*, August 13, 2012, accessed November 29, 2012, http://www.npr.org/2012/08/13/158679182/the-last-word-in-business.
2. Center for American Women and Politics, "The Gender Gap: Voting Choices in Presidential Elections," Eagleton Institute of Politics, Rutgers University, 2008, accessed March 22, 2012, http://www.cawp.rutgers.edu/fast_facts/voters/documents/GGPresVote.pdf.
3. *The Terminal*, directed by Steven Spielberg (Universal City, California: DreamWorks, 2004).
4. *Forrest Gump*, directed by Robert Zemeckis (Hollywood, California: Paramount Pictures, 1994).
5. Eric Lincoln, "Old Plans Revived for Category 5 Hurricane Protection," *Riverside* U.S. Army Corps of Engineers, New Orleans District 16 (2004): 5, accessed July 5, 2012, http://www.mvn.usace.army.mil/pao/Riverside/Sept-Oct_04_Riv.pdf.
6. Lindsay M. Howden and Julie A. Meyer, "Age and Sex Composition: 2010," *U.S. Census Bureau*, 2011, accessed February 6, 2012, http://www.census.gov/prod/cen2010/briefs/c2010br-03.pdf.
7. John McQuaid and Mark Schleifstein, "Washing Away," *The Times-Picayune*, June 23-27, 2002, accessed July 5, 2012, http://www.nola.com/hurricane/content.ssf?/washingaway/index.html.
8. Bill Chappell, "U.S. Military's Suicide Rate Surpassed Combat Deaths in 2012," *National Public Radio*, January 14, 2013, accessed June 11, 2013, http://www.npr.org/blogs/thetwo-way/2013/01/14/169364733/u-s-militarys-suicide-rate-surpassed-combat-deaths-in-2012.
9. Freedom House, *Freedom in the World 2012*, accessed October 11, 2012, http://freedomhouse.org/report/freedom-world/freedom-world-2012.
10. "State Religion," *Wikipedia*, 2012, accessed October 9, 2012, http://en.wikipedia.org/wiki/State_religion.

Chapter 10

1. Abu Dhabi Gallup Center, *Muslims in India: Confident in Democracy Despite Economic and Educational Challenges,* November 2011, accessed June 12, 2013, http://www.gallup.com/strategic consulting/153635/BRIEF-ENGLISH-Muslims-India-Confident-Democracy-Despite-Economic-Educational-Cha.aspx.
2. Pew Research Center for the People and the Press, "Majority Views NSA Phone Tracking as Acceptable Anti-Terror Tactic: Public Says Investigate Terrorism, Even If It Intrudes on Privacy," June 10, 2013, accessed June 11, 2013, http://www.people-press.org/2013/06/10/majority-views-nsa-phone-tracking-as-acceptable-anti-terror-tactic/1/.
3. Gregory Acs, Kenneth Braswell, Elaine Sorensen, and Margery Austin Turner, "The Moynihan Report Revisited," *Urban Institute,* June 2013, accessed June 14, 2013, http://www.urban.org/publications/412839.html.

Chapter 11

1. Ryan Murphy, "10 Most Superstitious Athletes," *Men's Fitness* 2012, accessed October 18, 2012, http://www.mensfitness.com/leisure/sports/10-most-superstitious-athletes.
2. Darrell Huff, *How to Lie with Statistics* (New York: Norton, [1954] 1982).
3. Daryl J. Bem, "Feeling the Future: Experimental Evidence for Anomalous Retroactive Influences on Cognition and Affect," *The Journal of Personality and Social Psychology* 100 (2001): 407–25.
4. Benjamin Radcliffe, "Controversial ESP Study Fails Yet Again," *Discovery News,* September 12, 2012, accessed October 23, 2012, http://news.discovery.com/human/controversial-esp-study-fails-yet-again-120912.html.
5. Liz Neporent, "Walking Speed Predicts Who Will Live Longer," ABC News, January 5, 2011, accessed October 23, 2012, http://abcnews.go.com/Health/Wellness/walking-speed-predicts-longevity-elderly/story?id=12539377#.UcGYoZyE5cU.
6. Sam Hananel, "Fact Check: Are Federal Employees Overpaid?" *Star Tribune,* April 7, 2011, accessed October 18, 2012, http://www.startribune.com/printarticle/?id=119385569.
7. Global Public Square, "Global Poverty is Falling, So What's the Problem?" *CNN,* May 1, 2013, accessed June 15, 2013, http://globalpublicsquare.blogs.cnn.com/2013/05/01/global-poverty-is-falling-so-whats-the-problem/?hpt=hp_bn2.
8. "Poverty and Equity Data," *The World Bank* 2013, accessed June 15, 2013, http://povertydata.worldbank.org/poverty/home/.
9. Hananel, "Fact Check."

Chapter 12

1. Steven D. Levitt and Stephen J. Dubner, *Freakonomics: A Rogue Economist Explores the Hidden Side of Everything* (New York: HarperCollins, 2009).
2. Caroline Lambert, "French Women in Politics: The Long Road to Parity," *Brookings,* May 1, 2001, accessed June 19, 2013, http://www.brookings.edu/research/articles/2001/05/france-lambert.
3. Transparency International, "The Report," *Corruption Perceptions Index 2011: In Detail.* 2011, accessed November 20, 2012, http://cpi.transparency.org/cpi2011/in_detail/.
4. Monty G. Marshall and Benjamin R. Cole, *Polity IV Project* 2012, accessed August 11, 2012, http://www.systemicpeace.org/polity/polity4.htm.

Chapter 13

1. Freedom House, *Freedom in the World 2012,* accessed October 11, 2012, http://freedomhouse.org/report/freedom-world/freedom-world-2012.
2. Fund for Peace, *The Failed States Index 2012: The Indicators* 2012, accessed November 17, 2012, http://ffp.statesindex.org/indicators.
3. Jon Terbush, "Chicago's Murder Rate isn't Nearly as Bad as You Think," *The Week,* June 11, 2013, accessed June 19, 2013, http://theweek.com/article/index/245447/chicagos-murder-rate-isnt-nearly-as-bad-as-you-think.

4. Monty G. Marshall and Benjamin R. Cole, *Polity IV Project* 2012, accessed August 11, 2012, http://www.systemicpeace.org/polity/polity4.htm.
5. Transparency International, "The Report," *Corruption Perceptions Index 2011: In Detail* 2011, accessed November 20, 2012, http://cpi.transparency.org/cpi2011/in_detail/.
6. "GDP per Capita PPP," *GapMinder* 2011, accessed November 17, 2012, http://www.gapminder.org/data/.
7. CIA, "Literacy," *The World Factbook* 2012, accessed November 17, 2012, https://www.cia.gov/library/publications/the-world-factbook/fields/2103.html.
8. "Income Share of Richest 10%," *GapMinder* 2012, accessed November 17, 2012, http://www.gapminder.org/data/.
9. Freedom House, *Freedom in the World 2012*.
10. Politico, "2012 Election Central," 2012, accessed November 19, 2012, http://www.politico.com/2012-election/map/#/President/2012/.
11. U.S. Census Bureau, *The 2012 Statistical Abstract: The National Data Book* 2012, accessed November 19, 2012, http://www.census.gov/compendia/statab/cats/population.html.

Chapter 14

1. "Dealing with Doubt," *Radiolab,* March 26, 2013, accessed June 21, 2013, http://www.radiolab.org/2013/mar/26/dealing-doubt/.
2. Jim Rutenberg, "Data You Can Believe in," *The New York Times,* June 20, 2013, accessed June 21, 2013, http://www.nytimes.com/2013/06/23/magazine/the-obama-campaigns-digital-masterminds-cash-in.html?pagewanted=1&tntemail0=y&_r=1&emc=tnt.
3. Elizabeth Mendes and Joy Wilke, "Americans' Confidence in Congress Falls to Lowest on Record," *Gallup Politics,* June 13, 2013, accessed July 16, 2013, http://www.gallup.com/poll/163052/americans-confidence-congress-falls-lowest-record.aspx.
4. Scott Adams, *Dilbert,* January 3, 1991, accessed June 20, 2013, http://www.dilbert.com/strips/comic/1991-01-03/.

Glossary

Absolute zero: When a variable cannot have a negative value because zero means the total absence of the characteristic.

Aggregation bias: When a relationship found in aggregate data does not reflect a relationship in individual-level data.

Analysis of variance (ANOVA): A statistical test used when the dependent variable is on a continuous scale (it has an interval level of measurement) and the independent variable has a limited number of categories.

Antecedent variable: A variable that comes before both the independent and dependent variables in the causal chain.

Associational (correlational) validity: The claim of validity based on the measure being correlated with other measures that are used to operationalize a particular concept.

Betas: The standardized coefficients in a regression.

Bivariate statistics: Statistics that analyze the relationship between two variables.

Box-and-whiskers plot: A graphic presentation of the five-point summary of data.

Case: The individuals (person, group, country, etc.) that are being identified in order to study their attributes; also identified as unit of analysis.

Chi-square statistic: Measure of statistical significance that compares observed frequency to expected value of cells in a contingency table.

Coding: The process of translating information into numbers.

Coding sheet: A form for entering data as they are collected.

Coefficient: The constants in a regression equation.

Collinearity: When two or more independent variables are so highly correlated that regression is unable to distinguish the effects of each controlling for the other.

Conceptual definition: The meaning of a term in theoretical language.

Concordant pair: Two cases that give evidence of a positive relationship.

Consensual validity: The claim of validity based on expert agreement.

Content analysis: The systematic study of human communication, especially texts.

Contingent: When the probability of an event is dependent on the outcome of another event.

Cramer's V: Measure of association appropriate for a non-ordered relationship between categorical variables.

Cross sectional surveys: The same instrument used at different points in time to question different groups of people to see how public opinion has changed.

Data cleaning: The process of correcting errors made during data entry.

Dependent variable: The effect to be explained.

Descriptive statistics: The numerical characteristics of a full population.

Dichotomous variable: A variable that has only two possible values.

Discordant pairs: Two cases that give evidence of a negative relationship.

Dummy variable: A dichotomous variable that has been coded so that those cases with an attribute are coded "1" and those without it are coded "0."

Ecological fallacy: When you incorrectly infer a relationship exists at the individual level because it exists at an aggregate level.

Empirical: Capable of verification by experiment or observation of the physical world.

Eta: A measure of association found through ANOVA.

Eta-squared: A PRE measure of association found through ANOVA; identifies the proportion of variance explained by the independent variable.

Exhaustive: When listing the alternative categories of an attribute, each case must belong in a category.

Expected value: The number of cases that should be in a cell if the two variables are independent.

Experiment: A controlled test in which the purported cause is varied in order to measure the resulting change in the effect.

Exponential relationship: A relationship that curves upward at ever-increasing amounts, looking like a "J."

Face validity: The claim of validity based on logical evaluation.

Falsifiable: A statement that is not merely true by definition but subject to evidence that could confirm or disconfirm it.

Figure: A pictorial presentation of information.

Frequency: A tabulation of the number of cases in each of the categories of a variable.

F-statistic: Measures how different the observed relationship is from the null hypothesis; the F-statistic is used to conduct an F-test, which will yield a measure of statistical significance.

Gamma: A measure of association appropriate for ordinal-level variables.

Gauss-Markov assumptions: The assumptions about the data that are necessary for OLS to give the best linear unbiased estimate of a relationship.

Hawthorne effect: The finding that the process of being studied affects behavior.

Heteroscedastic: When distribution of the residuals varies across the values of the dependent variable, looking like a funnel cloud.

Homoscedastic: When the residuals are distributed in a constant pattern around the values of the dependent variable.

Hypothesis: The statement of the relationship between the dependent and independent variables.

Independent: When the probabilities of two events are unrelated.

Independent variable: The hypothesized cause of the effect to be explained.

Inferential statistics: Analysis that allows you to generalize from a sample to a population.

Indicator: A characteristic that gets at one aspect of a concept.

Interquartile range: The range within which the middle two-fourths of the cases lie.

Intersection: When all of a number of events occur concurrently.

Interval: Data that have both order and a uniformly sized scale.

Intervening variable: A variable that comes between the independent and dependent variables in the causal chain.

Kendall's tau: A measure of association for ordinal-level data.

Lambda: A measure of association appropriate for nominal-level variables.

Level of measurement: The amount of information contained in the quantification of a variable in terms of categorization, order, and scale.

Linear relationship: A relationship that, when graphed, approximates a straight line.

Log linear relationship: When a relationship increases at decreasing amounts, looking like a lower case "r."

Marginals: The total percent and frequency for each row or column in a cross tabulation.

Mean: The arithmetic average.

Median: The middle value when a set of data is arranged in order.

Measure of central tendency: The three statistics that measure the average value of a variable.

Mode: The most frequently observed value.

Model: A word picture of concepts connected by arrows indicating the direction of relationships.

Multiple regression: Estimate of a linear relationship controlling simultaneously for more than one independent variable.

Mutually exclusive: When no one case can belong in more than one alternative category of an attribute.

Negative relationship: When increases in the independent variable lead to decreases in the dependent variable; also known as an inverse relationship.

Nominal: Data that have no order or scale, only categories.

Null hypothesis: The statement that there is no relationship between the dependent and independent variables.

Observation: The act of noting and recording phenomena in a systematic way.

One-tailed: A range that excludes one tail of the distribution.

Operational definition: The way in which a concept is measured.

Ordinal: Data that have order, but not a uniform scale.

OLS: Ordinary least squares regression.

Outlier: A case that has a value far from the others in the distribution.

Panel study: The same instrument used at different points in time to question the same group of people to see how their views have changed.

Parameter: Statistics based on measures of a population.

Pearson's r: A measure of association appropriate for interval-level variables; also called the correlation.

Population: The entire group of cases being studied.

Positive relationship: When increases in the independent variable lead to increases in the dependent variable; also known as a direct relationship.

Predictive validity: The claim of validity based on its ability to predict an event the concept is expected to cause.

PRE measures: Measures of association measuring the predictive value of the independent variable; stands for Proportional Reduction in Error.

Public record: Any information gathered and maintained by a governmental body that is openly available.

P values: Another name for measures of statistical significance, as a reminder that they are probabilities.

Qualitative data: Information that is collected as descriptions and explanations rather than as numbers.

Quantitative data: Information that is collected and analyzed as numbers.

Range: The number given by subtracting the lowest value of an attribute from the highest.

Ratio: Data that have order, a uniformly sized scale, and an absolute zero.

Reify: To treat an abstraction as if it is real.

Reliability: The degree to which a measure yields consistent results.

Residual: The difference between the value of Y and its estimate $\hat{Y}$.

Robust: When a statistical measure is not overly affected by minor violations of its assumptions.

R-square: A PRE measure of association found through regression; identifies the proportion of variance explained by the regression model.

Sample: A subset of cases from which generalizations are made.

Scientific method: A method of research designed to answer a question about how the world works by stating a hypothesis that is then tested by empirical evidence.

Specification: The independent variables included in a model.

Spurious relationship: When two variables appear to be correlated, but the causal relationship actually lies with a third variable with which both variables are correlated.

Squared relationship: When data are curved in either a valley or a mountain shape.

Standard deviation: The square root of the average squared distance from the mean.

Standard error of the mean: The standard deviation of sample estimates around the population mean.

Standardized value: When a raw score for a case is translated into the number of standard deviations that case is from the mean.

Statistical significance: When it is very unlikely that observed differences are due to chance; measures of statistical significance are sometimes called "p values."

Substantive significance: The strength of the relationship between the dependent and independent variables.

Survey: A questionnaire used to collect the self-reported attitudes of people.

Symmetrical: When the shape of one side of a distribution is a mirror image of the other side.

Symmetrical measure of association: The results do not depend on which variable is identified as the dependent variable.

Table: Numbers or words presented in rows and columns.

Theory: A story of how the world works; used to explain why some phenomenon occurred the way it did.

Two-tailed: A range that crosses the mean but does not include the two tails of the distribution.

Type I error: A false positive finding.

Type II error: A false negative finding.

Unimodal: Having a single mode or peak.

Unit of analysis: The individuals (person, group, country, etc.) that are being identified in order to study their attributes; also identified as a case.

Univariate statistics: Statistics that analyze a single variable.

Validity: The degree to which a measure corresponds with its concept.

Variable: A characteristic that, when it is measured for different cases, varies.

Variance: The average squared distance from the mean.

z-score: The number of standard deviations a case is from the mean.

Index

A
Abortion
   legalized abortion–political ideology correlation, 317 (table)
   legalized abortion–race correlation, 269 (table), 269–272, 270 (table), 271 (table)
   religious fundamentalism–abortion–race correlation, 324 (table), 324–328, 325 (table), 325–328, 326 (table), 330–332, 331 (table)
   religious fundamentalism–abortion–rape correlation, 333–338, 334 (figure), 335 (figure), 336 (figure), 337 (figure), 338–339 (table)
   teen abortion rate–birth rate correlation, 359 (table), 359–363, 360 (table), 362 (table)
Absolute zero, 23
Abu Dhabi Gallup Center, 453
A Civil Action (film), 145, 451
Acs, Gregory, 453
Adams, Scott, 454
Adjusted R-squared, 394
African nation study, 149 (table), 149–151, 152 (table), 155–157, 176 (table)
Aggregation bias, 234
Altman, David, 450
American Cancer Society, 147
"American Cancer Society Responds to Changes to USPSTF Mammography Guidelines," 451
American Journal of Political Science, 16
American National Election Study (ANES), 3, 207, 261 (table), 317 (table), 383, 383 (table), 413 (table), 414 (table), 452
Analysis of variance (ANOVA)
   analytical purpose, 425 (table), 426
   Apply It Yourself assignments, 208, 210–211
   basic concepts, 185–188, 186 (figure), 187 (figure), 188 (table), 189 (table), 190–193, 193 (table), 195–196
   levels of measurement, 276 (table)
   political examples, 196–197 (table), 196–197, 198 (table), 199, 202–208, 203 (figure), 204 (figure), 205–206 (figure)
   professional writing tips, 432–433
   Statistical Package for the Social Sciences (SPSS), 199–208, 200 (box), 201–202 (box)
   use guidelines, 440
   Your Turn exercises, 208, 209 (table)
Antecedent variables, 321, 321 (figure)
Apply It Yourself assignments
   basic statistics, 16
   bivariate regression, 374–376
   chi-square statistic, 272
   continuous probability, 143–144
   Cramér's V, 272
   hypothesis testing, 208, 210–211
   means testing, 177–178
   measurement, 49–50
   measures of association, 317–318
   measures of central tendency, 74–75
   measures of dispersion, 108–110
   multiple regression analysis, 414–416
   multivariate relationships, 340–342
   visual presentations, 243–244
Arab League Internet usage–state fragility correlation, 303–304, 304 (table), 305 (table), 306, 307 (table), 308 (table), 314 (figure)
Artificial normal distributions, 113
Aschwanden, Christie, 451
Assange, Julian, 120 (box)
Assembly line production, 179
Associational (correlational) validity, 26
Attribute categories, 29–30
Aurora, Colorado, 143
Autism–vaccination correlation, 21
Averages
   see Measures of central tendency

B

Baker Library Historical Collections, 452
Baseball statistics
   batting records, 51, 52 (table), 53
   measures of central tendency, 51, 52 (table), 53
   measures of dispersion, 76, 77 (table)
   normal curves, 112
   salary distribution, 76, 77 (table)
   statistical analysis, 17
Basic statistics
   Apply It Yourself assignments, 16
   barriers, 9–11
   equation-centered formats, 10
   importance, 11
   introductory statistics, 7–9
   political examples, 13–15
   statistical packages, 11–12
   Your Turn exercises, 15–16
Bates, Timothy, 168, 451
Batting records, 51, 52 (table), 53
BBC News, 451
Bell curves, 113
   see also Normal curves
Bem, Daryl J., 319, 453
Bernhard, Michael, 450
"Best Law Schools," 78 (table), 80 (table), 106
Best Linear Unbiased Estimate (BLUE), 381–382, 395, 399
Betas, 384
Between-group sum of squares (BSS), 186–189, 192, 195–197, 199
Bill and Melinda Gates Foundation, 181 (box)
Birth rate–teen abortion rate correlation, 359 (table), 359–363, 360 (table), 362 (table)
Bivariate regression
   Apply It Yourself assignments, 374–376
   dichotomous independent variables, 356 (table), 356–357
   fitting a line, 346–350, 347 (figure), 349 (table), 357–358
   income–gender correlation example, 356 (table), 356–357
   levels of measurement, 276 (table)
   murder rate example, 354–356, 355 (figure), 355 (table)
   ordinary least squares (OLS), 346–350, 428–429
   relationship representations, 344 (figure), 344–346, 345 (figure)
   R-square, 353, 359
   state government expenditures by population example, 345 (figure), 346–353, 349 (table), 351 (table), 366–371, 368 (figure), 369 (figure), 370 (figure), 371 (table)
   Statistical Package for the Social Sciences (SPSS), 363 (box), 363–370, 364 (box), 366 (box), 367–368 (figure), 369 (figure), 370 (figure), 371 (table)
   statistical significance, 350–353, 351 (table), 358–359
   teen abortion rate–birth rate correlation example, 359 (table), 359–363, 360 (table), 362 (table)
   time series data, 354–356, 355 (figure), 355 (table)
   Your Turn exercises, 371 (table), 371–374, 373 (table), 374 (table)
   see also Regression analysis
Bivariate scatter plots, 402 (box), 402–403, 406 (figure)
Bivariate statistics, 7
Blaser, Vince, 451
Board of Education, Brown v. (1954), 5
Box-and-whiskers plot, 80, 81, 82 (figure), 92 (figure)
Bradley Effect, 27
Bradley, Tom, 27
Braswell, Kenneth, 453
Brodke, Michell Haff, 450
Broockman, David E., 4, 449
Brown v. Board of Education (1954), 5
Brutus, Stephane, 450
Budget discussion and approval, 229 (box)
Bull, Andy, 451
Bureau of the Census
   see U.S. Census Bureau
Bush, George W.
   election prediction polls, 162–163
   public opinion polls, 6 (box)
Business Software Alliance, 196–197 (table), 198 (table)
Butler, Daniel M., 4, 449

C

California, Miller v. (1974), 449
Card counting, 111
"Career Leaders for Slugging Average," 450
Carroll, Lewis, 449

Carter, Jimmy, 106
Cases, 18–19, 438
Casinos, 111
Catholicism, 333–338
Cause-and-effect relationships, 9, 319–320
Census History Staff, 449
Center for American Women and Politics, 452
Center for Systematic Peace, 3, 304 (table),
    316 (table), 379 (table), 380 (table),
    395 (table), 396 (table)
Centers for Disease Control and Prevention
    (CDC), 2, 400, 400 (table)
Central Intelligence Agency (CIA)
    see CIA World Factbook
Central tendency
    see Measures of central tendency
Chadwick, Henry, 17
Chance, 319
Chappell, Bill, 452
Chicago, Illinois, 393 (box)
"Chinese Olympic Swimmer Ye Shiwen Denies
    Doping," 451
Chi-square statistic
    analytical purpose, 424, 425 (table)
    calculations, 252–258, 256 (table), 257 (table),
        258 (box), 260
    chi-square table, 447–448
    definition, 245–246
    political examples, 261 (table), 261–263,
        262 (table)
    Statistical Package for the Social Sciences
        (SPSS), 264 (box), 264–268, 266 (figure)
Christie, Chris, 366
CIA World Factbook, 379 (table), 380 (table),
    389 (table), 390 (table), 395 (table),
    396 (table), 454
Civil Action, A (film), 145, 451
Clark, Kenneth and Mamie, 5
Clinton, Bill, 6 (box)
Coded data, 29–30
Coding, 29
Coding sheets, 29–30, 41 (table), 42 (table)
Coefficients, 346
Coffey, Daniel J., 5, 6, 449, 450
Cole, Benjamin R., 176 (table), 304 (table),
    316 (table), 379 (table), 380 (table),
    395 (table), 396 (table), 452, 453, 454
Collapsed data, 225 (box), 225–226
Collinearity, 394–397, 395 (table)

Conceptual definition, 20–21, 24–29
Concordant pairs, 287 (table), 287–291,
    288 (table), 289 (table), 291–292 (table),
    294–295 (table), 296, 306–308,
    308 (table), 426
Confidence intervals, 159–163, 160 (figure),
    161 (figure), 169–170
Congressional district reapportionment,
    373 (table)
Congressional district votes, 127 (table),
    127–132, 130 (figure), 131 (figure),
    132 (figure), 142 (table)
Congressional Quarterly, 127 (table), 142 (table)
Connelly, Marjorie, 451
Consensual validity, 25–26
Content analysis, 5
Context analysis, 7–8
Contingency tables
    background information, 220–221
    collapsed data, 225 (box), 225–228,
        227 (table)
    data presentation, 213, 231–232
    pattern analysis, 224–225, 422–423
    political examples, 226–228, 227 (table),
        232–235, 233 (table), 234 (table)
    professional writing tips, 433–434
    relationship representations, 286 (table),
        286–287, 287 (table), 288 (table)
    setup and creation, 221 (table), 221–223,
        222 (table), 223 (table), 224 (box)
    Statistical Package for the Social Sciences
        (SPSS), 235–238, 236 (box), 237 (box),
        239–240 (figure), 241–242, 441
    three-way contingency tables, 324–332,
        325 (table), 326 (table), 328 (box),
        331 (table), 433–434
Contingent probabilities, 249–251, 250 (table),
    251 (table), 252 (table)
Continuous probability
    Apply It Yourself assignments, 143–144
    basic concepts, 111–112
    normal curves, 112–114, 113 (figure), 114 (figure)
    political examples, 126–135, 127 (table),
        130 (figure), 131 (figure), 132 (figure),
        133 (figure)
    probability estimates, 117–120, 118 (figure),
        119 (figure), 124 (figure), 124–126,
        125 (figure), 128–132, 130 (figure),
        131 (figure), 132 (figure)

Statistical Package for the Social Sciences
(SPSS), 135 (box), 135–142,
136–137 (box), 138 (figure), 139 (figure),
140 (figure), 141 (figure)
value calculations, 132–135, 133 (figure)
Your Turn exercises, 142 (table), 142–143
z-scores, 114–132, 115 (figure), 116 (figure),
117 (figure), 118 (figure), 119 (figure),
126–132, 127 (table)
Coppedge, Michael, 450
Correctional validity, 26
Correlational evidence, 145–146
Correlation matrix, 313 (figure), 314 (figure),
404, 404–405 (figure), 441
Corruption Perceptions Index 2011, 166 (table),
188 (table), 189 (table), 316 (table),
379 (table), 380 (table), 395 (table),
396 (table)
Countervailing relationships, 323, 323 (figure)
"Countries," 41 (table), 42 (table), 47 (table), 48
Cramér's V
analytical purpose, 424, 425 (table)
calculations, 258–259, 259 (box), 261
definition, 245–246
levels of measurement, 276 (table)
political examples, 261 (table), 261–263
Statistical Package for the Social Sciences
(SPSS), 264 (box), 264–268, 266 (figure)
Crime data
continuous probability, 137–141, 138 (figure),
139 (figure), 140 (figure), 141 (figure),
143–144
gun violence rates, 393 (box)
regression analysis, 354–356, 355 (figure),
355 (table)
Cross sectional surveys, 28
Cross tabulation
see Contingency tables
Cuban missile crisis, 5
"Current Knesset Members," 63 (table)

D

Dark Knight Rises, The (film), 143
Data cleaning, 36–38, 38 (box)
Data coding, 29–30
Data collection, 4–6
Data files
data cleaning, 36–37, 38 (box)
frequency tables, 38–39, 39 (box)
importing files, 34–36, 36 (box), 436–437
opening files, 35–36, 37 (figure)
"Data in Gapminder World," 450
Daum, Diane L., 450
"Dealing with Doubt," 454
Deceptive statistics, 1–2
Degrees of freedom, 155
Dependent variables
antecedent variables, 321, 321 (figure)
contingency tables, 286 (table), 422–423
hypothesis testing, 181–183, 421–422
interaction effects, 323 (figure), 323–324
intervening variables, 321–322, 322 (figure)
measures of association, 424, 426
regression analysis, 346–350
relationship correlations, 8
scientific method, 3–4
spurious non-relationships, 322–323, 323 (figure)
visual representations, 213–214, 278–281,
279–280 (figure), 281 (figure)
see also Contingency tables; Multiple
regression
Descriptive statistics, 7–8, 148
Dichotomous variables, 86–89, 97, 277,
287 (table), 287–290, 288 (table),
289 (table), 356 (table), 356–357
"Digest of Education Statistics," 93–94 (table),
96 (table)
Dilbert, 428
Direct relationship, 182
Discordant pairs, 287 (table), 287–291,
288 (table), 289 (table), 291–292 (table),
294–295 (table), 296, 306–308,
308 (table), 426
Discrete probability, 246–252
Dishonesty, 27–28
Dispersion measures
see Measures of dispersion
Disraeli, Benjamin, 1
Dogbert, 428
Domestic war–political competition relationship
chi-square statistic, 252–257, 257 (table)
contingent probabilities, 221 (table), 221–223,
222 (table), 223 (table), 249–251,
250 (table), 251 (table), 252 (table)
expected values, 256 (table)
Dominant religion–level of freedom correlation,
309 (table), 309–310
"Don't Ask, Don't Tell" (DADT) policy, 5

"Driving while black" study, 168–169
Dropout rate, 93–94 (table), 95 (table), 96 (table)
Dubner, Stephen J., 343, 453
Duke, Annie, 417
Dummy variables, 157–159, 356, 383–384, 401–402

E
Eastpoint, Michigan, 168
Ecological fallacy, 234
Economic Policy Institute, 94 (box)
Election prediction polls, 154, 162–163, 169–170
Emanuel, Rahm, 393 (box)
Empirical tests, 4, 180
Employment distribution example, 157 (table), 157–159, 159 (table)
Equal Employment Opportunity Commission (EEOC), 3, 157
Equal probability calculations, 246–247
Eriksson, Mikael, 221 (table), 222 (table), 223 (table), 250 (table), 251 (table), 252 (table), 256 (table)
Eta/eta-squared (η^2), 192–193, 201, 206 (figure), 425 (table), 426
European Union application example, 39–48, 41 (table), 42 (table), 43 (figure), 44 (figure), 45–46 (figure), 47 (figure), 47 (table)
Excel data files, 34–36, 36 (box), 37 (figure), 43 (figure), 44 (figure), 436–437
Exhaustive, 30
Expected values, 251, 253–254, 256 (table)
Experiments, 4
Exponential relationships, 387–389, 388 (figure), 389 (table), 390 (figure), 390 (table)
Extreme poverty, 322 (box)

F
Face validity, 25
Fahey, Jonathon, 450
Failed States Index, 385 (figure), 386 (figure), 387 (table)
False positive/false negative predictions, 146–147, 185
False statistics, 1–2
Falsifiable, 183
Federal Bureau of Investigation (FBI), 3, 137, 142, 355 (table), 451

Female literacy
    life expectancy correlations, 237–238
    state stability factors, 379 (table), 380 (table), 381, 395 (table), 396 (table), 411 (figure)
Female literacy–life expectancy relationship, 241 (table), 241–242
Fenno, Richard F., 4, 6, 449
Field of Dreams (film), 51, 450
Figures, 212–213
    see also Graphs
Fish, Steven, 450
Five-point summary, 80, 81, 90–91, 421
Flake, Jeff, 16
Flash Crash, 28
Fleischer, Ari, 6 (box)
Fleischman, Howard L., 65 (table), 72–73 (table)
Forbes, 159
Ford, Gerald R., 106
Foreign oil dependency, 68 (box)
Former British colonies study, 149 (table), 149–151, 152 (table), 155–157, 176 (table)
Forrest Gump (film), 247, 452
Freakonomics (Levitt and Dubner), 343, 359, 360
Freedom House, 209 (table), 230 (table), 231 (figure), 243 (table), 268, 284 (table), 309 (table), 314 (figure), 386 (figure), 387 (table), 452, 453, 454
Freedom House Political Rights Score, 230 (table), 386 (figure)
French constitutional amendment, 354 (box)
Frequency distributions, 31, 32 (box), 33
Frequency tables, 38–39, 39 (box), 46, 47 (figure), 47 (table), 61–62, 91 (table), 431–432
F-statistic, 190–191, 423, 425 (table)
Fund for Peace, 387 (table), 453

G
Gallup, George, 6 (box)
Gallup polls, 6 (box), 154, 162–163, 169–170
Gambling, 111–112
Gamma
    analytical purpose, 425 (table), 426
    Arab League Internet usage–state fragility correlation example, 307 (table), 307–308, 308 (table)
    calculations, 290–291, 291–292 (table), 292 (box), 294–295 (table), 306–308
    definition, 274

dichotomous variables, 287 (table), 287–290, 288 (table), 289 (table)
gun control–gun ownership correlation example, 289 (table), 289–290
gun control–political ideology correlation example, 292–294, 293 (table), 294–295 (table), 296
levels of measurement, 276 (table), 277–278
relationship correlations, 287
Gapminder, 69, 72, 149 (table), 152 (table), 241 (table), 242, 379 (table), 380 (table), 392 (table), 393 (table), 395 (table), 396 (table)
Gas prices, 273 (figure), 273–274, 274 (figure)
Gates Foundation, 181 (box)
Gates, Robert M., 5
Gauss-Markov assumptions
  basic principles, 381–382
  collinearity, 394–397, 395 (table)
  correctly specified models, 393–394, 427
  exponential relationships, 387–389, 388 (figure), 389 (table), 390 (figure), 390 (table)
  homoscedastic errors, 397–398, 398 (figure)
  interval-level variables, 382–384, 383 (table)
  linear relationships, 384–393, 385 (figure), 427
  log linear relationships, 390–393, 391 (figure), 392 (figure), 392 (table), 393 (table)
  means of zero, 397
  political examples, 395 (table), 395–397, 396 (table)
  squared relationships, 385 (figure), 385–386, 387 (table)
"GDP per Capita PPP," 454
Gee-Whiz Graphs, 215
Geisel, Theodor, 449
General Social Survey, 5, 269 (table), 271 (table), 289 (table), 293 (table), 298 (table), 301 (table), 302 (table), 324 (table), 325 (table), 326 (table), 331 (table), 338–339 (table), 340 (table), 356 (table), 374 (table)
"George Carlin Quotes," 450
Gerring, John, 450
Giffords, Gabrielle, 54
Gleditsch, Nils Petter, 221 (table), 222 (table), 223 (table), 250 (table), 251 (table), 252 (table), 256 (table)
Global Enteric Multicenter Study (GEMS), 181 (box)

Global Public Square, 453
Gosset, William Sealy, 154
Grade point average (GPA), 79–80, 120–121
Graphs
  political examples, 229–231, 231 (figure)
  relationship representations, 213–220, 214 (figure), 215 (figure), 216–217 (figure), 228–231, 344 (figure), 344–346, 345 (figure)
  scale, 219 (figure), 219–220, 220 (figure)
  trend patterns, 217, 218 (figure), 219, 219 (figure), 220 (figure)
  visual representations, 213 (figure)
Green, John C., 450
Green, Joshua, 449
Gross domestic product (GDP) per capita
  African nation study, 155–157
  hypothesis testing, 182–183
  life expectancy correlations, 390–393, 391 (figure), 392 (figure), 392 (table), 393 (table)
  population–sample comparisons, 149 (table), 149–151, 152 (table)
  state instability, 314 (figure), 379 (table), 380 (table), 395 (table), 395–397, 396 (table), 410 (figure)
Gross, Neil, 449
Guided missile accuracy, 76, 78
Gun control–gun ownership correlation, 289 (table), 289–290
Gun control–political ideology correlation, 292–294, 293 (table), 294–295 (table), 296
Gun violence rates, 393 (box)

H
Halperin, Rick, 55–56 (table), 57 (table), 60 (table), 73–74 (table)
Hananel, Sam, 453
Hanks, Tom, 246, 247
Harr, Jonathan, 451
Harvard clerk appointments, 66 (table), 66–67, 67 (table)
Hawthorne Effect, 179
Height of Presidents, 106–108, 107 (table), 108 (table)
"Heights of Presidents and Presidential Candidates of the United States," 107 (table)
Hensley, Scott, 449
Heteroscedastic data, 397–398

Hicken, Allen, 450
Hindu population, 278 (box)
Hogan, Steve, 143
Holder, Eric, 137
Hollerith cards, 19
Hollerith, Herman, 19
Holmes, James, 143
Homoscedastic data, 397–398, 398 (figure)
Honesty, 27–28
Hopstock, Paul J., 65 (table), 72–73 (table)
Hout, Michael, 233 (table), 234 (table)
Howden, Lindsay M., 248–249 (table), 452
Huff, Darrell, 1–2, 9, 319, 449, 453
Hypothesis testing
    analysis of variance, 185–199, 186 (figure), 187 (figure), 188 (table), 189 (table), 193 (table), 196–197 (table), 198 (table), 276 (table)
    Apply It Yourself assignments, 208, 210–211
    basic concepts, 421–422
    hypotheses, 4, 181–183, 194–195
    null hypothesis, 8–9, 183–185, 252–253, 288 (table), 422
    political examples, 194–195, 202–208, 203 (figure), 204 (figure), 205–206 (figure)
    Statistical Package for the Social Sciences (SPSS), 199–208, 200 (box), 201–202 (box)
    theory development, 180
    Your Turn exercises, 208, 209 (table)

I
Importing files, 36 (box), 436–437
Income
    income equality–state instability correlation, 379 (table), 380 (table), 380–381, 395 (table), 396 (table), 411 (figure)
    income equality–unemployment rate correlation, 397–398, 398 (figure)
    income–gender correlation, 355 (table), 356 (table), 356–357
    income in developing countries, 69–72, 71 (figure)
    income–race correlation, 374 (table)
"Income Share of Richest 10%," 379 (table), 380 (table), 395 (table), 396 (table), 411 (figure), 454
Independent events, 247 (table), 247–248, 248–249 (table)
Independent variables
    antecedent variables, 321, 321 (figure)
    contingency tables, 286 (table), 422–423
    hypothesis testing, 181–183, 421–422
    interaction effects, 323 (figure), 323–324
    intervening variables, 321–322, 322 (figure)
    measures of association, 424, 426
    regression analysis, 346–350, 356 (table), 356–357
    relationship correlations, 8
    scientific method, 3–4
    spurious non-relationships, 322–323, 323 (figure)
    visual representations, 213–214, 278–281, 279–280 (figure), 281 (figure)
    see also Contingency tables; Multiple regression
India, 278 (box)
Indicators, 22
Industrial Revolution, 179
Infant mortality rate–rural population correlation, 388 (figure), 388–389, 389 (table), 390 (figure), 390 (table)
Inferential statistics, 7–8
Interaction effects, 323 (figure), 323–324
"Internet Statistics: Users (Per Capita)," 304 (table), 307 (table)
Interquartile range, 79–80, 90–91, 421
Intersection, 250
Interval-level measures
    analytical purpose, 425 (table)
    appropriate measures, 274–278, 275 (figure), 276 (table), 422–424, 426
    basic concepts, 23, 23 (figure), 24 (figure), 420
    contingency tables, 225–226, 231, 235, 311
    Gauss-Markov assumptions, 382–384, 383 (table)
    graphs, 213–215, 228–229
    Pearson's r, 274, 302–303, 310, 343, 401
    regression analysis, 357, 398, 399
    scatter plots, 402
Intervening variables, 321–322, 322 (figure)
Introductory statistics, 7–9
Inverse relationship, 182
Iraq, 243 (table)
Israeli Knesset, 62–63, 63 (table)

J
Jackson, Joe, 51, 52 (table)
Jacobellis v. Ohio (1964), 449

Jacobson, Gary C., 4, 449
"Joe Jackson," 450
Jones, Jeffrey M., 451
Journal of Personality and Social Psychology, 319
Journal of the American Medical Association, 319

K
Keith, Tamara, 452
Kendall's tau, 276 (table), 296
Kennedy, John F., 5
Khrushchev, Nikita, 5
Kim, Jibum, 233 (table), 234 (table)
Kinsella, W. P., 450
Kiviat, Barbara, 450
Kremlinology, 5, 6
Kroenig, Matthew, 450

L
Lambda
　analytical purpose, 424, 425 (table), 426
　calculations, 297–302, 300 (box), 308–310, 309 (table)
　definition, 274
　levels of measurement, 276 (table), 278
　limitations, 300–302
　political examples, 298 (table), 298–302, 301 (table), 302 (table), 309 (table), 309–310, 317 (table)
Lambert, Caroline, 453
Landrieu, Mitch, 137, 142
Lao-Tzu, 2, 449
Law school tuition, 78 (table), 78–82, 80 (table), 82 (figure), 101, 102–103 (figure), 103–106, 104 (figure), 105 (figure)
Lederer, Howard, 417
Legalization of abortion
　see Abortion
Levels of measurement, 22–24, 23 (figure), 24 (figure), 274–277, 275 (figure), 276 (table), 419–420
Levitt, Steven D., 343, 453
Life expectancy
　female literacy, 237–238, 241 (table), 241–242
　gross domestic product (GDP) per capita, 390–393, 391 (figure), 392 (figure), 392 (table), 393 (table)
"Life expectancy at birth," 392 (table), 393 (table)
Lincoln, Eric, 452

Lindberg, Staffan I., 450
Linear relationships, 384–393, 385 (figure)
"List of Law Clerks of the Supreme Court of the United States," 66 (table)
"List of Legislatures by Country," 452
"List of Sovereign States and Dependent Territories in Europe," 41 (table), 42 (table), 47 (table), 48
"Local Area Unemployment Statistics: Unemployment Rates for States," 82 (table), 83 (table), 84 (table), 85 (table), 91 (table), 95 (table)
Logit, 382, 428
Log linear relationships, 390–393, 391 (figure), 392 (figure), 392 (table), 393 (table)
Loughner, Jared, 54
LSAT scores, 79–80, 113
Luketic, Robert, 451

M
"Major League Baseball Salaries," 77 (table)
Malloy, Dannell, 366
Mammograms, 146–147
Manmade normal distributions, 113
Marginals, 250–251
Margin of error, 160 (figure), 160–163, 170–171
Marsden, Peter, 233 (table), 234 (table)
Marshall, Monty G., 176 (table), 304 (table), 316 (table), 379 (table), 380 (table), 395 (table), 396 (table), 452, 453, 454
Marx, Karl, 179, 180
Maternal mortality rate increase study, 177–178
"Math Model Helps Predict Olympic Medal Winners," 452
Math scores, 64–66, 65 (table)
Maximum likelihood estimate (MLE), 428–429
McCain, John, 127
McCormick, Sheila A., 450
McMann, Kelly, 450
McNeil, Donald, Jr., 452
McQuaid, John, 452
Mean
　definition, 54, 420
　political examples, 64–67, 65 (table), 66 (table), 67 (table)
　raw data calculations, 54–55, 55 (table)
　tabular data calculations, 55–56 (table), 55–58, 57 (table)

Meaningful numbers, 427–428
Means testing
   Apply It Yourself assignments, 177–178
   confidence intervals, 159–163, 160 (figure), 161 (figure), 169–170
   definition, 148
   employment distribution example, 157 (table), 157–159, 159 (table)
   pattern analysis, 424
   political examples, 165–171, 166 (table), 173–176, 174 (figure), 175 (figure)
   population–sample comparisons, 148–151, 149 (table), 150 (table), 151 (figure), 152 (table), 164–171
   proportions, 157 (table), 157–159, 159 (table), 167–169
   sample size, 163–164, 164 (table), 170–171
   standard error of the mean, 152–157, 161–162
   Statistical Package for the Social Sciences (SPSS), 171–176, 172 (box), 173 (box), 174 (figure), 175 (figure)
   symbols, 150, 150 (table), 165
   t-distribution, 154–157, 159–161
   use guidelines, 439–440
   Your Turn exercises, 176 (table), 176–177
Measurement
   Apply It Yourself assignments, 49–50
   attribute measures, 20–25, 23 (figure), 24 (figure), 33, 33–34 (box), 418–419
   census enumeration, 18–19, 20
   conceptual versus operational definitions, 20–21, 24–29
   influencing factors, 418–419
   variable measures, 33, 33–34 (box)
   Your Turn exercises, 48–49
Measures of association
   Apply It Yourself assignments, 317–318
   basic principles, 274–278, 275 (figure), 276 (table), 277 (table), 302–303
   data interpretation, 193 (table), 259 (table)
   levels of measurement, 24
   political examples, 312
   Statistical Package for the Social Sciences (SPSS), 310–314, 311 (box), 313 (figure), 314 (figure)
   statistical significance, 9, 424
   substantive significance, 191–193, 193 (table)
   Your Turn exercises, 314–316, 316 (table), 317 (table)
   see also Cramér's V; Gamma; Lambda; Pearson's r
Measures of central tendency
   analytical purpose, 425 (table)
   Apply It Yourself assignments, 74–75
   average value calculations, 64–72, 69 (box), 70 (figure), 71 (figure)
   definition, 53
   levels of measurement, 419–420
   mean, 54–58, 55–56 (table), 57 (table), 64–67, 65 (table), 66 (table), 67 (table), 420
   median, 58–62, 60 (table), 61 (table), 64–67, 65 (table), 66 (table), 67 (table), 420
   mode, 62–64, 63 (table), 64–67, 65 (table), 66 (table), 67 (table), 420
   Statistical Package for the Social Sciences (SPSS), 68–72, 69 (box), 70 (figure), 71 (figure)
   Your Turn exercises, 72–74 (table), 72–73
Measures of dispersion
   analytical purpose, 425 (table)
   Apply It Yourself assignments, 108–110
   importance, 76, 78
   levels of measurement, 419–420
   political examples, 90–97, 91 (table), 92 (figure), 93–94 (table)
   ranges, 78 (table), 78–82, 80 (table), 82 (figure), 90–91
   Statistical Package for the Social Sciences (SPSS), 98 (box), 98–106, 100 (box)
   Your Turn exercises, 106–108, 108 (table)
   see also Standard deviation
Media expenditures, 417–418
Median
   definition, 58, 420
   political examples, 64–67, 65 (table), 66 (table), 67 (table)
   raw data calculations, 58–59
   tabular data calculations, 59–62, 60 (table), 61 (table)
"Member States of the European Union," 41 (table), 42 (table), 47 (table), 48
Memmott, Mark, 450
Memo-writing guidelines, 12–13 (box), 12–15, 430–431
Mendes, Elizabeth, 454
Meyer, Julie A., 248–249 (table), 452

Middle Eastern oil, 68 (box)
Military personnel suicide rate, 253 (box)
Millennium Development Goals (MDGs), 322 (box)
Miller, Bennett, 450
Miller Test, 20–21
Miller v. California (1974), 449
Minority discrimination, 157 (table), 157–159, 159 (table), 168–169
Minority population, 86 (table), 87 (table), 89 (table)
Mode
  definition, 62, 420
  political examples, 64–67, 65 (table), 66 (table), 67 (table)
  raw data calculations, 62
  tabular data calculations, 62–63, 63 (table)
Models, 181–183, 378
Moderate to severe diarrhea (MSD), 181 (box)
Moneyball (film), 450
Monson, J. Quin, 450
Moody, Chris, 450
Mortenson, Greg, 237
Multicollinearity, 395
Multiple regression
  Apply It Yourself assignments, 414–416
  basic concepts, 377–378, 398–399
  congressional district reapportionment, 373 (table)
  correlation matrix, 404–405 (figure)
  data assumptions, 381–382, 399
  ordinary least squares (OLS), 399, 428–429
  R-square, 391
  state instability example, 404, 405 (figure), 406–409, 407 (figure), 408 (figure), 409 (figure), 410–412 (figure)
  Statistical Package for the Social Sciences (SPSS), 401–404, 404–405 (figure), 406 (figure), 406–409, 407 (figure), 408 (figure), 409 (figure), 410–412 (figure)
  teen pregnancy rate example, 373 (table), 400 (table), 400–404
  variable analysis, 378–381, 379 (table), 380 (table)
  Your Turn exercises, 412–413, 413 (table), 414 (table)
  see also Gauss-Markov assumptions
Multivariate relationships
  analytical purpose, 425 (table)
  Apply It Yourself assignments, 340–342

interaction effects, 323 (figure), 323–324
spurious non-relationships, 322–323, 323 (figure)
spurious relationships, 320–322, 321 (figure), 322 (figure)
Statistical Package for the Social Sciences (SPSS), 332 (box), 332–333, 334 (figure), 335 (figure), 336 (figure), 337 (figure)
three-step process, 421–422
three-way contingency tables, 324–330, 325 (table), 326 (table), 328 (box), 331 (table)
Your Turn exercises, 339–340, 340 (table)
Murder rates
  see Crime data
Murphy, Ryan, 453
Muslim population, 278 (box)
Mutually exclusive, 30
Myers, Dee Dee, 6 (box)

N
National Association of State Budget Officers, 349 (table), 351 (table), 371 (table)
National Campaign to Prevent Teen and Unplanned Pregnancy, 359 (table), 360 (table), 373 (table)
National Center for Education Statistics (NCES), 93–94 (table), 96 (table), 161, 451
National Opinion Research Center, 3, 5, 269 (table), 271 (table), 289 (table), 293 (table), 298 (table), 301 (table), 302 (table), 324 (table), 325 (table), 326 (table), 331 (table), 338–339 (table), 340 (table), 356 (table), 374 (table)
National Science Foundation (NSF), 16
National Security Agency (NSA), 312
Nation Master, 304 (table), 307 (table), 314, 389 (table), 390 (table)
Nazi regime, 2
Negative relationships
  contingency tables, 286 (table), 286–287, 287 (table), 288 (table)
  hypothesis testing, 182–183
  relationship correlations, 273–274, 278–281, 279–280 (figure), 281 (figure), 306–308, 308 (table), 344, 344 (figure)
  visual representations, 213–214, 214 (figure), 285–286

Negative z-scores, 117, 117 (figure)
Neporent, Liz, 453
Newport, Frank, 451
New Start Treaty, 19
New York Times, 13, 14, 78, 101, 153
Nielsen ratings, 418
1919 World Series, 52 (table)
Nominal-level measures, 22, 23 (figure), 24 (figure), 274–275, 275 (figure), 276 (table), 277, 419
Normal curves, 112–114, 113 (figure), 114 (figure), 148 (figure)
  see also z-scores
Null hypothesis, 8–9, 183–185, 252–253, 288 (table), 422
Nunez, Michael, 452

O
Obama, Barack, 5, 127, 169–170, 212, 452
Obama campaign, 417–418
Obscenity tests, 20–21
Observation, scientific, 4
Odds makers, 111–112
Ohio, Jacobellis v. (1964), 449
Oil production/oil consumption, 68 (box)
One-tailed probabilities, 118 (figure), 118–120, 119 (figure), 147, 148 (figure)
Operational definition, 20–21, 24–29
Operational measures
  levels of measurement, 22–24, 23 (figure), 24 (figure)
  number assignments, 21–22
Opinion polls, 6 (box)
Ordinal-level measures, 22, 23 (figure), 24 (figure), 214, 274–277, 275 (figure), 276 (table), 306–308, 419–420
Ordinary least squares (OLS), 346–350, 381–382, 397–398, 399, 428–429
Organization for Economic Cooperation and Development (OECD), 64
Outliers, 58
"Overview of Race and Hispanic Origin: 2010," 86 (table), 87 (table)

P
Pakistan, 278 (box)
Panel study, 28
Parameters, 149

Partisanship study
  analysis of variance (ANOVA), 202–208, 203 (figure), 204 (figure), 205–206 (figure), 210–211
  contingency tables, 226–228, 227 (table), 232–235, 233 (table), 234 (table)
  lambda calculations, 298 (table), 298–302, 301 (table), 302 (table), 317 (table)
Patterson, Kelly D., 450
Paxton, Pamela, 450
Pearson's r
  analytical purpose, 425 (table), 426
  calculations, 281–283, 282 (box), 303
  definition, 274
  levels of measurement, 276 (table), 276–277
  political examples, 283–285, 284 (table), 285 (table), 303–304, 304 (table), 305 (table), 306
  relationship correlations, 278–281, 279–280 (figure), 281 (figure), 379 (table)
  Statistical Package for the Social Sciences (SPSS), 313 (figure)
Pelczar, Maria P., 65 (table), 72–73 (table)
Peters, Gerhard, 215 (table), 217 (figure), 218 (figure), 219 (figure), 220 (figure)
Pew Research Center for the People and the Press, 3, 453
Phony statistics, 1–2
Polish Solidarity Movement, 229–231, 231 (figure)
Political campaigns, 417–418
Political rights
  Iraq, 243 (table)
  political rights–civil liberties correlation, 283–285, 284 (table), 285 (table)
Politico, 454
Polity Project IV, 61, 61 (table), 176 (table), 221 (table), 222 (table), 223 (table), 250 (table), 251 (table), 252 (table), 256 (table)
Poor Student Graduation Rates, 317–318
Pope, Jeremy C., 450
Population
  definition, 7
  pattern analysis, 8, 11
  population–sample comparisons, 148–151, 149 (table), 150 (table), 151 (figure), 152 (table), 164–171

Positive relationships
  contingency tables, 286 (table), 286–287, 287 (table), 288 (table)
  hypothesis testing, 182–183
  relationship correlations, 273–274, 278–281, 279–280 (figure), 281 (figure), 306–308, 308 (table), 344, 344 (figure)
  visual representations, 213 (figure), 213–214, 285–286
Post-Soviet corruption study, 165–166, 166 (table), 187, 188 (table), 189 (table), 190, 209 (table)
Poulson, Kevin, 451
Poverty, 322 (box)
"Poverty and Equity Data," 453
Power of statistics, 2–3
Predictive validity, 26
PRE measures, 193, 297, 353, 425 (table), 426–427
Presidential height, 106–108, 107 (table), 108 (table)
Probability estimation
  contingent events, 249–251, 250 (table), 251 (table), 252 (table)
  cross tabulations, 247 (table), 248
  discrete events, 246–252
  equal probability calculations, 246–247
  independent events, 247 (table), 247–248, 248–249 (table)
  relationship correlations, 8
  statistical significance, 423–424
  Type I and Type II errors, 146–147
  see also Continuous probability
Probit, 382, 428
Professional writing tips, 430–435
Progressive Insurance, 7–8
Proportional Reduction in Error (PRE) measures, 193, 297, 353, 425 (table), 426–427
Proportions
  confidence intervals, 162–163
  means testing, 157 (table), 157–159, 159 (table), 167–169
  standard deviation, 86 (table), 86–89, 87 (table), 88 (table), 89 (table), 162–164, 164 (table)
Public opinion polls, 6 (box)
Public records, 4
Putnam, Robert D., 5, 449
p values, 184

Q
Qualitative data, 6, 21–22
Quantitative data, 6, 21–22
Questionnaires, 5

R
Race
  discrimination, 157 (table), 157–159, 159 (table), 168–169
  legalized abortion–race correlation, 269 (table), 269–272, 270 (table), 271 (table)
  multiple regression analysis, 383 (table), 383–384
  race–ideology relationship, 261 (table), 261–263, 262 (table)
  race–income correlation, 374 (table)
  religious fundamentalism–abortion–race correlation, 324 (table), 324–328, 325 (table), 326 (table), 330–332, 331 (table)
Radcliffe, Benjamin, 453
Random error, 8–9, 146–147
Ranges, 78 (table), 78–82, 80 (table), 82 (figure), 90–91
Rape, 333–338
Ratio-level measures, 23, 23 (figure), 24 (figure), 274–275, 275 (figure), 276 (table), 420
Reading scores, 72–73 (table), 72–73
Regression analysis
  analytical purpose, 425 (table), 426–427
  Apply It Yourself assignments, 374–376, 414–416
  basic concepts, 377–378
  congressional district reapportionment, 373 (table)
  dichotomous independent variables, 356 (table), 356–357
  fitting a line, 346–350, 347 (figure), 349 (table), 357–358
  income–gender correlation example, 356 (table), 356–357
  levels of measurement, 276 (table)
  murder rate example, 354–356, 355 (figure), 355 (table)
  ordinary least squares (OLS), 346–350, 428–429
  professional regression tables, 434–435
  relationship representations, 344 (figure), 344–346, 345 (figure)

R-square, 353, 359, 391
  state government expenditures by population example, 345 (figure), 346–353, 349 (table), 351 (table), 366–371, 368 (figure), 369 (figure), 370 (figure), 371 (table)
  state instability example, 404, 405 (figure), 406–409, 407 (figure), 408 (figure), 409 (figure), 410–412 (figure)
  statistical significance, 350–353, 351 (table), 358–359
  teen abortion rate–birth rate correlation example, 359 (table), 359–363, 360 (table), 362 (table)
  teen pregnancy rate example, 373 (table), 400 (table), 400–401
  time series data, 354–356, 355 (figure), 355 (table)
  use guidelines, 442
  variable analysis, 378–381, 379 (table), 380 (table)
  Your Turn exercises, 371 (table), 371–374, 373 (table), 374 (table), 412–413, 413 (table), 414 (table)
  see also Bivariate regression; Gauss-Markov assumptions; Multiple regression; Statistical Package for the Social Sciences (SPSS)
Regression tables, 434–435
Reify, 22
Relationships
  see Contingency tables; Graphs; R-square
Reliability, 26–28
Religion
  religious freedom example, 265–268, 266 (figure), 267 (figure)
  religious fundamentalism–abortion–race correlation, 325 (table), 325–328, 326 (table), 330–332, 331 (table)
  religious fundamentalism–abortion–rape correlation, 333–338, 334 (figure), 335 (figure), 336 (figure), 337 (figure), 338–339 (table)
Rentrak, 418
Research question, 3–4
Residuals, 347, 362 (table)
Residual scatter plots, 403, 403 (box)
Rhodes, Richard, 452
Ripley, Tom, 449

Robinson, Phil, 51, 450
Robust, 399
Roe v. Wade (1973), 343, 359
Romney, Mitt, 154
Roosevelt, Franklin D., 6 (box)
Ross, Dennis, 320
Roth, G., 449
Routine mammograms, 146–147
R-square, 353, 359, 427
Rutenberg, Jim, 454
Ryan, Ann Marie, 450

S
Saad, Lydia, 451
Salsburg, David, 451
Samples
  pattern analysis, 8–9
  population–sample comparisons, 148–151, 149 (table), 150 (table), 151 (figure), 152 (table), 164–171
Sample size, 163–164, 164 (table), 170–171
Santorum, Rick, 13–15
Scale, 219 (figure), 219–220, 220 (figure)
Scatter plots, 441–442
Schleifstein, Mark, 452
Schmit, Mark J., 450
Schwarz, Alan, 449
Scientific method, 3–4, 8–9, 21
Scientific observation, 4
Segal, David, 450
Self-reports, 28
Semetko, Holli A., 450
Shelley, Brooke E., 65 (table), 72–73 (table)
Shoeless Joe (book), 51
Shoeless Joe Jackson, 51, 52 (table)
Significance
  see Statistical significance; Substantive significance
Silver, Nate, 153, 227 (table)
60 Minutes, 111, 451
Skaaning, Svend-Erik, 450
SLAPS test, 21
Slavery, 20
Smith, Tom W., 233 (table), 234 (table)
Smoking–lung cancer correlation, 145–146
Social capital, 5
Software piracy, 194–197, 196–197 (table), 198 (table)
Solidarity Movement, 229–231, 231 (figure)

Sollenberg, Margareta, 221 (table), 222 (table), 223 (table), 250 (table), 251 (table), 252 (table), 256 (table)
Sorensen, Elaine, 453
South Sudan, 243
Soviet corruption study, 165–166, 166 (table), 187, 188 (table), 189 (table), 190, 209 (table)
Specification, 393–394
Spielberg, Steven, 452
"Sports Betting: Billy Walters," 451
Sports gambling, 111–112
Spurious non-relationships, 322–323, 323 (figure)
Spurious relationships, 9, 320–322, 321 (figure), 322 (figure)
Squared relationships, 385 (figure), 385–386, 387 (table)
Stableford, Dylan, 450
Standard deviation
  analytical purpose, 424
  definition, 420–421
  dichotomous variables, 86–89, 97
  normal distributions, 113–114
  population–sample comparisons, 150–151, 151 (figure), 153–154
  proportions, 86 (table), 86–89, 87 (table), 88 (table), 89 (table), 162–164, 164 (table)
  raw data calculations, 82 (table), 82–84, 83 (table), 92–94, 93–94 (table)
  tabular data calculations, 84 (table), 84–89, 85 (table), 86 (table), 87 (table), 88 (table), 89 (table), 95 (table), 95–97, 96 (table)
Standard error of the mean, 152–157, 161–162
Standard error of the slope, 346–353, 358–359
Standardized coefficient, 356
Standardized values, 135
  see also z-scores
Stango, Nick, 451
"State and County QuickFacts," 89 (table)
State Death Penalty Executions
  mean calculations, 54–58, 55–56 (table), 57 (table)
  measures of central tendency, 73, 73–74 (table)
  median calculations, 58–61, 60 (table)
State Fragility Index, 304 (table), 314, 314 (figure), 316 (table), 378–380, 379 (table), 380 (table), 385 (figure), 395 (table), 396 (table)

State government expenditures by population, 345 (figure), 346–353, 349 (table), 351 (table), 366–371, 368 (figure), 369 (figure), 370 (figure), 371 (table)
State instability example, 401–404, 405 (figure), 406–409, 407 (figure), 408 (figure), 409 (figure), 410–412 (figure)
"State Religion," 452
Statistical average
  see Measures of central tendency
Statistical evidence, 145–146
Statistical Package for the Social Sciences (SPSS)
  analysis of variance (ANOVA), 199–208, 200 (box), 201–202 (box)
  basic principles, 34
  bivariate regression analysis, 363 (box), 363–370, 364 (box), 366 (box), 367–368 (figure), 369 (figure), 370 (figure), 371 (table)
  bivariate scatter plots, 402 (box), 402–403, 406 (figure)
  case selection, 438
  chi-square statistic, 264 (box), 264–268, 266 (figure)
  collapsing variables, 99–101, 100 (box), 437–438
  contingency tables, 235–238, 236 (box), 237 (box), 239–240 (figure), 241–242, 265 (box), 441
  continuous probability, 135 (box), 135–142, 136–137 (box), 138 (figure), 139 (figure), 140 (figure), 141 (figure)
  correlation matrix, 313 (figure), 314 (figure), 404, 404–405 (figure), 441
  Cramér's V, 264 (box), 264–268, 266 (figure)
  cross tabluations, 264 (box), 265–268, 267 (figure), 441
  data cleaning, 36–38, 38 (box)
  data files, 34–39, 36 (box)
  European Union application example, 39–48, 41 (table), 42 (table), 43 (figure), 44 (figure), 45–46 (figure), 47 (figure), 47 (table)
  Excel data files, 34–36, 36 (box), 37 (figure), 43 (figure), 436–437
  frequency tables, 38–39, 39 (box), 46, 47 (figure), 47 (table)
  hypothesis testing, 199–208, 200 (box), 201–202 (box)

law school tuition application, 101,
    102–103 (figure), 103–106, 104 (figure),
    105 (figure)
means testing, 171–176, 172 (box), 173 (box),
    174 (figure), 175 (figure)
measures of association, 310–314, 311 (box),
    313 (figure), 314 (figure)
measures of central tendency, 68–72,
    69 (box), 70 (figure), 71 (figure)
measures of dispersion, 98 (box),
    98–106, 100 (box)
multiple regression analysis, 401–404,
    404–405 (figure), 406 (figure), 406–409,
    407 (figure), 408 (figure), 409 (figure),
    410–412 (figure)
multivariate relationships, 332 (box),
    332–333, 334 (figure), 335 (figure),
    336 (figure), 337 (figure)
Pearson's r, 313 (figure)
religious freedom example, 265–268,
    266 (figure), 267 (figure)
residual scatter plots, 403, 403 (box)
scatter plots, 441–442
three-way contingency tables, 332 (box),
    332–333, 334 (figure), 335 (figure),
    336 (figure), 337 (figure)
use guidelines, 11–12, 436–442
variable computations, 438
variable measures, 44 (figure), 44–45,
    45–46 (figure)
Statistical significance
  analytical purpose, 425 (table)
  definition, 171
  F-statistic, 190–191
  null hypothesis, 184
  pattern analysis, 9, 423–424
  regression analysis, 350–353, 351 (table),
    358–359
  Type I and Type II errors, 146, 147
  see also Chi-square statistic
Staton, Jeffrey, 450
Stewart, Potter, 20
Strand, Håvard, 221 (table), 222 (table),
    223 (table), 250 (table), 251 (table),
    252 (table), 256 (table)
Strauss, Valerie, 450
Student's t-distribution, 154–157, 159–161
Substantive significance, 9, 191–193, 245, 424,
    425 (table), 426–427

Sum of squares (SS), 84, 85 (table), 93–94 (table),
    96 (table), 187, 188 (table), 282–283, 420
Superstitious behaviors, 319
Supply-and-demand economics, 273 (figure),
    273–274, 274 (figure)
Surveys, 5
Swedish census, 7
Symbols, 150, 150 (table), 165
Symmetrical curves, 113
Symmetrical measure of association, 283

T
Tabellverket, 7
Tables, 213
  see also Contingency tables
t-distribution, 154–157, 159–161, 423,
    425 (table), 445–446
  see also z-scores
Teen abortion rate–birth rate correlation,
    359 (table), 359–363, 360 (table),
    362 (table)
Teen pregnancy rate, 373 (table), 400 (table),
    400–401
Teorell, Jan, 450
Terbush, Jon, 453
Terminal, The (film), 246–247, 452
Testable hypothesis, 181, 183–184
Test score comparisons, 94 (box)
Thee-Brenan, Megan, 451
Theory, 180
The Terminal (film), 246–247, 452
Third variables
  see Multivariate relationships
Three-way contingency tables
  basic concepts, 324–328, 325 (table),
    326 (table)
  political examples, 330–338, 331 (table),
    334 (figure), 335 (figure), 336 (figure),
    337 (figure), 338–339 (table)
  professional writing tips, 433–434
  setup and creation, 328 (box), 329–330
  Statistical Package for the Social Sciences
    (SPSS), 332 (box), 332–333, 334 (figure),
    335 (figure), 336 (figure), 337 (figure)
Through the Looking-Glass (Carroll), 20
Time series data, 354–356, 355 (figure),
    355 (table)
Total sum of squares (TSS), 185, 187–188, 192,
    195–197, 199

Transparency Index, 188 (table), 189 (table), 190, 314 (figure), 379 (table), 380 (table), 380–381, 395 (table), 396 (table), 410 (figure)
Transparency International, 166, 166 (table), 188 (table), 189 (table), 316 (table), 379 (table), 380 (table), 395 (table), 396 (table), 453, 454
Turner, Margery Austin, 453
Twain, Mark, 1, 449
21 (film), 111, 451
Two-tailed probabilities, 120–122, 121 (figure), 159–161, 160 (figure)
2009 EEO-1 National Aggregate Report, 157 (table), 451
Type I and Type II errors, 146–147, 164–165, 185

U

Uncertainty, 417, 428–429
Unemployment rates, 29 (box), 82 (table), 83 (table), 84 (table), 85 (table), 91 (table), 92 (figure)
Uniform Crime Reports, 3, 137, 142, 355 (table)
Unimodal curves, 113
United Nations Mission in South Sudan (UNMISS), 243
United Nations (UN), 3
Unit of analysis, 19
Univariate statistics, 7, 148, 419–421, 425 (table)
Unreliability, 27–28
Unstandardized coefficient, 356
U.S. Army Corps of Engineers, 247
USA Today Salaries Database, 77
U.S. Bureau of Labor Statistics, 29 (box), 82 (table), 83 (table), 84 (table), 85 (table), 91 (table), 95 (table), 450
U.S. Census Bureau, 3, 18–19, 86 (table), 87 (table), 89 (table), 157, 227 (table), 248–249 (table), 373 (table), 449, 450, 454
U.S. Department of Defense (DOD), 5, 6
U.S. News & World Report, 78 (table), 80 (table), 106
U.S. Preventative Services Task Force (USPSTF), 146–147

V

Vaccination–autism correlation, 21
Validity, 25–26
Vardi, Nathan, 451

Variables, 22, 33, 33–34 (box), 346, 378–381, 419–422, 438–439
  see also Dependent variables; Independent variables
Variance, 84–86, 85 (table), 93–94 (table), 96 (table), 420
  see also Analysis of variance (ANOVA)
Vedantam, Shankar, 450
Violations of assumptions, 427
Visual presentations
  Apply It Yourself assignments, 243–244
  contingency tables, 213, 220–228, 221 (table), 222 (table), 223 (table), 224 (box), 225 (box), 227 (table), 231–235, 233 (table), 234 (table), 422–423
  graphs, 213 (figure), 213–220, 214 (figure), 215 (figure), 216–217 (figure), 218 (figure), 219 (figure), 220 (figure), 228–231, 231 (figure)
  selection guidelines, 213
  Your Turn exercises, 242–243, 243 (table)
Voter choice, 127 (table), 127–132, 130 (figure), 131 (figure), 132 (figure), 142 (table)
Voter turnout study, 214–215, 215 (figure), 215 (table), 216–217 (figure), 218 (figure), 219 (figure), 220 (figure), 302 (table)

W

Wakefield, Andrew, 21
Walesa, Lech, 229
Wallensteen, Peter, 221 (table), 222 (table), 223 (table), 250 (table), 251 (table), 252 (table), 256 (table)
Walters, Billy, 111–112
Weber, Max, 3, 449
Westlake, Texas, 159–160
Wheelwright, Kentucky, 69, 72
White House, 452
WikiLeaks, 120 (box)
Wilke, Joy, 454
Wilson, Rick, 16
Within-group sum of squares (WSS), 186–189, 192, 195–197, 199
Wittich, C., 449
Wooley, John, 215 (table), 217 (figure), 218 (figure), 219 (figure), 220 (figure)
World Bank, 3, 322 (box), 450, 453
World Series of Poker Tournament of Champions, 417

Y

Ye Shiwen, 145, 146
Your Turn exercises
    basic statistics, 15–16
    bivariate regression analysis, 371 (table), 371–374, 373 (table), 374 (table)
    chi-square statistic, 269–272
    continuous probability, 142 (table), 142–143
    Cramér's V, 269–272
    hypothesis testing, 208, 209 (table)
    means testing, 176 (table), 176–177
    measurement, 48–49
    measures of association, 314–316, 316 (table), 317 (table)
    measures of central tendency, 72–74 (table), 72–73
    measures of dispersion, 106–108, 108 (table)
    multiple regression analysis, 412–413, 413 (table), 414 (table)
    multivariate relationships, 339–340, 340 (table)
    visual presentations, 242–243, 243 (table)

Z

Zaillian, Steven, 451
Zemeckis, Robert, 452
z-scores
    calculations, 122–123, 126–128, 127 (table), 439
    definition, 114
    negative z-scores, 117, 117 (figure)
    one-tailed probabilities, 118 (figure), 118–120, 119 (figure)
    political examples, 126–132, 127 (table), 130 (figure), 131 (figure), 132 (figure)
    probability estimates, 117–120, 118 (figure), 119 (figure), 124 (figure), 124–126, 125 (figure), 128–132, 130 (figure), 131 (figure), 132 (figure)
    relationship correlations, 278–281
    Statistical Package for the Social Sciences (SPSS), 135 (box), 138 (figure), 139 (figure), 140 (figure), 141 (figure)
    two-tailed probabilities, 120–122, 121 (figure)
    z table distributions, 114–115, 115 (figure), 116 (figure), 443–444
    see also Normal curves; t-distribution

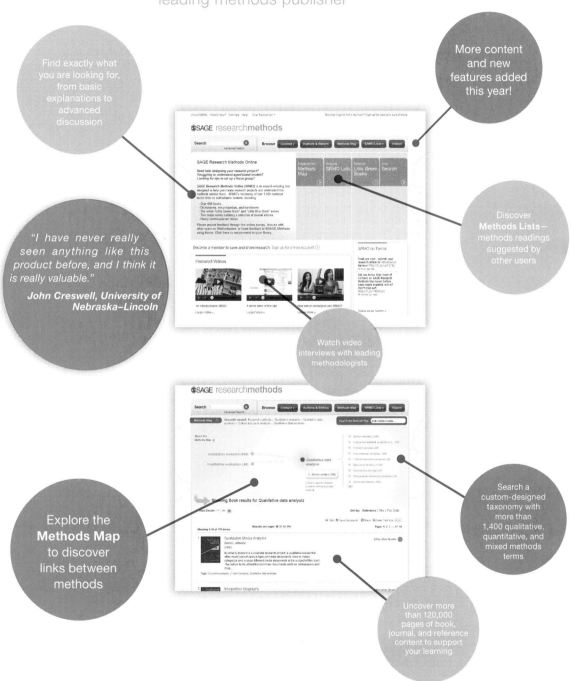